Nanostructured Carbon Electron Emitters and Their Applications

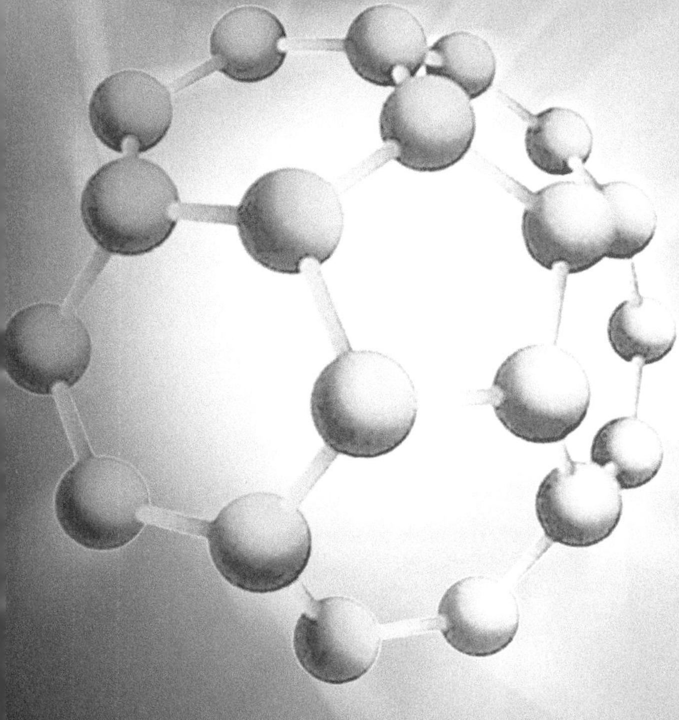

Nanostructured Carbon Electron Emitters and Their Applications

edited by

Yahachi Saito

JENNY STANFORD
PUBLISHING

Published by

Jenny Stanford Publishing Pte. Ltd.
Level 34, Centennial Tower
3 Temasek Avenue
Singapore 039190

Email: editorial@jennystanford.com
Web: www.jennystanford.com

British Library Cataloguing-in-Publication Data
A catalogue record for this book is available from the British Library.

Nanostructured Carbon Electron Emitters and Their Applications

ISBN 978-981-4877-62-6 (Hardcover)
ISBN 978-1-003-14199-0 (eBook)

Contents

Preface

Carbon forms a variety of allotropes (structures) due to the diverse hybridization of *s*- and *p*-electron orbitals; besides the well-known graphite (two-dimensional structure) and diamond (three-dimensional network), new forms such as C_{60} fullerene (cage structure), nanotube (quasi-one-dimensional structure), graphene (truly two-dimensional network), and carbyne (truly one-dimensional chain) have also been discovered. A new family of carbon isotopes (fullerene, carbon nanotube (CNT), graphene, carbyne), called "nanostructured carbon" or "nanocarbon," has attracted great interests both in academia and industrial circles since the successful macroscopic production of C_{60} in 1990. These allotropes exhibit extreme properties such as ultrahigh mechanical strength, ultrahigh charge–carrier mobility, and high thermal conductivity, attracting considerable attention for their electronic and mechanical applications as well as for exploring new physics and chemistry in the field of basic materials science.

Electron sources are important in a wide range of areas, from basic physics and scientific instruments to medical and industrial applications. CNT and graphene behave as excellent electron-field emitters owing to their exceptional properties and offer several benefits compared to traditional cathodes (e.g., thermionic and photo-cathodes). Field emission (FE) produces very intense electron currents from a small surface area with a narrow energy spread, providing a highly coherent electron beam. This combination of high current, small source size, and narrow energy distribution not only provides us with the brightest electron sources but also explores a new field of electron beam–related research.

This book presents the enthusiastic research and development of CNT-based FE devices and focuses on the fundamental aspects of FE from nanocarbon materials, including CNT and graphene, and the latest research findings related to it. The book consists of 17 chapters contributed by proficient researchers leading the related scientific areas. Chapters 1 to 5 describe the fundamental characteristics of CNT electron emitters, Chapters 6 to 11 deal with FE from graphene,

Chapters 12 and 13 discuss applications of FE to X-ray and UV generation, Chapters 14 and 15 review electron sources in vacuum devices and space applications, and Chapters 16 and 17 report on the new forms of carbon produced via FE from CNT.

This book project started in May 2018, and it is my pleasure to bring it to its successful completion even after several interruptions caused due to the COVID-19 pandemic that affected the publication process. I hope the book would provide students and researchers an insight into the field of nanoscience and nanotechnology and contribute to the development of related disciplines.

I am thankful to the authors who contributed excellent chapters reviewing the latest status of their research fields. I express my special thanks to Prof. Richard Forbes for his kind and careful critical reading of the book and also to the staff at Jenny Stanford Publishing for their suggestions and support during this project.

Yahachi Saito
Nagakute, Japan
August 2021

Chapter 1

FEM and FIM of Carbon Nanotubes

Yahachi Saito
Toyota Physical and Chemical Research Institute,
Yokomichi 41-1, Nagakute, Aichi 480-1192, Japan
ysaito@toyotariken.jp

Carbon nanotube (CNT) possesses beneficial properties as field electron emitters. In this chapter, structural, electrical, mechanical, and thermal properties of CNT are briefly described, and their field-emission phenomena observed by field-emission microscopy (FEM) and field ion microscopy (FIM) are reviewed.

1.1 Structure of CNT and Electron Emission Properties

CNT is made of seamlessly rolled graphene (honeycomb lattice of carbon atoms) with outer diameters ranging from about 1 nm to over 50 nm, depending on the number of walls comprising the nanotube. CNTs composed of one sheet of graphene are called single-wall CNT (SWCNT) [1–3], and those made of more than two sheets, multiwall CNT (MWCNT) [4]. Among MWCNTs, those made of two walls and three walls are called double-wall CNT (DWCNT) and triple-wall CNT (TWCNT), respectively. The length of nanotubes exceeds 10 μm, and the longest ones are reportedly on the order of a centimeter [5,

Nanostructured Carbon Electron Emitters and Their Applications
Edited by Yahachi Saito

ISBN 978-981-4877-62-6 (Hardcover), 978-1-003-14199-0 (eBook)
www.jennystanford.com

6]. Thus, CNT exhibits quasi-one-dimensional properties; van Hove singularities (steep maxima) in electronic density of states (DOS) are induced, which brings about characteristic optical absorption and photoluminescence [7]. In addition, ballistic transport of electrons (i.e., no scattering of carriers) and quantum conductance are observed [8].

For mass production of CNTs, catalyst-assisted chemical vapor deposition (CVD) method is mainly employed now, though arc-discharge and laser evaporation of carbon were used at a pioneering stage of CNT studies. The arc-discharge technique is a unique one in the production of high-quality MWCNTs without metal catalysts. These methods of CNT preparations are briefly reviewed in Ref. [9].

CNTs possess various extreme properties as summarized in Tables 1.1 and 1.2 in addition to its geometrical shape with extremely high aspect ratio (length to diameter). Especially, the following properties make CNTs most favorable electron field emitters: (1) needle-like shape with a sharp tip, (2) high electrical and thermal conductivity, (3) high chemical and thermal stability, (4) high mechanical strength, and (5) low surface diffusion of carbon atoms. When an electric potential is applied to a conductor with a sharp tip, electric field concentrates at the sharp point. The field strength (F) at the tip surface is inversely proportional to the radius of curvature (r) of the tip:

$$F = V/kr \tag{1.1}$$

where V is an electric potential of the emitter and k is a constant depending on the emitter shape being approximately 5 in the presence of a cylindrical or conical shank [10]. Therefore, the sharper the tip of the emitter is, the stronger the field strength at the tip is even with the same applied voltage. The first advantage of CNTs, needle-like shape with a tip of an extremely small radius of curvature, realizes easily high electric field required for field emission (FE). Owing to the second advantage, surfaces of CNTs are inert and stable against reactions with residual gas molecules in vacuum. Therefore, the CNT emitters operate stably even at a vacuum of 10^{-4} Pa, in contrast to metal emitters such as tungsten and molybdenum needles that need ultrahigh vacuum to operate stably. High thermal conductivity facilitates to dissipate the Joule heat generated by current flowing along a CNT during FE operation to the cathode substrate. The third advantage of CNT emitters enables them to endure the high tensile stress due to electrostatic force (Maxwell tension). Together with this robustness, the last property of CNTs (low surface diffusion) keeps their original shapes (slender and sharp tip) even under a high field for prolonged usage. Needless to say, since CNTs (especially MWCNTs formed by arc discharge) are a good electrical conductor, they can transfer and emit high density of electron current (10^8 A/cm^2) through their length.

Table 1.1 Electrical properties of CNT

	Value	**Reference**
Work function		
SWCNT	4.7 eV (contact pot. difference)	X. Cui et al., *Nano Lett.* **3** (2003), 783
	4.8 eV (photoelectron emission)	S. Suzuki et al., *Appl. Phys. Lett.* **76** (2000), 4007
MWCNT	4.3 eV (photoelectron emission)	H. Ago et al., *J. Phys. Chem. B*, **103** (1999), 8116
	4.6–4.8 eV (contact pot. difference)	R. Gao et al., *Appl. Phys. Lett.* **78** (2001), 1757
	4.51–4.78 eV (field emission)	Z. Xu et al. *Appl. Phys. Lett.* **87** (2005), 163106
SWCNT, DWCNT, MWCNT bundles	4.7–4.9 eV (thermionic emission from side wall)	P. Liu et al., *Nano Lett.* **8** (2008), 647
Electrical resistance		
Individual SWCNT	~7 kΩ/μm (bridged single tube)	S. Li et al., *Nano Lett.* **4** (2004), 2003
Electrical resistivity		
Aligned SWCNT film	$\lesssim 5.5 \times 10^{-4}$ Ω-cm (// alignment)	J. E. Fisher et al., *J. Appl. Phys.* **93** (2003), 2157
	$\lesssim 3.4 \times 10^{-3}$ Ω-cm (⊥)	
Aligned SWCNT rope	0.825×10^{-3} Ω-cm (// rope) at RT	J. Hone et al., *Appl. Phys. Lett.* **77** (2000), 667
	5.0×10^{-3} Ω-cm (⊥) at RT	

(*continued*)

Table 1.1 *(continued)*

	Value	**Reference**
Individual MWCNT	Longitudinal (// *c*-plane)	
	Resistivity $\rho = 10^{-5}$–10^{-4} Ω-cm	T. W. Ebbesen et al., *Nature* **382** (1996), 54
	Transversal (⊥ c-plane)	
	Intershell conductance 4.5 mS/μm for d = 33 nm	A. Stetter et al., *Phys. Rev. B* **82** (2010), 115451
	13–19 mS/μm for d = 13–18 nm (270–350 mS/μm^2)	Y. Shinomiya et al., *Surf. Interface Anal.* **48** (2016), 1206
	Intershell resistivity	
	≈ 3 Ω-cm for d = 30 nm	A. Stetter et al., *Appl. Phys. Lett.* **93** (2008)172103
	= (0.8–1.1) × 10^{-3} Ω-cm at high temp.	Y. Shinomiya et al., unpublished
cf. Graphite	In-plane resistivity ~5 × 10^{-5} Ω-cm	K. Matsubara et al., *Phys. Rev. B* **41** (1990), 969
	c-axis resistivity ~10^{-3} Ω-cm for natural and synthetic single crystal	
	~10^{-2}–100 Ω-cm for HOPG	

	Value	Reference
Max. current		
SWCNT in vacuum	>20 μA (10^9 A/cm^2) on SiO_2/Si	Z. Yao et al., *Phys. Rev. Lett.* **84** (2000), 2941
MWCNT in vacuum	>10^9 A/cm^2 on SiO_2/Si	B. Q. Wei et al., *Appl. Phys. Lett.* **79** (2001), 1172
	~2 × 10^9 A/cm^2 on Si_3N_4 membrane	G. E. Begtrup et al., *Phys. Rev. Lett.* **99** (2007), 155901
	~0.7–0.35 × 10^9 A/cm^2 suspended	K. Yamauchi et al., unpublished
	(decrease with the increase of layer number from 6 to 13)	
Carrier mobility (FET) of SWCNT	≈100 cm^2/Vs (isolated on SiO_2)	F. Nihey et al., *Jpn. J. Appl. Phys.*, **42** (2003), L128
	~10^5 cm^2/Vs at 50 K (isolated on SiO_2)	X. Zhou et al., *Phys. Rev. Lett.* **95** (2005), 146805
	~1000 cm^2/Vs (aligned array)	S. J. Kang et al., *Nat. Nanotech.* **2** (2003), 230
	≈80 cm^2/Vs (random network)	C. Cao et al., *Nature* **454** (2008), 495
cf. Graphite (in-plane)	Electron mobility 20,000 cm^2/Vs	M. S. Dresselhaus et al., *Science of Fullerenes and Carbon Nanotubes* (Academic Press, 1995) Chap. 2
	Hole mobility 1500 cm^2/Vs	

Table 1.2 Mechanical and thermal properties of CNT

Property	Value	Reference
Young's modulus Individual SWCNT	~ 1.2 TPa (test in AFM)	T. W. Tombler et al., *Nature* **405** (2000), 769
Individual MWCNT	0.40–4.15 (1.8 in average) TPa (thermal vibration in TEM)	M. M. J. Treacy et al., *Nature* **381** (1996), 678
	1.28 (±0.59) TPa (test in AFM)	E. W. Wong et al., *Science* **277** (1997), 1971
	0.1–1 TPa (resonance vibration in TEM, decrease with the increase of diameter)	P. Poncharal et al. *Science* **283** (1999), 1513
	0.27–0.95 TPa (test in AFM)	M. F. Yu et al., *Science* **287** (2000), 637
	0.91 (±0.18) TPa (tensile test in TEM)	B. G. Demczyk et al., *Mater. Sci. Eng. A* **334** (2002), 173
Tensile strength		
Individual SWCNT	~200 GPa (test in AFM)	J. Kong et al., *Nature* **395** (1998), 878
Individual MWCNT	11–63 GPa (AFM meas.)	M. F. Yu et al., *Science* **287** (2000), 637
	150 (±45) GPa (tensile test in TEM)	B. G. Demczyk et al., *Mater. Sci. Eng. A* **334** (2002), 173

Property	Value	Reference
Thermal conduct. at RT		
Individual SWCNT	~3500 W/Km (diam., *d* =1.7 nm)	E. Pop et al., *Nano Lett.* **6** (2006), 96
Aligned SWCNTrope	~2000–7000 W/Km (*d* = 1–3 nm)	C. Yu et al., *Nano Lett.* **5** (2005), 1842
	~220 W/Km	J. Hone et al., *Appl. Phys. Lett.* **77** (2000), 667
Aligned SWCNT film	60 W/Km (// alignment)	J. E. Fisher et al., *J. Appl. Phys.* **93** (2003), 2157
	11 W/Km (⊥)	
Dense-mat of SWCNT	35 W/Km	J. Hone et al., *Phys. Rev. B* **59** (1999), R2514
Individual MWCNT	≳3000 W/Km	P. Kim et al., *Phys. Rev. Lett.* **87** (2001), 215502
	~2000 W/Km (*d* = 9.8 nm)	M. Fujii et al., *Phys. Rev. Lett.* **95** (2005), 065502
	~500 W/Km (*d* = 28 nm)	
On-set temperature of sublimation	3200 K on Si_3N_4 membrane	G. E. Begtrup et al., *Phys. Rev. Lett.* **99** (2007), 155901
(MWCNT in vacuum)	2200–2800 K suspended	H. Maruyama et al., *Appl. Phys. Exp.* **3** (2010), 025101
	3100–3130 K, suspended	H. Nakahara et al., unpublished

1.2 FEM of CNT

FEM is a powerful technique to reveal not only surface structure and work function distribution but also dynamic surface phenomena such as adsorption and desorption. This technique is mainly applied to refractory metals such as tungsten and molybdenum so far, because they are amenable to be shaped into a sharp needle with a tip radius less than a few tens nm by electrochemical etching and are also mechanically strong to withstand in a high electric field.

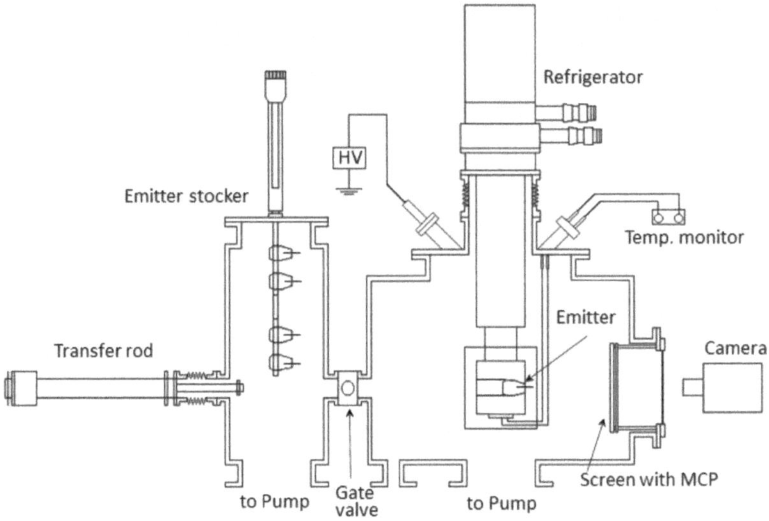

Figure 1.1 Schematic of an FEM apparatus.

FEM apparatuses are usually designed by individual researchers for their own research purposes. One example of a typical FEM apparatus is shown schematically in Fig. 1.1, which can also be used for FIM as described in Section 1.4. The emitter is usually mounted on a heating loop (e.g., hairpin-shaped tungsten wire with diameter of 0.15 mm) for cleaning the emitter surface by evaporating adsorbates. For FEM study, it would be desirable to use a single, isolated CNT fixed at the apex of a metal needle. Such emitters can be prepared by using micromanipulators in a scanning electron microscope (SEM) [11], but the emitters thus prepared are usually covered with contaminants (hydrocarbon deposit) during the SEM observation. Electrophoresis is an alternative method to attach a thin bundle of CNTs at the tip of a metal needle [12], though the control of the number of CNTs in an attached bundle is difficult. The simple and easy method is to glue a bulk bundle of as-grown CNTs to the tip of a heating loop by using a conductive paste [13]. The last method retains CNT surfaces clean, but an enormous number of CNTs protrude out at the end of the bundle.

The tip of CNT emitter is placed at about 30–40 mm in front of an anode plate made of a phosphor screen, or sometimes a microchannel plate (MCP), on which field-emission patterns are observed. The base pressure of the FEM vacuum chamber is 10^{-7} to 10^{-8} Pa. The negative voltage of 0.5 to 2 kV is applied to the emitter relative to the screen.

Various types of CNTs including MWCNT with closed and open tips, SWCNT, VGCF (vapor-grown carbon fiber), and so on have been studied to reveal electron emission properties using FEM so far. In the following subsections, typical FEM images from various CNTs are individually described and their features are commented.

1.2.1 MWCNT with a Closed Cap

FEM images of MWCNT with highly graphitized (i.e., well-developed graphite) layers often display interesting features, as shown in Fig. 1.2. These FEM images were obtained using MWCNTs produced by arc-discharge technique [14]. The FEM images are featured as follows: First, each image consists of pentagonal rings, and second, interference fringes are observed between adjacent rings [15]. We called an image with these features a "pentagon pattern." The number of pentagonal rings observed from one CNT emitter is usually five or six; the maximum number is six. The tips of arc-grown MWCNTs are capped by graphite layers with polyhedral shapes as shown in Fig. 1.3. In order to give a positive curvature to a graphene sheet, pentagons (five-membered carbon rings) have to be introduced to the sheet; six pentagons are required to have a curvature of 2π steradian (i.e., a hemispherical cap). The portion where a pentagon is located extrudes like a vertex of a polyhedron, while the other flat regions are made of hexagons. The shape of a CNT cap depends on how the six pentagons are arranged at the end. FEM images in Fig. 1.2a,b exhibit five-fold and six-fold symmetries, respectively, while that in Fig. 1.2c have a lower symmetry. In Fig. 1.4, structure models with high symmetries (five-fold and six-fold symmetries) and a low symmetry are shown, corresponding to FEM patterns in Fig. 1.2, respectively.

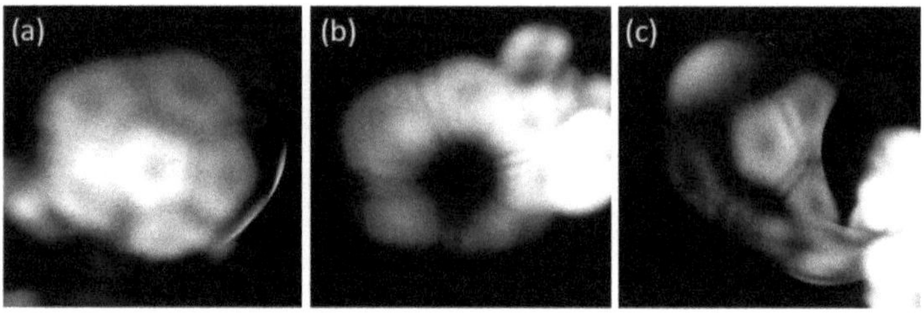

Figure 1.2 Typical FEM images of MWCNTs exhibiting (a) five-fold, (b) six-fold, and (c) low symmetries; Figs. (a) and (b) from [15].

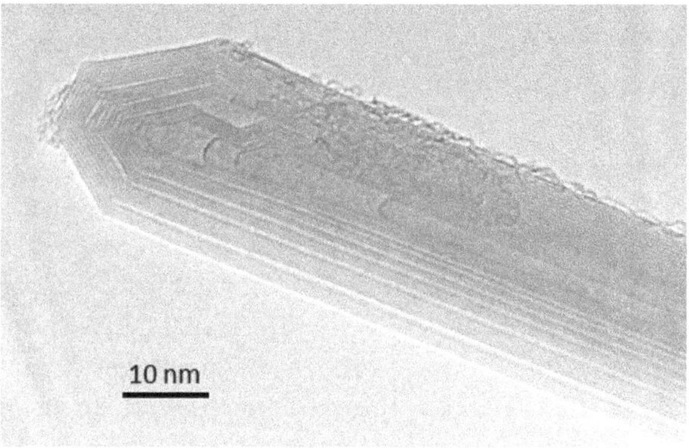

Figure 1.3 TEM image of a tip of MWCNT produced by arc-discharge method.

Since the pentagon sites are locally pointed, the electric field around them would be stronger than that on other flat regions. In addition, it is theoretically indicated that the pentagon site has a higher density of states near the Fermi level EF than the flat hexagon regions [16]. Therefore, the electron tunneling through the pentagons is expected to occur dominantly. It should also be noted that interference fringes are observed between the neighboring pentagons, indicating high coherency of electron waves emitted from adjacent pentagons.

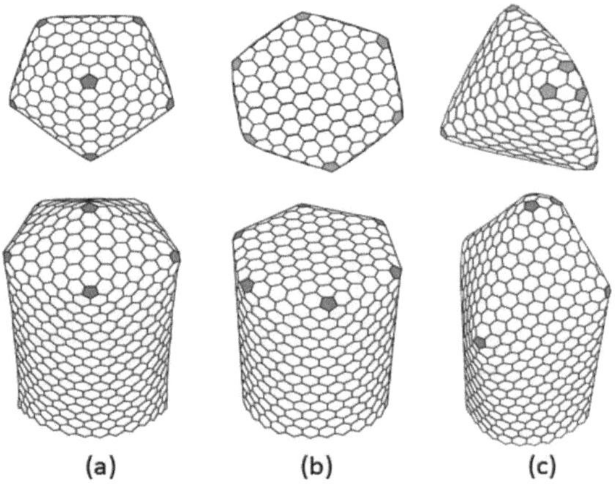

Figure 1.4 Structure models with (a) five-fold, (b) six-fold, and (c) low symmetries; Figs. (a) and (b) from [15].

Clear-cut pentagon patterns (e.g., Fig. 1.2) are observed for the well-graphitized CNT with clean surfaces. When residual gas molecules are adsorbed on the CNT cap, these molecules are observed as bright spots. Clean surface of a CNT tip is acquired by heating at about 1000 °C for few minutes in ultrahigh vacuum (e.g., $<10^{-7}$ Pa) by which adsorbates on CNT surfaces are desorbed. Adsorption of a gas molecule occurs preferentially on a pentagon site, where the electric field is the strongest and the field gradient is also highest. The polarization force acting on the adsorbate moves it in the direction of higher field, and thus the adsorbate is attracted at a pentagon site. Adsorption of a single molecule onto a clean pentagon brings about a bright spot in FEM images with a stepwise increase in emission current. When a molecule is desorbed, the FEM image and the emission current are recovered to those before the adsorption [17]. Adsorption and desorption of residual molecules on a surface of electrons emitter are the origin of current fluctuation. Dynamic behaviors of various gas molecules (hydrogen, nitrogen, oxygen, carbon monoxide, carbon dioxide, methane) adsorbed on a CNT cap and their effects on FEM images and emission current have been reported in Refs. [18, 19]. Interestingly, FEM images of adsorbates, N_2 [20] and CO_2 [21], show not only molecular structures with atomic resolution but also molecular motions on a CNT cap, though most of adsorbed molecules in the FEM images appear simple, bright spots (structure-less).

Theoretical studies on molecular adsorption effects on FE [22, 23], taking into accounts of electronegativity, work function, local field enhancement, electronic states, etc., predict emission current increase or decrease, contradictions between different theoretical predictions being found. Experimentally, however, all kinds of molecules brought about emission enhancement upon their adsorption on a CNT. Emission enhancement due to molecular adsorptions may be caused by a local field enhancement (an adsorbate is a kind of hump on a hemispherical cap) and/or by a resonance tunneling through a molecular level.

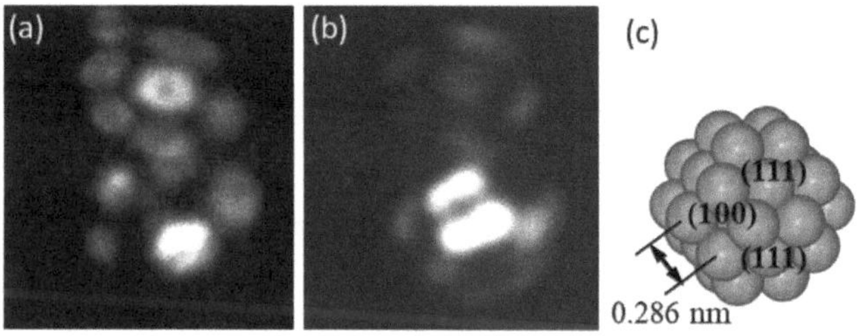

Figure 1.5 FEM images of (a) clean MWCNT caps, (b) an Al cluster on a cap of MWCNT, and (c) a structure model of the Al cluster with a cubo-octahedral form. Two caps of MWCNTs are observed in (a), and the Al cluster is deposited on the cap of the lower MWCNT [18].

Besides molecular adsorption, atomic clusters of metal (e.g., Al) deposited on a CNT cap exhibited interesting FEM images as shown in Fig. 1.5, which also suggests possibility of atomic resolution in FEM using CNT emitters [18]. An Al cluster shows a high symmetry image indicating a shape of cubooctahedron faceted by (111) and (100) surfaces, which is a crystal form characteristic of face-centered cubic metals [24]. A structure model of the Al cluster is shown in Fig. 1.5c. Atoms at the corners and edges look bright, while flat surfaces of (111) and (100) are dark in the FEM image. The distance between the bright spots at the corner of the cluster is estimated to be in the range from 0.28 to 0.31 nm, using the size of the carbon pentagon (approximately 0.25 nm in diameter) as a measure of magnification of the FEM image, under an assumption that the pentagon image originates from five carbon atoms comprising a pentagon. The distance between the bright spots in the FEM image is in good agreement with the nearest neighbor distance on the Al

(100) surface. In this case the magnification enhancement due to a small protrusion on a CNT cap seems not to be so large.

1.2.2 Thin MWCNT with a Cone-Shaped Tip

Slender MWCNTs (diameter ~10 nm) with a cone-shaped tip, as shown in Fig. 1.6, were produced by RF (radio frequency)-plasma heating of graphite in argon gas [25]. The cone angle is 19.2 degree, indicating the presence of five pentagons at the apex. The cone-shaped CNTs gave two types of FEM patterns: one is the "dim" (blurred) pattern and the other is the "pentagon" pattern consisting of clear pentagon rings as shown respectively in Fig. 1.7a,b. The latter type is typically observed for MWCNTs, as explained in the previous subsection, though the number of pentagonal rings is five in this case. Interference fringes are also observed in boundary regions between the neighboring pentagons. On the other hand, the "dim" pattern is similar to that observed for SWCNTs (see the next subsection). FEM patterns from SWCNTs resemble scanning tunneling microscope (STM) images of C_{60} fullerenes as pointed out by Dean and Chalamala [26], suggesting that the pattern reflects the local DOS of electrons on the tip of SWCNTs.

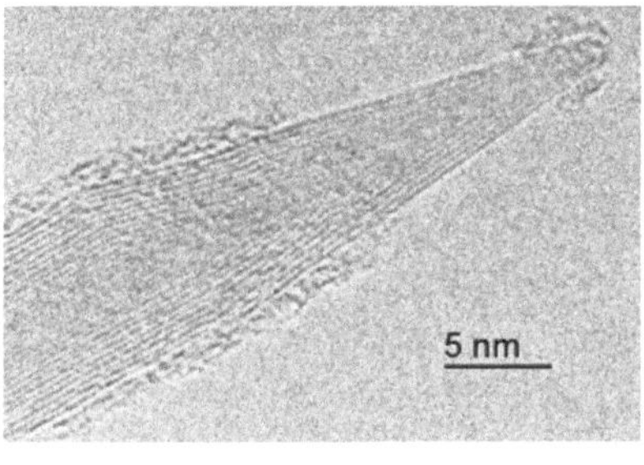

Figure 1.6 TEM image of a slender MWCNT with a cone-shaped tip [27].

A comparative TEM and FEM study on the cone-shaped MWCNTs has shown that the "pentagon" patterns originate from CNTs with apex radii R larger than ca. 2 nm, while the "dim" patterns originate from those with the smaller R [27]. This conclusion is consistent with the previous FEM studies for ordinal MWCNT and SWCNT; the diameters of ordinal MWCNTs are in the range from about 5 to 50 nm, and those of SWCNTs are in the range from about 1 to 2 nm.

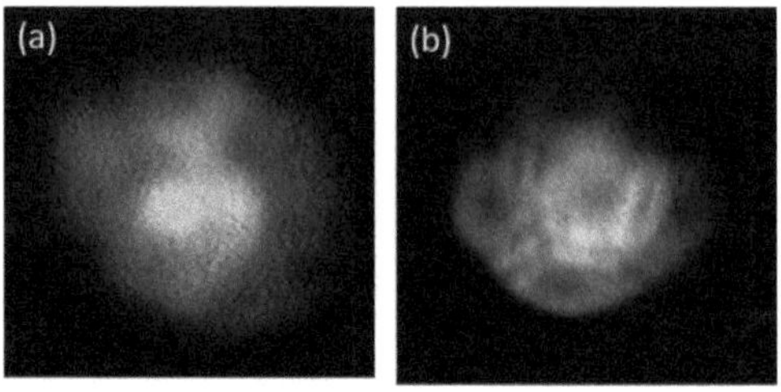

Figure 1.7 FEM images from cone-shaped MWCNTs showing (a) "dim" (blurred) pattern and (b) "pentagon" pattern [27].

When a CNT cap is approximated by a hemisphere and six pentagons are evenly distributed on the cap, the distances between adjacent pentagons become shorter for the narrower CNT. A closed tip (i.e., a curved graphene) of a CNT is considered as a part of a fullerene cage. When 12 pentagons are evenly distributed over a fullerene cage like C_{60}, the shape of which is a truncated icosahedron, the pentagon–pentagon separation $s_{\text{p-p}}$ is approximately equal to the radius of curvature of a closed graphene cage. In the case of a CNT cap, $s_{\text{p-p}}$ is on average approximated to the radius of the CNT body (the cylinder part), because the radius of curvature of a CNT cap is equal to the radius of the CNT body. Thus, the value of $s_{\text{p-p}}$ that differentiates the two types of FEM pattern is considered to be approximately 2 nm. That is to say, carbon pentagons are separately observed by FEM when the $s_{\text{p-p}}$ is larger than about 2 nm, while they are merged and blurred when the $s_{\text{p-p}}$ is smaller.

According to the argument on the spatial resolution of FEM [28, 29], resolutions of 0.2 and 0.35 nm are possible for emitters with tip radius $R = 1$ and 4 nm, respectively. The resolution of our FEM images from sharp tips ($R \simeq 1$ nm) approaches the atomic scale, but is not enough to resolve individual atoms on the CNT caps. Electron emission intensity in FEM depends on the two principal factors: one is the local field strength (tunneling probability) and the other is the local electron density of states at the Fermi level. When the tip radius is small, say $R \lesssim 2$ nm, CNT caps have a round shape as a whole like C_{60} fullerene, indicating that the electric field may appear evenly over the pentagon and hexagon sites on the caps. Therefore, FEM images of CNTs with small tip radii arguably reflect the surface electronic states of the CNT caps, as discussed by Dean and Chalamala [26].

When the tip radius is large, say $R \gtrsim 2$ nm, on the other hand, the separations between the adjacent pentagons $s_{\text{p-p}}$ are also large, which makes the topological character of pentagons strikingly clear; geometrical strain in a curved hexagon sheet is concentrated to the pentagons, so the pentagon sites extrude like vertices of a polyhedron as has been illustrated in literature [15, 30]. Electric field at the local protrusions is higher than that in the other flat regions on the CNT cap, and thus tunneling probability of electrons at the pentagon sites is higher than that in the other flat hexagon regions. Therefore, the electrons are preferentially emitted from pentagons on a CNT cap. Since the $s_{\text{p-p}}$ for the large tip radii ($R \gtrsim 2$ nm) is well within the reach of FEM resolution, the preferential emission from locally protruding pentagons makes them clearly visible. If an eigenstate of electrons is distributed over adjacent pentagons, electrons even though being emitted from different pentagons are coherent and interfere with each other and thus interference fringes are formed between the adjacent pentagons.

When the radius of CNT cap is too large, i.e., $R \gtrsim 50$ nm ($s_{\text{p-p}} \gtrsim$ 50 nm), on the contrary, interference fringes between the pentagons are rarely observed. Such a thick CNT is obtained as vapor-grown carbon fibers (VGCFs, Showa Denko Co.) as shown in Fig. 1.8a. A typical FEM image from a VGCF is shown in Fig. 1.8b, in which one fringe (indicated by an arrow) is observed between two pentagons which are considered to be closely located with each other.

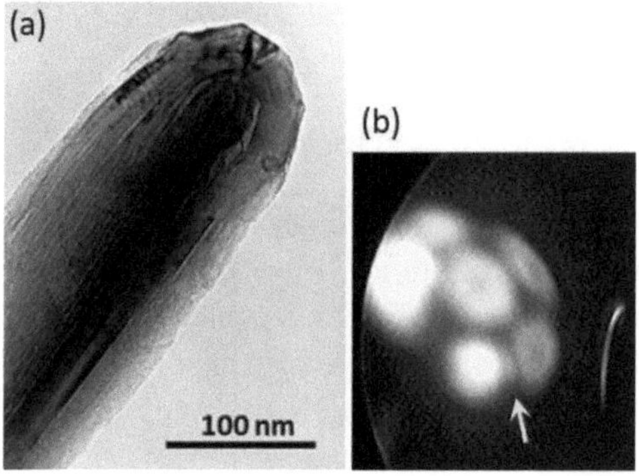

Figure 1.8 (a) TEM image of a vapor-grown carbon fiber (VGCF) and (b) FEM image from a VGCF.

1.2.3 SWCNTs

Metal catalysts are indispensable for the production of SWCNTs by CVD, arc discharge, and laser evaporation; SWCNTs grow from seed metal particles. Grown SWCNTs stick with each other and form bundles by van der Waals forces. Metal particles can be removed by acid treatment following oxidation (etching) of carbon layers coating metal particles. After the wet-chemical purification, SWCNTs are strongly bundled as shown in Fig. 1.9a. SWCNTs employed for FEM were prepared by arc discharge, and purified by the wet method [31]. A lump of SWCNTs consisting of numerous CNTs was fixed on top of a tungsten filament for FE experiment. An example of FEM images from a purified SWCNT bundle is shown in Fig. 1.9b, which exhibits a blurred image, being totally different from the clear pentagon pattern described above. Even though the emitter is a bundle of SWCNTs, only the most extruding CNT will emit electrons.

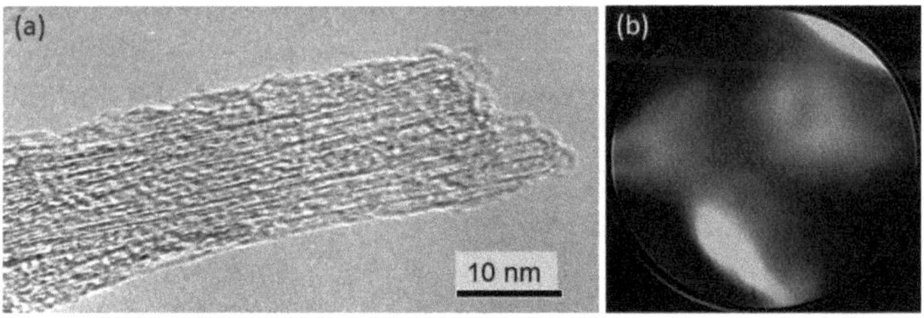

Figure 1.9 (a) TEM of a bundle of SWCNTs after wet purification processes and (b) FEM image from a bundle of purified SWCNTs.

1.2.4 MWCNT with an Open Tip

Open-ended MWCNTs were prepared by either wet oxidation with the aid of $CuCl_2$ intercalation [32] or dry oxidation (heating in dry air) [33] of as-prepared MWCNTs. A typical TEM picture of MWCNTs with open tips, prepared by oxidation of the former process, is shown in Fig. 1.10a. FEM images from the open-ended MWCNTs exhibit "doughnut-like" annular rings [34], as shown in Fig. 1.10b. A black region in the center of the ring corresponds to a hollow cavity of the CNT. The annular ring is continuous, i.e., no fine structure is observed. The continuous annular ring pattern is characteristic of the open MWCNT with a thick wall (thickness $\gtrsim$ 5 nm) and a large opening at the end, while a thin MWCNT consisting of a few layers with an open end shows a radially striped "daisy-like" FEM pattern as reported by Nakayama et al. [35]. As will be described in the later section, the "daisy-like" patterns are observed even for open-ended thick MWCNTs by using field ion microscopy (FIM). The origin of the striped "daisy-like" FEM patterns is presumably same as those observed for graphene emitters (see Chapter 7).

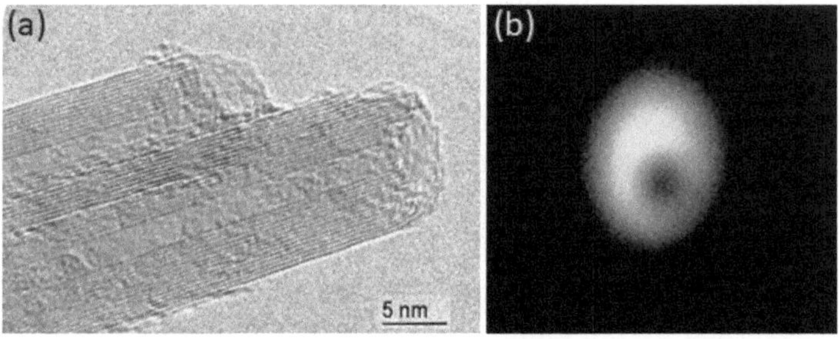

Figure 1.10 (a) TEM picture of open-ended MWCNTs, and (b) FEM image from open-ended MWCNTs.

In Table 1.3, CNT structures (cap structure and radius) and the corresponding FEM patterns are summarized.

Table 1.3 FEM patterns depending on cap radius and tip structure of CNT

CNT	**FEM pattern**
Capped CNT	
SWCNT, MWCNT with cap radius $R \lesssim 2$ nm	Dim pattern
MWCNT: ~ 2 nm < $R \lesssim 50$ nm	Pentagon pattern with interference fringes
MWCNT: $R \gtrsim 50$ nm	Separated pentagons without fringes or with only limited fringes
Open tip CNT	
thin MWCNT (wall thickness $\lesssim$ 5 nm)	Daisy-like pattern
thick MWCNT (wall thickness $\gtrsim$ 5 nm)	Doughnut-like pattern

1.3 Energy Spectra of Emitted Electrons

The energy spread of electrons emitted from FE sources is much narrower than that from thermionic electron sources (hot cathodes). The narrow energy spread is advantage for forming a finely focused electron beam, a high coherent electron wave, and so on. In a metal (e.g., W) field emitter, where emitted electrons come from electrons in the conduction band, the width of the energy distribution is defined by the occupied electronic density of states (DOS) around the Fermi level and the tunneling barrier (i.e., tunneling probability). Above the Fermi level, the tunneling probability increases but the population of electrons decreases very sharply. Below the Fermi level, the DOS increases but the tunneling probability decreases steeply. The resulting field-emission energy distribution (FEED) of the electrons has a characteristic shape predicted by the Fowler–Nordheim (F–N) theory [10]. The FEED has a peak at the Fermi level with exponential tails on both sides of the peak; the exponents are governed by the temperature of the metal on the high energy side and by the tunneling barrier width on the low energy side. The F–N theory is valid for field emission from a flat surface of an ideal metal. Therefore, deviations from the F–N energy distribution are encountered in emitters with atomic-scale protrusions and/or nonmetallic emitters [36].

The full width at half maximum (FWHM) of the FEED for a W cold emitter is typically about 0.2 eV. At an early stage of CNT field-emission studies, not only narrow energy distribution (0.21 eV [37], ca. 0.3 eV [38]) but also the presence of fine structures in FEED [38, 39] have been reported. These early FEED measurements are not guaranteed in respect that they were obtained from clean CNT surfaces (i.e., free from adsorbates). Oshima et al. carried out FEED measurements from clean MWCNT caps and from an adsorbate on the cap [40]. As described in the previous section, FEM patterns from a clean MWCNT cap exhibit pentagon rings. A FEED from a pentagon is shown in Fig. 1.11a. The tip temperature is room temperature and the voltage applied to the MWCNT is −780 V. A few features different from ordinary metal emitters are found. First, on the lower energy side of the main peak, a hump is observed at about 0.5 eV. Such fine structures below the Fermi level have been commonly observed for metal emitters with emission area of atomic scale (called "nanotips"

[41]), showing that the emission area of the present nanotube emitter is also in an atomic scale (i.e., size of a pentagon). Second, the descending slop of the main peak on the higher energy side is dull compared to the W cold emitter (Fig. 1.11c). Third, the main peak is rather broad, 0.3 eV (FWHM), compared to W cold emitters (about 0.2 eV). All these features were observed from all the clean pentagons. The origins of these features are not yet clear but the electronic DOS consisting of π-electrons and semimetallic nature of carbon nanotubes would be possible causes.

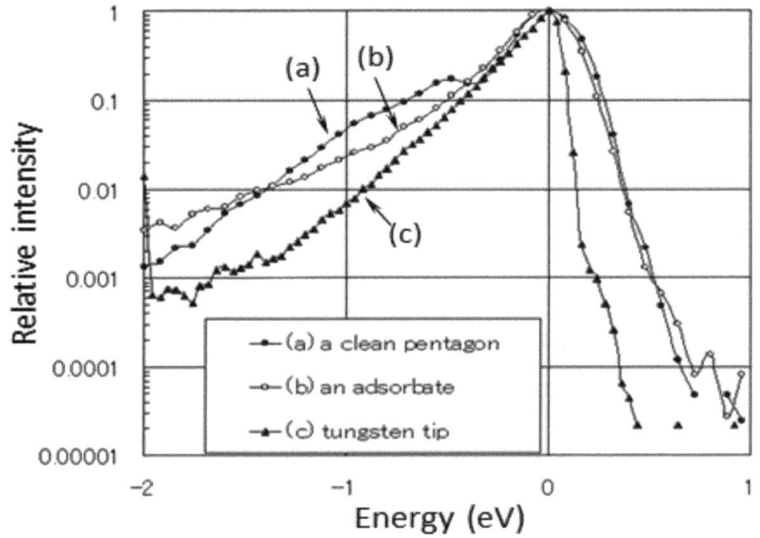

Figure 1.11 Field-emission energy distributions (FEEDs) from (a) a clean pentagon, (b) an adsorbate on a MWCNT cap, and (c) a conventional tungsten emitter. The origin of the abscissas (0 eV) is taken to the Fermi level [40].

In energy distributions of electrons emitted through an adsorbed molecule, the humps on the lower energy side disappeared though the other two features (i.e., broad main maximum and a gentle slope at the higher energy side) observed in the clean pentagon still retained (Fig. 1.11b). The molecular species adsorbed, not specified in this experiment, is inferred to be carbon monoxide because CO was one of the major residual gas species in the ultrahigh vacuum (UHV) chamber and the behavior of the adsorbate on the CNT is similar to that of CO [19].

1.4 Field-Emission Study of CNT in TEM

Since the first measurements demonstrating the excellent field-emission properties of CNT, a variety of CNT-based FE devices such as flat panel displays, vacuum fluorescence lamps, electron sources for X-ray tubes, microwave device, electron microscopes, and so on have been experimentally manufactured. For understanding their operation and improving their performance, it is important to directly observe the behavior of individual CNT emitter in electric field and to measure the electron emission properties with simultaneous observation of their morphology and structures.

Pioneering studies for microscopic observation of CNTs behavior in an electric field were carried out by Wang et al. [42, 43] using TEM and by Wei et al. [44] using SEM. Both of them used MWCNTs as field emitters. Under moderate electric fields, MWCNTs bent and reoriented reversibly, but under high-field conditions sufficient to extract large FE currents, MWCNTs were suffered from significant structural damages; for example, peeling of the outer layers of the MWCNT and subsequent evaporation likely due to the Joule heating, resulting in a thinned and shortened MWCNT. Jin et al. [45] observed that a single MWCNT could emit a current up to a few tens of microampere but prolonged large emission led to sudden emission failure resulting from a drastic structure damage. Other studies by in situ TEM of CNT emitters include electron holography carried out by Cumings et al. [46] that revealed the magnitude and spatial distribution of the electric field surrounding an individual field-emitting MWCNT. The effect of tip structures on electron emission was investigated by Wang et al. [47], and the relationship between the gap distance (from a CNT tip to an anode surface) and the emission current was also shown by Asaka et al. [48]. Since the aforementioned TEM studies are reviewed in Ref. [49], these are not rewritten here. Other intriguing phenomena of CNT field emitters observed by in situ TEM, self-oscillation of CNT during field emission, are briefly described here. A more detailed and comprehensive review is given in Chapter 2 by Ayari et al.

The self-oscillation phenomena in field emission are encountered for MWCNTs [12, 50, 51] and SiC nanowires [52]. A series of TEM images in Fig. 1.12a–f show the self-oscillation of a MWCNT from a start to a cessation during the increase and decrease of the applied voltage. The corresponding FE current (I_e) and applied voltage (V_a) are shown as a function of time in Fig. 1.12g. As increasing the applied bias voltage, the self-oscillation suddenly occurs above a certain threshold voltage, and the oscillation ceases by decreasing the bias voltage at a certain voltage. The threshold voltages where the oscillation starts and stops do not necessarily coincide; the starting voltage is often higher than the ceasing one, and sometimes the starting voltage is lower or nearly same as the ceasing one. This vague behavior (ill reproducibility) may be due to the large fluctuation of FE current. Long MWCNTs were apt to show self-oscillation: MWCNTs longer than 3 μm always exhibited the self-oscillation. Furthermore, field emission of electrons is requisite for the generation of self-oscillation. When the bias-voltage was reversed, i.e., positive voltage was applied to a MWCNT, no oscillation was observed.

The natural frequency (eigenfrequency) of a cantilever beam with circular cross-section under free vibration is written as:

$$f_n = \frac{k_n^2}{2\pi l^2}\sqrt{\frac{EI}{\rho A}}\ [\text{Hz}] \tag{1.2}$$

where k_n = 1.875 for the first mode vibration, E is the Young's modulus of the beam material, I is the moment of inertia of the beam cross-section, A is the cross-section area, l is the length, and ρ is the material density. Mechanical eigenfrequencies of a MWCNT with diameter 10–20 nm are in the range from about 1 to 25 MHz, increasing with the decrease of length from 10 to 3 μm. Compared to this high frequency of the MWCNT free vibration, the dominant frequency component in FE current oscillation is rather low, e.g., ca. 50 kHz for a MWCNT of 3.8 μm in length and 20 nm in diameter [53].

A CNT emitter under an electric field experiences electrostatic force and a mechanical restoring force. Deflection of the CNT tip is derived from a balance between an electrostatic and an elastic force. When an unbalance between the electric force and the restoring force occurred suddenly by, for example, fluctuation in FE current, electromechanical vibration of the CNT would be induced. Based on mechanical vibration coupled with an electric circuit contribution, a few mechanisms explaining the self-oscillation of CNT and nanowire in an electric field are proposed [51, 52, 54]. Nonlinear nature of the self-oscillation is predicted, but there still remain some phenomena that cannot be explained by the proposed mechanisms.

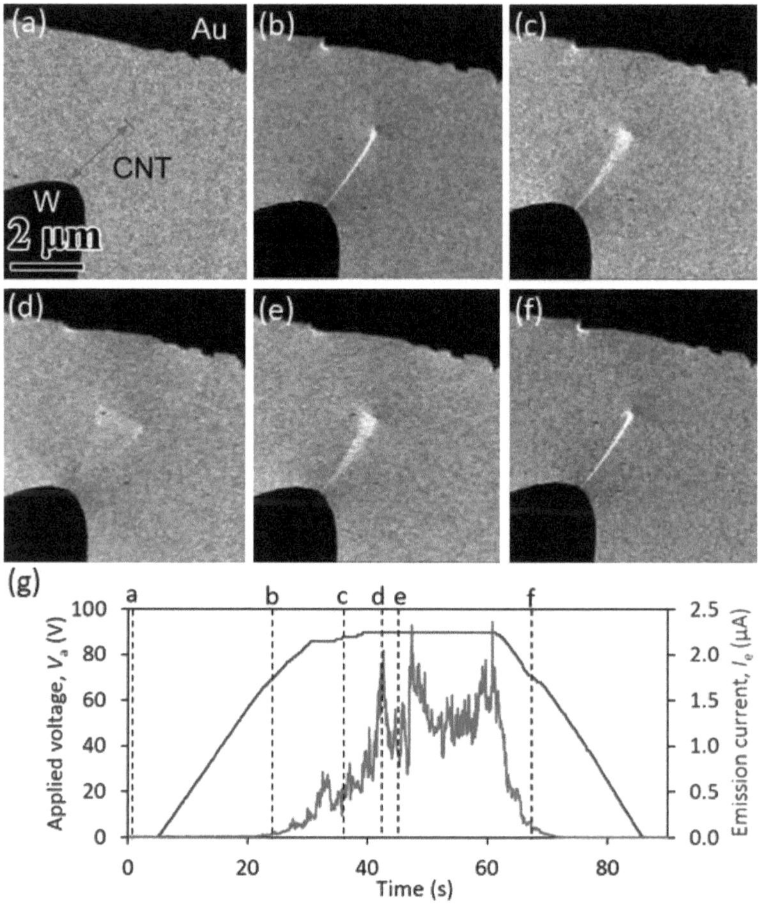

Figure 1.12 (a)–(f) TEM images showing an MWCNT on the tip of W needle before, during, and after the self-oscillation. (g) FE current (I_e) and applied voltage (V_a) as a function of time. Letters "a" to "f" at the upper side of the frame in (g) show the times when TEM images (a) to (f) were taken, respectively.

The self-oscillation of CNT emitters is detrimental to electron sources such as electron microscopes and X-ray tubes, which need stable and constant emission, but the phenomena may render CNT emitters applicable to other active elements (autonomous electromagnetic device) such as high-frequency radio generators [55], micromass sensors [56, 57], and so on.

1.5 FIM of CNT

There have been a number of FEM studies showing "pentagon patterns" that indicate the presence of carbon pentagon rings on a cap of CNT and the special role in the field emission of electrons, as described in the previous sections. Compared with FEM studies, FIM of CNT has not been widely studied. Reasons for the limited study may be partially the difficulty of FIM experiments and the additional requirement for apparatuses (ultrahigh vacuum, gas-inlet, cryogenic system, etc.). In this section, FIM studies on MWCNTs with capped and open-ended (or broken) tips and SWCNT as well as DWCNT bundles are reviewed.

The CNT emitter and a microchannel plate with a phosphor screen for observation of FIM images, being placed about 40 mm in front the CNT emitter, are installed in an ultrahigh vacuum chamber with a base pressure less than 10^{-6} Pa. The CNT emitter fixed on a hairpin-shaped tungsten filament (0.15 mm in diameter) was cooled down to 50–80 K by a helium refrigerator or liquid nitrogen. The emitter was executed the flash heating at about 1000 °C for several seconds to a few tens seconds, and then imaging gases (e.g., Ne, He, a mixture of He and H_2) were introduced up to a desired pressure in the range from 3×10^{-4} to 1×10^{-2} Pa. Electric voltages applied to the emitter was varied typically from 1.0 to 4 kV, depending on the emitters (curvature of a CNT cap) and the pressure of the imaging gas.

1.5.1 Capped MWCNT

In Fig. 1.13a,b, an FEM image of a capped MWCNT produced by arc-discharge and the corresponding FIM image of the same MWCNT are shown, respectively. The FEM image displays a typical pentagon pattern, though the number of observed pentagons is five in this image. In the corresponding FIM image, five clusters consisting of fine spots are observed as indicated by arrows. These fine spots in each cluster are considered to correspond with carbon atoms at a pentagon ring, where the electric field is strongest than the other flat hexagon regions (because the corner atoms are protruded) and thus imaging gases are effectively ionized. Another example of FIM images, demonstrating that these fine spots originate from carbon atoms at pentagons, is shown in Fig. 1.14 [58], which is measured from a capped MWCNT. In Fig. 1.14a, a cluster of bright dots arranged in a pentagon is observed at the lower-right part of the picture, corresponding to a five-membered ring of carbon atoms. Another cluster of bright spots, in which only three spots are observed because the rim of the observation screen partially shadowed the cluster, is recognized at the upper-right of the picture. After taking this FIM picture at an emitter voltage of 3.0 kV, the voltage was raised to 3.6 kV. Then, the FIM image changed as shown in Fig. 1.14b, in which large bright spots appear instead of fine spots. The image resolution is not as high as that in Fig. 1.14a, though each large spot has a variation of contrast within the spot. The resolution and contrast of FIM sensitively depend on imaging conditions especially the emitter voltage, as exemplified in Fig. 1.15.

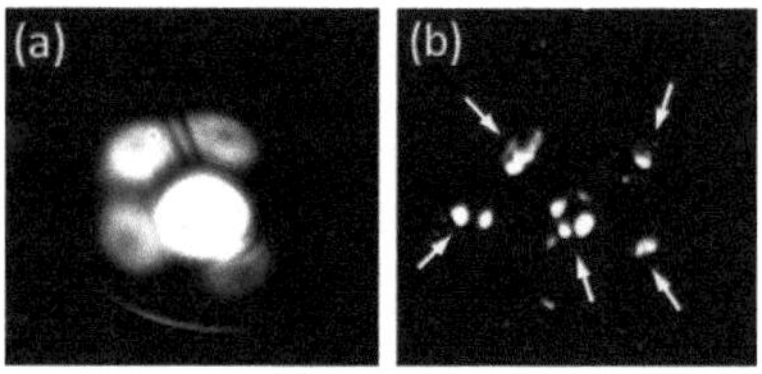

Figure 1.13 (a) FEM image (emitter voltage V_E = −700 V, at room temperature) of a capped MWCNT produced by arc-discharge technique and (b) the corresponding FIM image (V_E = 2.7 kV, Ne = 8.4×10^{-4} Pa, at liq. N_2 temperature).

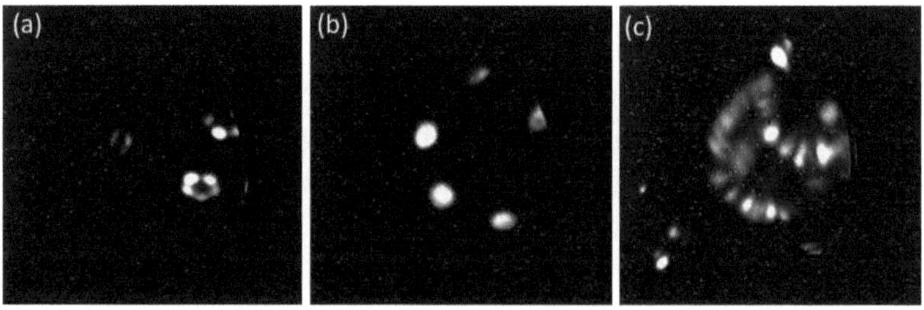

Figure 1.14 FIM images of a capped MWCNT at different emitter voltages: (a) 2.7 kV, (b) 2.8 kV, and (c) 3.2 kV. He = 1.8×10^{-5} Torr and H_2 = 1.6×10^{-5} Torr, at 70 K [58].

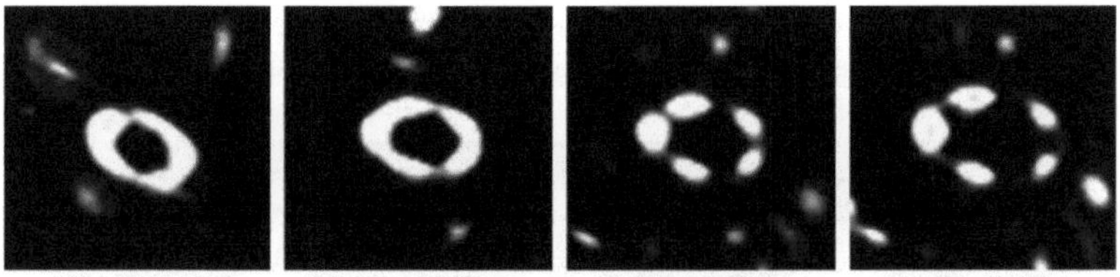

Figure 1.15 FIM images of a carbon pentagon at the emitter voltage V_E from 3.0 to 3.6 kV. Ne 3.3×10^{-4} Pa, at liq. N_2 temperature.

1.5.2 Broken Tip and Open-Ended MWCNT

At a higher emitter voltage, the FIM image utterly changed to the one as shown in Fig. 1.14c, spots being arranged annularly and prolonged radially. This "daisy-like" pattern is frequently encountered at the final stage with the high applied voltages. During the FIM experiment an extremely high electric field on the order of 10 V/nm is generated on the top of a CNT, which brings about the field evaporation of carbon atoms and clusters (see Chapter 18). The daisy-like pattern indicates a breakage of the CNT cap and the formation of holes on graphite layers after the field evaporation. This conclusion is supported by the FIM observations of open-ended MWCNTs that were subjected to the oxidation treatment prior to the FIM measurement. As revealed by TEM, which is already shown in the previous section, the caps of MWCNTs are etched and have open structures with free edges of graphene layers. Figure 1.16 shows FIM images obtained from two different open-ended MWCNTs that were subjected to the oxidation treatment. The FIM image of Fig. 1.16a exhibits the daisy pattern, being the same as that in Fig. 1.14c. In the other FIM image of Fig. 1.16b, the streaking of spots in the radial direction is exaggerated, and double spots (dipole spots) are also observed as indicated by arrows. Noticeable streaking and dipole spots are commonly observed in FIM of graphene emitters, showing that these characters are responsible from graphene edges. Details of the origin of the daisy pattern are presented in Chapter 7.

Related to the breakage of CNT caps under very high electric field, monoatomic and polyatomic carbon ions are emitted via field evaporation. Mass analysis of field-evaporated ions from CNT emitters showed that C_{20}^+ cluster ions are emitted with prominent intensity, details of which are described in Chapter 18.

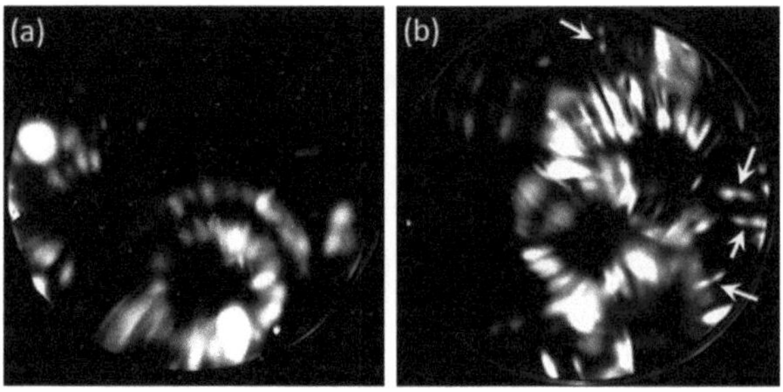

Figure 1.16 FIM images of open-ended MWCNTs: (a) V_E = 2.1 kV, H_2 = 1.0 × 10^{-4} Torr, at 70 K and (b) V_E = 4.0 kV, He = 6.0 × 10^{-6} Torr and H_2 = 5.4 × 10^{-5} Torr, at 70 K.

1.5.3 SWCNT and DWCNT Bundles

Figure 1.17 shows FIM images of SWCNTs at two different emitter voltages. The SWCNTs were prepared using chemical purification processes, and they were bundled (Fig. 1.9). A few bright and several faint spots are observed in Fig. 1.17a taken at a lower emitter voltage (1.8 kV). In Fig. 1.17b, taken at a higher voltage (2.1 kV), double spots (dipole spots) appear as indicated by arrows. It is notable that the double spots are also observed for open-ended MWCNTs (Fig. 1.16b), suggesting the breakage of SWCNT caps at high applied voltage.

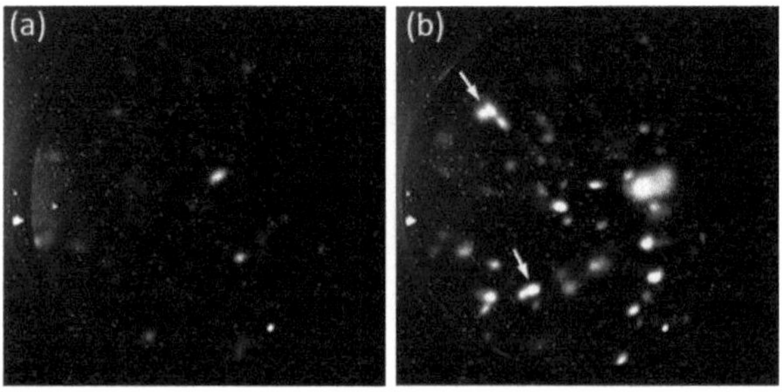

Figure 1.17 FIM images from a bundle of SWCNTs at different emitter voltages: (a) 1.8 kV and (b) 2.1 kV. H_2 = 8.8 × 10^{-5} Torr, at 70 K.

An FIM image from a bundled DWCNTs, which were prepared by arc discharge with metal catalysts [59] and were in an as-grown form (i.e., without purification), is shown in Fig. 1.18. The FIM image of DWCNTs exhibits arced spots or a small circle as indicated by arrows, in addition to small spots and streaks. The circular pattern seemingly reflects opening of a DWCNT cap.

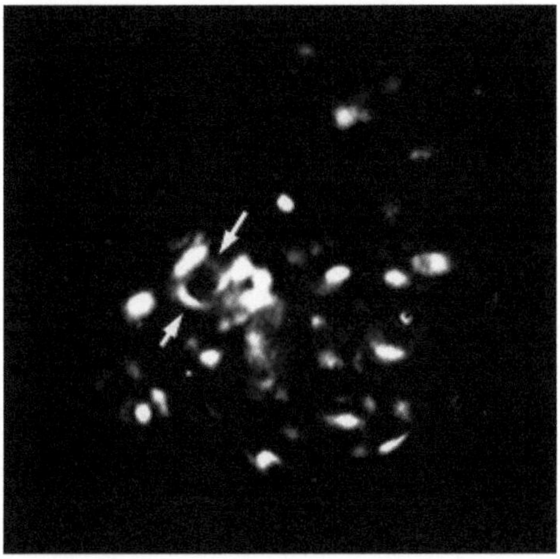

Figure 1.18 FIM image from a bundle of DWCNTs. V_E = 3.6 kV, Ne 1.4 × 10^{-4} Pa, at liq. N_2 temp.

1.6 Summary and Conclusion

Field electron and ion emission microscope images from various types of carbon nanotubes (single-walled, double-walled, and multiwall CNTs with capped or open ends) are presented, and the origins of the images are discussed in correlation with CNT structures. As the pentagon images of capped CNTs suggest, FEM of a nanoscale tip, like CNT emitter, shows high spatial resolution, possibly in atomic scale. Possible atomic images of adsorbates and a deposited Al cluster on CNT emitters are presented and discussed. FIM from capped CNTs shows atomically resolved carbon pentagons. During FIM measurement, the FIM image of carbon pentagons suddenly changed to a daisy-like pattern, indicating a breakage and opening of a CNT cap under high electric field. Arrays of fine streaks in the daisy pattern show features commonly found for FIM images of graphene edges (see Chapter 7).

Acknowledgments

The author's research on field emission from nanocarbon materials has been financially supported by JSPS KAKENHI Grant Numbers JP18H01882 and JP23656060.

References

1. Iijima, S. and Ichihashi, T. (1993). Single-shell carbon nanotubes of 1-nm diameter, *Nature*, **363**, pp. 603–605.
2. Bethune, D. S., Kiang, C. H., de Vries, M. S., Gorman, G., Savoy, R., Vazquez, J. and Beyers, R. (1993). Cobalt-catalysed growth of carbon nanotubes with single-atomic-layer walls, *Nature*, **363**, pp. 605–607.
3. Saito, Y., Yoshikawa, T., Okuda, M., Fujimoto, N., Sumiyama, K., Suzuki, K., Kasuya, A. and Nishina, Y. (1993). Carbon nanocapsules encaging metals and carbides, *J. Phys. Chem. Solids*, **54**, pp. 1849–1860.
4. Iijima, S. (1991). Helical microtubules of graphitic carbon, *Nature*, **354**, pp. 56–58.
5. Zheng, L. X., O'Connel, M. J., Doorn, S. K., Liao, X. Z., Zhao, Y. H., Akhadov, E. A., Hoffbauer, M. A., Roop, B. J., Jia, Q. X., Dye, R. C., Peterson, D. E., Huang, S. W. and Liu, J. (2004). Ultralong single-wall carbon nanotubes, *Nat. Mater.*, **3**, pp. 673–676.
6. Yun, Y., Shanov, V., Tu, Y., Subramaniam, S. and Schulz, M. J. (2006). Growth mechanism of long aligned multiwall carbon nanotube arrays by water-assisted chemical vapor deposition, *J. Phys. Chem. B*, **110**, pp. 23920–23925.
7. Bachilo, S. M., Strano, M. S., Kittrell, C., Hauge, R. H., Smalley, R. E. and Weisman, R. B. (2002). Structure-assigned optical spectra of single-walled carbon nanotubes, *Science*, **298**, pp. 2361–2366.
8. Bachtold, A., Fuhrer, M. S., Plyasunov, S., Forero, M., Anderson, E. H., Zettl, A. and McEuen, P. L. (2000). Scanned probe microscopy of electronic transport in carbon nanotubes, *Phys. Rev. Lett.*, **84**, pp. 6082–6085.
9. Saito, Y. (2010). *Carbon Nanotube and Related Field Emitters, Fundamental and Applications*, ed. Saito, Y., Chapter 1 "Structures and Synthesis of Carbon Nanotubes," (Wiley-VCH, Weinheim), pp. 3–14.

10. Gomer, R. (1961). *Field Emission and Field Ionization*, Chapter 1 "Theory of Field Emission," (Harvard University Press, Cambridge, MA), pp. 1–31.

11. Nakahara, H., Ichikawa, S., Ochiai, T., Kusano, Y. and Saito, Y. (2011). Carbon nanotube electron source for field emission scanning electron microscopy, *e-J. Surf. Sci. Nanotechnol.*, **9**, pp. 400–403.

12. Saito, Y., Seko, K. and Kinoshita, J. (2005). Dynamic behavior of carbon nanotube field emitters observed by in situ transmission electron microscopy, *Diam. Relat. Mater.*, **14**, pp. 1843–1847.

13. Saito, Y., Hamaguchi, K., Hata, K., Tohji, K., Kasuya, A., Nishina, Y., Uchida, K., Tasaka, Y., Ikazaki, F. and Yumura, M. (1998). Field emission from carbon nanotubes; Purified single-walled and multi-walled tubes, *Ultramicroscopy*, **73**, pp. 1–6.

14. Saito, Y., Yoshikawa, T., Inagaki, M., Tomita, M. and Hayashi, T. (1993). Growth and structure of graphitic tubules and polyhedral particles in arc-discharge, *Chem. Phys. Lett.*, **204**, pp. 277–282.

15. Saito, Y., Hata, K. and Murata, T. (2000). Field emission patterns originating from pentagons at the tip of a carbon nanotube, *Jpn. J. Appl. Phys.*, **39**, pp. L271–L272.

16. Tamura, R. and Tsukada, M. (1995). Electronic states of the cap structure in the carbon nanotube, *Phys. Rev. B*, **52**, pp. 6015–6026.

17. Hata, K., Takakura, A. and Saito, Y. (2001). Field emission microscopy of adsorption and desorption of residual gas molecules on a carbon nanotube tip, *Surface Sci.*, **490**, pp. 296–300.

18. Saito, Y. (2009). Atomic detail observation of adsorbed molecules and metal clusters on carbon nanotube electron emitters, *Coord. Chem. Rev.*, **253**, pp. 2912–2919.

19. Hata, K., Takakura, A. and Saito, Y. (2003). Field emission from multiwall carbon nanotubes in controlled ambient gases, H_2, CO, N_2 and O_2, *Ultramicroscopy*, **95**, pp. 107–112.

20. Kawakita, K., Hata, K., Sato, H. and Saito, Y. (2006). Development of microfocused X-ray source by using carbon nanotube field emitter, *J. Vac. Sci. Tech. B*, **24**, pp. 950–952.

21. Kishimoto, Y. and Hata, K. (2008). Behaviors of single CO_2 molecule on pentagon at carbon nanotube tip observed by field emission microscopy, *Surf. Interface Anal.*, **40**, pp. 1669–1672.

22. Park, N., Han, S. and Ihm, J. (2001). Effects of oxygen adsorption on carbon nanotube field emitters, *Phys. Rev. B*, **64**, 125401.

23. Li, Z. and Wang, C.-Y. (2006). First-principles study of field emission properties of gas adsorption on the carbon nanotubes, *Chem. Phys.*, **330**, pp. 417–422.

24. Kimoto, K. and Nishida, I. (1977). The crystal habits of small particles of aluminium and silver prepared by evaporation in clean atmosphere of argon, *Jpn. J. Appl. Phys.*, **16**, pp. 941–948.

25. Koshio, A., Yudasaka, M. and Iijima, S. (2002).Metal-free production of high-quality multi-wall carbon nanotubes, in which the innermost nanotubes have a diameter of 0.4 nm, *Chem. Phys. Lett.*, **356**, pp. 595–600.

26. Dean, K. A. and Chalamala, B. R. (1999). Field emission microscopy of carbon nanotube caps, *J. Appl. Phys.*, **85**, pp. 3832–3836.

27. Saito, Y., Tsujimoto, Y., Koshio, A. and Kokai, F. (2007). Field emission patterns from multiwall carbon nanotubes with a cone-shaped tip, *Appl. Phys. Lett.*, **90**, 213108.

28. Rose, D. J. (1956). On the magnification and resolution of the field emission electron microscope, *J. Appl. Phys.*, **27**, pp. 215–220.

29. Brodie, I. (1978). The visibility of atomic objects in the field electron emission microscope, *Surface Sci.*, **70**, pp. 186–196.

30. Kobayashi, K. (2000). Superstructure induced by a topological defect in graphitic cones, *Phys. Rev. B*, **61**, pp. 8496–8500.

31. Tohji, K., Goto, T., Takahashi, H., Shinoda, Y., Shimizu, N., Jeyadevan, B., Matsuoka, I., Saito, Y., Kasuya, A., Ohsuna, T., Hiraga, K. and Nishina, Y. (1996). Purifying single-walled nanotubes, *Nature*, **383**, p. 679.

32. Ikazaki, F., Ohshima, S., Uchida, K., Kuriki, Y., Hayakawa, H., Yumura, M., Takahashi, K. and Tojima, K. (1994). Chemical purification of carbon nanotubes by use of graphite intercalation compounds, *Carbon*, **32**, pp. 1539–1542.

33. Ebbesen, T. W., Ajayan, P. M., Hiura, H. and Tanigaki, K. (1994). Purification of nanotubes, *Nature*, **367**, p. 519.

34. Saito, Y., Hamaguchi, K., Hata, K., Uchida, K., Tasaka, Y., Ikazaki, F., Yumura, M., Kasuya, A. and Nishina, Y. (1997). *Nature*, **389**, pp. 554–555.

35. Tanaka, H., Akita, S., Pan, L. and Nakayama, Y. (2004). Daisylike field-emission images from standalone open-ended carbon nanotube, *Jpn. J. Appl. Phys.*, **43**, pp. L197–L199.

36. Purcell, S. T., Binh, V. T. and Garcia, N. (1995). 64 meV measured energy dispersion from cold field emission nanotips, *Appl. Phys. Lett.*, **67**, pp. 436–438.

37. de Heer, W. A., Bonard, J. -M., Fauth, K., Chatelain, A., Forro, L. and Ugarte, D. (1997). Electron field emitters based on carbon nanotube films, *Adv. Mater.*, **9**, pp. 87–89.

38. Fransen, M. J., von Rooy, T. L. and Kruit, P. (1999). Field emission energy distributions from individual multiwalled carbon nanotubes, *Appl. Surface Sci.*, **146**, pp. 312–327.

39. Lovall, D., Buss, M., Graugnard, E., Andres, R. P. and Reifenberger, R. (2000). Electron emission and structural characterization of a rope of single-walled carbon nanotubes, *Phys. Rev. B*, **61**, pp. 5683–5691.

40. Takakura, A., Hata, K., Saito, Y., Mastuda, K., Kona, T. and Oshima, C. (2003). Energy distributions of field emitted electrons from a multi-wall carbon nanotube, *Ultramicroscopy*, **95**, pp. 139–143.

41. Binh, V. T., Purcell, S. T., Garcia, N. and Doglioni, J. (1992). Field-emission electron spectroscopy of single-atom tips, *Phys. Rev. Lett.*, **69**, pp. 2527–2530.

42. Wang, Z. L., Poncharal, P. and de Heer, W. A. (2000). Measuring physical and mechanical properties of individual carbon nanotubes by in situ TEM, *J. Phys. Chem. Solid*, **61**, pp. 1025–1030.

43. Wang, Z. L., Gao, R. P., de Heer, W. A. and Poncharal, P. (2002). In situ imaging of field emission from individual carbon nanotubes and their structural damage, *Appl. Phys. Lett.*, **80**, pp. 856–858.

44. Wei, Y., Xie, C., Dean, K. A. and Coll, B. F. (2001). Stability of carbon nanotubes under electric field studied by scanning electron microscopy, *Appl. Phys. Lett.*, **79**, pp. 4527–4529.

45. Jin, C., Wang, J., Wang, M., Su, J. and Peng, L.-M. (2005). In-situ studies of electron field emission of single carbon nanotubes inside the TEM, *Carbon*, **43**, pp. 1026–1031.

46. Cumings, J., Zettl, A., McCartney, M. R. and Spence, J. C. H. (2002). Electron holography of field-emitting carbon nanotubes, *Phys. Rev. Lett.*, **88**, pp. 056804.

47. Wang, M. S., Wang, J. Y. and Peng, L.-M. (2006). Engineering the cap structure of individual carbon nanotubes and corresponding electron field emission characteristics, *Appl. Phys. Lett.*, **88**, pp. 243108.

48. Asaka, K., Nakahara, H. and Saito, Y. (2008). Nanowelding of a multiwalled carbon nanotube to metal surface and its electron field emission properties, *Appl. Phys. Lett.*, **92**, pp. 023114.

49. Asaka, K. and Saito, Y. (2010). *Carbon Nanotube and Related Field Emitters: Fundamentals and Applications*, ed. Saito, Y., Chapter 9 "In-

situ Transmission Electron Microscopy of CNT Emitters," (Wiley-VCH, Weinheim), pp. 109–117.

50. Purell, S. T., Vincent, P., Journet, C. and Binh, V. T. (2002). Tuning of nanotube mechanical resonances by electric field pulling, *Phys. Rev. Lett.*, **89**, pp. 276103.

51. Weldon, J. A., Aleman, B., Sussman, A., Gannett, W. and Zettl, A. K. (2010). Sustained mechanical self-oscillations in carbon nanotubes, *Nano Lett.*, **10**, pp. 1728–1733.

52. Barois, T., Perisanu, S., Vincent, P., Purcell, S. T. and Ayari, A. (2013). Role of fluctuations and nonlinearities on field emission nanomechanical self-oscillators, *Phys. Rev. B*, **88**, pp. 195428.

53. Kato, T. (2017). In situ TEM on oscillation phenomena of carbon nanotubes during field emission, *Master degree thesis of Nagoya University*.

54. Kleshch, V. I., Obraztsov, A. N. and Obraztsova, E. D. (2010). Electromechanical self-oscillations of carbon nanotube field emitter, *Carbon*, **48**, pp. 3895–3900.

55. Jensen, K., Weldon, J., Garcia, H. and Zettl, A. (2007). Nanotube radio, *Nano Lett.*, **7**, pp. 3508–3511.

56. Jensen, K., Kim, K. and Zettl, A. (2008). An atomic-resolution nanomechanical mass sensor, *Nat. Nanotechnol.*, **3**, pp. 533–537.

57. Nishio, M., Sawaya, S., Akita, S. and Nakayama, Y. (2005). Carbon nanotube oscillators toward zeptogram detection, *Appl. Phys. Lett.*, **86**, pp. 133111.

58. Saito, Y., Mizushima, R. and Hata, K. (2002). Field ion microscopy of multiwall carbon nanotubes: Observation of pentagons and cap breakage under high electric field, *Surface Sci.*, **499**, pp. L119–L123.

59. Saito, Y., Nakahira, T. and Uemura, S. (2003). Growth conditions of double-walled carbon nanotubes in arc discharge, *J. Phys. Chem. B*, **107**, pp. 931–934.

Chapter 2

Electromechanical Self-Oscillations of Carbon Field Emitters

A. Ayari, P. Vincent, S. Perisanu, P. Poncharal, and S. T. Purcell
Université Claude Bernard Lyon 1, CNRS, Institut Lumière Matière, F-69622, Villeurbanne, France
anthony.ayari@univ-lyon1.fr

The spontaneous generation of AC currents and vibrations in field emitters based on carbon compounds is now a well-studied effect. We give in this chapter an overview of the different experimental realizations of these self-oscillations and of the different proposed mechanisms. We specially emphasize their state-of-the-art performances in terms of highest current, frequency, and stability for future applications.

2.1 Introduction

The invention of *triode* in 1906 marked the beginning of electronics and radio communications. This led around 1912 to the development of *radio receiver signal amplifiers*. Though it was just the simple

Nanostructured Carbon Electron Emitters and Their Applications
Edited by Yahachi Saito

ISBN 978-981-4877-62-6 (Hardcover), 978-1-003-14199-0 (eBook)
www.jennystanford.com

addition of a third electrode to a vacuum tube diode, this was at the origin of a momentous revolution for modern society. The fact that the current depends on two independent variables, the grid voltage and the cathode voltage, opened the path to manipulate electrical information and generate logical operations.

Though subsidiary at first, it appeared that in a triode circuit, electrical oscillations could be generated by a proper design but sometimes because of parasitic electrical elements as well. This spontaneous generation of an AC signal from a DC biased electrical circuit is the manifestation of a general phenomenon called *self-oscillation*. The importance of such phenomenon could not be better summarized by the following quotation from the bible of electronics by Horowitz and Hill: "A device without an oscillator either doesn't do anything or expects to be driven by something else (which probably contains an oscillator). It is not an exaggeration to say that an oscillator of some sort is as essential an ingredient in electronics as a regulated supply of DC power." [1].

In the past decades, the development of micro and then nano-*electromechanical* systems resulted in the emergence of field-emission devices in a diode configuration that shared the physical properties of triodes. In such devices the role of the grid was replaced by the presence of an additional degree of freedom: the displacement of the apex of the emitter. In this chapter, we will review the successful fabrication of self-oscillating electromechanical field emitter, describe the physical mechanism underlying this intriguing effect, and emphasize their performances as well as limitations.

2.2 Introduction to Electromechanical Self-Oscillators

2.2.1 Self-Oscillation

Before defining the term *self-oscillation* coined by Aleksandr Aleksandrovich Andronov [2], we will introduce complementary concepts that should not be confused between each other. The starting point of this paragraph is the idyllic model of the *harmonic oscillator*. Such a system is described by a simple linear second order differential equation and can oscillate indefinitely at a fixed frequency, amplitude, and phase given by some initial conditions. A more realistic model is the *damped oscillator* that takes into account the interaction of the oscillator with its environment leading to friction and nonconservation of energy. In order to maintain the amplitude of oscillation, the oscillator is usually driven by an external source (most of the time a self-oscillator). In reality, there exist other means to fight against damping. For instance, even in the field of nano- and micro-electromechanical systems, hundreds of them have been listed in this review [3]. One that is most often confused with self-oscillations is the *parametric instability* governed by the Mathieu equation [4]. The parametric oscillator is often used to describe the motion of a swing where the amplitude of motion can be amplified by applying a driving at twice the resonance frequency. Beside the fact that most kids do not parametrically pump a swing, this configuration has two major drawbacks compared to a self-oscillator: it needs an external oscillating driver and the phase of the oscillation is fixed by the driver.

A self-oscillator can be defined as a nonlinear dissipative system whose oscillations are maintained by drawing energy from a nonoscillating source. The amplitude fluctuates around a fixed nonzero value, the frequency is fixed, and the phase is free to drift. These three properties are essential for a self-oscillator. An AC generator, a clock, the heartbeat, a Gunn diode, the Belousov–Zhabotinsky reaction, and the laser are among the many examples of self-oscillators [5]. The self-oscillating triode is even considered by Willis Eugene Lamb Jr. (Nobel Prize in Physics in 1955) as the first MASER made by human [6]. A damped oscillator in contact with a thermal bath has a constant frequency and a free phase but lacks the amplitude condition. It is this difference that makes the superiority of a laser compared to a monochromatic, collimated arc-discharge lamp. The freedom of the phase is on the one hand a great advantage of a self-oscillator as it can lead, for instance, to complex collective behavior like synchronization. On the other hand, the freedom of the phase is the main drawback of a self-oscillator as this leads to an ordinary clock needing to be reset from time to time to a reference clock. For example, a quartz wristwatch keeps time to within ±15 seconds per month or better and a mechanical watch could be as bad as within ±15 seconds per day. The frequency stability of a self-oscillator changes over time and is characterized by what is called the Allan deviation, which is an estimate of the ratio of the frequency drift over the oscillation frequency for a fixed duration. The smaller the Allan deviation the more stable the self-oscillator. Just to fix the order magnitudes, a quartz in a computer has a relative frequency stability in the range 10^{-4}–10^{-5} and an atomic clock is better than 10^{-10}.

2.2.2 General Model of Self-Oscillations: The Van der Pol Oscillator

The simplest model of a self-oscillator is obtained by modifying the damping term in the damped oscillator equation and is called the Van der Pol oscillator [7]:

$$\frac{d^2x}{dt^2} - \varepsilon\,\omega(1 - x^2)\frac{dx}{dt} + \omega^2 x = 0 \tag{2.1}$$

where x is the dynamical variable describing the system (for instance the position or the current), ε is a small quantity responsible for a negative damping, ω is the resonance frequency, and t is the time.

A system with a negative damping is unstable. Its amplitude of motion increases as a function of time. The role of the additional nonlinear term in the damping is to saturate this mechanism once the amplitude is large enough. The interest of such a model is its simplicity. It allows to find analytical solutions in some limiting cases that describe rather well many systems. However, the presence of this negative damping seems mysterious and against the physical principle of energy conservation. The term ε is an effective parameter that hides the way an external source injects enough energy to compensate the intrinsic dissipation of the system. In practice, the sign and amplitude of ε depends on the coupling of the source to the system. For instance, the triode oscillator exhibits self-oscillations above a threshold applied voltage that corresponds to the value of the voltage where ε changes sign.

The negative damping model is only one among several mechanisms responsible for self-oscillations. Experimentally, it is not always easy to determine the correct mechanism at the origin of self-oscillations as will be shown in the following for the domain of self-oscillating field emitter. As was encountered in the triode case, many parasitic couplings can produce self-oscillations. The higher the amplification gain of the amplitude (related to ε in the case of the Van der Pol oscillator), the higher the risk of oscillation. This characteristic is dramatically enhanced for nano-electromechanical systems (NEMS) due to their high sensitivity to external perturbations.

One strategy to implement self-oscillations is to use a feedback loop with an external amplifier. This is the way an AC generator works. Typically, a quartz crystal with a well-defined resonance frequency is electrically coupled via the piezoelectric effect to an amplifier that acts as a negative resistance device. However, compared to the system developed in this review, such a circuit requires a bulky external electronic device and here only self-oscillators involving an *intrinsic* mechanism will be presented.

2.2.3 Specificity of the Single Clamped Field-Emission Geometry

The quest for nano-emitters with larger field enhancement factors pushed naturally the field toward long and thin nano-objects like nanotubes and nanowires. Such a geometry is also more prone to bending, which can also vary the enhancement factor due to the displacement of the emitter apex. So the field-emission current will depend on both the voltage and the position of the emitter. This type of nano-emitter can therefore be considered as a NEMS.

The field of NEMSs is expected to develop novel, low consumption, ultrasensitive devices in the continuity of the success of accelerometers based on micro-electromechanical systems. It faces, however, several challenges: firstly for detection as signals are generally very small and secondly because its physical robustness and movement are sensitive to unwanted external perturbations. Moreover, the extreme mechanical flexibility of these devices and the tight proximity of counter electrodes make them particularly vulnerable to the pull-in instability. This runaway instability, similar to door slamming, appears when the elastic restoring force is not capable anymore to counterbalance the diverging electrostatic attraction at short distance. In general, the NEMS end up definitely stuck to the counter electrode or damaged.

Despite these risks, several publications have demonstrated the impressive potentialities of field-emission NEMS as compact "readymade" nanodevices for radio applications [8, 9], mass sensing with atomic resolution [10], AC generators, or clocks [11, 12]. All of them require either intrinsic or extrinsic self-oscillations. A top down device where the counter electrode is vibrating and the field emitter is fixed has also been fabricated [13]. In parallel, various single clamped geometries have been patented by Nokia [14] for an oscillator, by the University of Berkeley [15] for a parametric amplifier, or by Toyota [16] for a signal processor.

2.3 Self-Oscillations of Individual Carbon Nano-emitters

Self-oscillations in NEMS based on carbon have been observed in several geometries but the single clamped geometry seems more favorable than the double clamped geometry that requires to be at low temperature [17, 18] or appears only in rare cases [19].

2.3.1 First Observations

The first convincing evidence of self-oscillation of carbon field emitters at the nanoscale was performed in a transmission electron microscope (TEM). In these pioneering studies, the main goal was to observe the gradual destruction of the emitter for large currents and self-oscillation was an unexpected curiosity. That is why the information concerning this phenomenon are rather scarce in these papers.

2.3.1.1 The “head-shaking” effect

In 2002, the group of Z.L. Wang at Georgia Tech performed field-emission experiments in a TEM on individual carbon MWCNT produced by arc discharge. They observed the static bending of the nanotube when a voltage difference is applied between the nanotube and a counter electrode. Their sample had a diameter around 20 nm, a length of a few microns, and a counter electrode at 2.5 μm. They observed spontaneous mechanical vibrations and called it a “head-shaking” effect [20]. Although no quantitative electromechanical parameters are given in the article, the applied voltage is probably in the 100 V range and the DC current in the μA range as for the other experiments described in their text. The authors propose that this effect might explain the current fluctuations observed in previous field-emission experiments on carbon emitters. Although a detailed model is not given in the article, they attributed the effect to a relaxation oscillator-type mechanism close to the Shishi-odoshi. The applied voltage bends the nanotube and charges accumulate on the apex surface until a sudden discharge takes place. The release of the electrons decreases the electrostatic force so that the nanotube relaxes to its rest position. Such mechanism is possible because of the high aspect ratio of the tube, which favors high amplitude of bending and large charges accumulation.

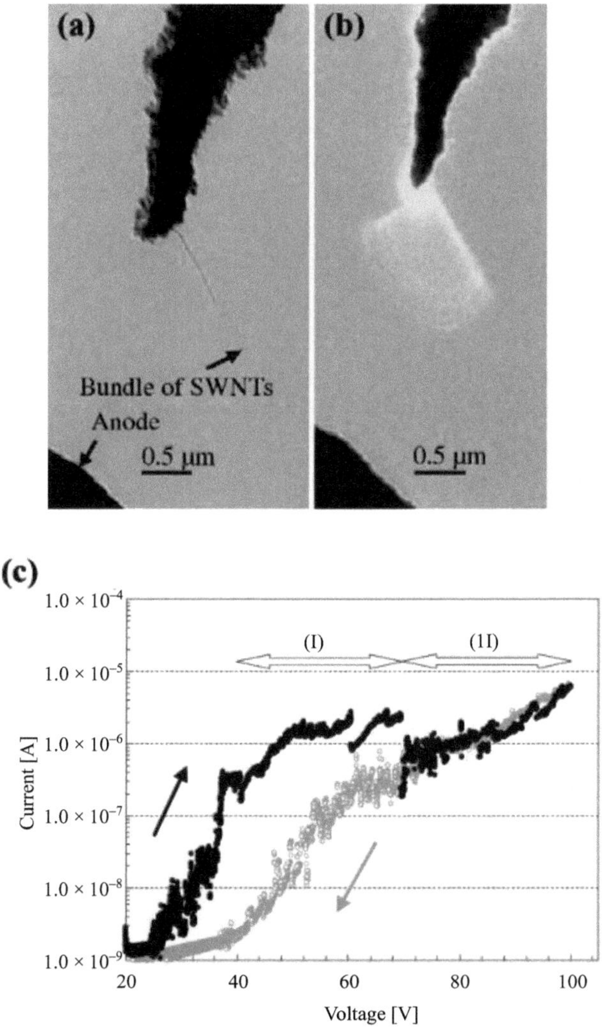

Figure 2.1 (a) TEM image of a SWCNT at 0 V, (b) TEM image of a SWCNT at 62 V, and (c) corresponding I–V characteristics [21].

2.3.1.2 Self-oscillations on SWCNT

Later similar experiments in a TEM by Saito et al. from Nagoya University reported oscillations on a bundle of SWCNTs produced by arc discharge [21]. The amplitude of oscillation appears to be very high as shown in Fig. 2.1 but most of the time they are barely visible.

The oscillations start at a voltage as low as 40 V for SWCNT length of 1 μm, a counter electrode a few microns away, and a DC current that can be above 1 μA. Compared to the previous work, I–V curves are presented that seem to indicate an abrupt change at the onset of oscillations (region I) although the data are rather noisy and gradual destruction (in region II) of the nanotube takes place at the same time as the self-oscillations. The origin of oscillation is not clarified by the author although they suspect it to be related to the resonance frequency.

2.3.2 Characterization of Nano-electromechanical Self-Oscillating Field Emitters

Probably, the most extensive characterization of the phenomenon of self-oscillation in nano-electromechanical field emitters has been performed with SiC nanowires [11, 12, 22]. Their low resonance frequency makes the detection easier and they are mechanically almost as robust as carbon nanotube to sustain high amplitude of oscillations. The characterization has been performed in electron microscopes and ultrahigh vacuum field-emission chambers.

2.3.2.1 Self-sustained oscillations in an electronic microscope

Experiments performed in a scanning electron microscope (SEM) or a TEM showed that these environments are not the most suitable to determine the intrinsic experimental parameters at the origin of self-oscillations. P. Vincent et al. from University of Lyon [22] showed that *even in the absence of a DC applied voltage*, self-sustained oscillations take place when the beam of an electron microscope impinges on a resistive 1D nanostructure under certain conditions. The SiC nanowires had lengths of several tens of microns and sometimes above a hundred microns with a radius of several tens of nanometers. It was demonstrated that the electron beam leaves some charges on the resonator that leads to a bending due to the asymmetry of the electrostatic geometry. Once charged, the resonator bends away from the electron beam—the additional charges leak to the ground at a time scale given by the product of the nanowire resistance and the capacitance of the electrostatic environment—so that the resonator returns back to its rest position to repeat the process. From the detection of the secondary electrons, the authors showed that the motion is periodic in time with a frequency close to the resonance frequency of the nanowire.

2.3.2.2 Self-oscillations in an ultrahigh vacuum environment

The secondary electron detector (SED) used in a microscope can also be used to detect the field-emission current when the electron beam of the microscope is blanked. With this method, field-emission currents in the pA range with a bandwidth in the MHz can be detected. In Ref. [11], SiC nanowires with similar dimensions as in Ref. [22] were attached to tungsten tips and used as field emitters. When a counter electrode was approached with a nanometric piezo-positioner at a distance of the order of 1 μm (see Fig. 2.2a), an AC current (Fig. 2.2c) was measured on the SED above a threshold DC voltage and an abrupt change in the DC current was measured with the electrometer connected to the tungsten tip (Fig. 2.2d). If these measurements were performed while imaging the nanowire with the microscope electron beam, it was possible to see a blurring of the image, indicating that the nanowire was vibrating (see Fig. 2.2b).

The authors proposed a mechanism for self-oscillations with two coupled equations, describing the mechanical and electrical dynamics of the system. We will present here a simplified linear model containing only the main ingredients. The displacement x of the apex can be described by the equation of a damped oscillator of resonance frequency $\omega/2\pi$, quality factor Q, driven by an electrostatic force proportional to the AC part of the voltage at the apex:

$$\frac{d^2x}{dt^2}+\frac{\omega}{Q}\frac{dx}{dt}+\omega^2 x=C'U \tag{2.2}$$

where the term C' is a parameter characterizing the actuation strength of the counter electrode. The voltage U is different from the applied DC voltage because of the delay induced by the resistance R and the capacitance C of the electrical circuit. This circuit can then be described by:

$$\tau\frac{dU}{dt}+U=-\beta' x \tag{2.3}$$

where $\tau = RC$ is the electrical time constant and β' is a term proportional to the spatial derivative of the field enhancement factor. It describes how the field-emission current changes when the position of the emitter changes. These two equations can be gathered into a single equation, which might have a negative damping and resemble the Van der Pol equation in the absence of the following nonlinear term:

$$\frac{d^2x}{dt^2}+\left(\frac{\omega}{Q}-\frac{C'\beta'\tau}{1+(\omega\tau)^2}\right)\frac{dx}{dt}+\left(\omega^2+\frac{C'\beta'}{1+(\omega\tau)^2}\right)x=0 \tag{2.4}$$

The more detailed model in Ref. [12] shows that a favorable condition for self-oscillation is when $\omega\tau \sim 1$, which is a rather general condition for self-oscillators [5]. This condition appears in a system that have dynamical variables that are characterized by two different time constants, here the period of oscillation for x and the relaxation time of the *RC* circuit for *U*. Self-oscillations may appear if a dephasing exist between the two degrees of freedom and this dephasing is more efficient if the time constants are not too far from each other. As the capacitance is rather small in nanodevices, it means that a high resistance is required or a high resonance frequency.

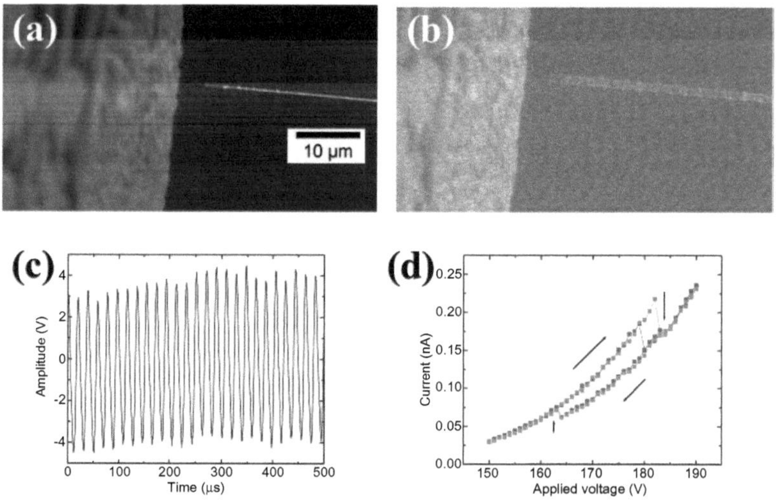

Figure 2.2 (a) SEM image of a SiC nanowire at rest, (b) the same nanowire in self-oscillation with an amplitude about 1 μm, (c) signal from the SED for a SiC nanowire with a DC voltage of 220 V and a DC field-emission current of 50 nA, and (d) emission current as a function of applied voltage for two successive voltage sweeps. The arrows show the direction of sweeping and current jumps [11].

In Ref. [12], the authors independently extracted all the experimental parameters for the model and obtained a reasonable agreement to predict the self-oscillation threshold voltage. Self-oscillations from the first and second mechanical mode have been observed for DC voltages close to 200 V. The AC current was measured at a frequency around 100 kHz and its amplitude was almost at 1 nA. It was shown that the ratio between the AC and DC current could be above 1 as long as the DC voltage was sufficiently above the threshold voltage. The Allan deviation was measured and its minimum value was 1.2×10^{-5}, which is a good value for a mechanical oscillator at room temperature without temperature stabilization.

2.3.3 Toward Large Scale Integration

In the perspective of industrial applications based on the effect of self-oscillation in field-emission nano-emitters, the group of A. Zettl in Berkley fabricated single clamped carbon nanotube self-oscillators [23] by a top-down approach (see Fig. 2.3). Optical and electron-beam lithography was used to fabricate the device and the nanotubes were deposited by spin casting. The nanotubes had a length of a few microns and a diameter probably between 5 and 10 nm. Besides the demonstration of the feasibility of the fabrication of a device, an interesting point of the paper is the movie presenting the onset of self-oscillation in a TEM.

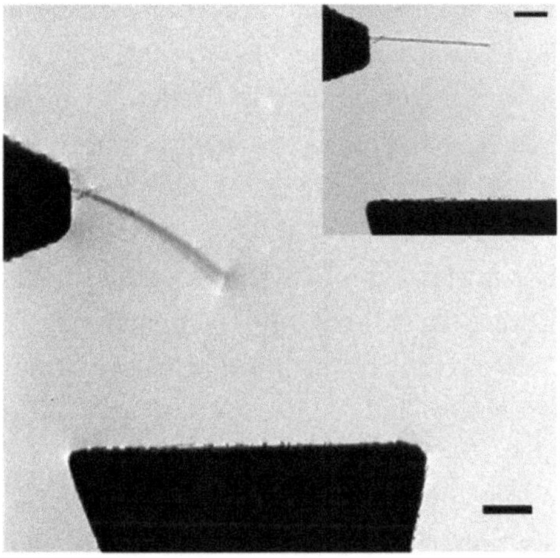

Figure 2.3 TEM image of a top-down carbon nanotube NEMS demonstrating sustained self-oscillations at 40 V; the inset shows the nanotube at 0 V (the scale bar is 1 μm) [23].

The performances of the device are comparable with the early work on nanotubes presented in the paragraph above, a DC current of 0.1–1 μA was achieved for a relatively low voltage between 50 and 60 V and an estimated vibration frequency of 4 MHz. However, the measurement of the AC current would have been more interesting for application purposes but was impossible due to capacitive leakage. The frequency was supposed to be identical to the measured resonance frequency although the movie gives the impression that the nanotube might be touching the counter electrode during oscillation which should affect the vibration frequency.

The authors emphasized that the design of their device was based on a predictive model. From this model they infer that there exists some room for improving their device, for instance to lower the onset voltage. The proposed model seems different from the model established in Refs. [11, 12]. Their explanation relies on the pull-in instability, which leads to an abrupt discharge followed by slower recharging. A thorough analysis of their study is difficult due to some missing details and some apparent inconsistencies. In particular, the criteria proposed by the author to determine the best conditions for self-oscillations, $\omega\tau \sim Q$, is several orders of magnitude different from the classical criteria $\omega\tau \sim 1$ without any justification. For the author, it is the mechanical relaxation time that needs to be comparable with the electrical relaxation time. The values of the capacitance and resistance that are crucial for the model are not explicitly given but a capacitance in the femto- or attoCoulomb can be expected which according to the RC constant they gave would lead to a nanotube resistance above 1 GΩ which is particularly high. No independent measurements of C, C' are performed to validate their finite element simulations of this parameter in their geometry. To conclude, it is not clear if the model proposed here is really different from the one in Refs. [11, 12] as it might be the same mechanism but in a different regime. This can be seen by returning to the Van der Pol model. There exist two regimes depending on the value of ε. For small ε, the solution is a sinusoid whereas for large ε, the solution is still periodic but with abrupt changes and this regime is called relaxation oscillation.

2.4 Self-Oscillation for Large Carbon Emitters

2.4.1 Self-Oscillations from Assembly of Nanotubes

Individual SWCNT or MWCNT emitters cannot sustain large currents without being irreversibly damaged or gradually shorted. The team of E. D. Obraztsova in Moscow State University has fabricated carbon self-oscillating field emitter made of large and compact assembly of nanotubes. In Ref. [24], they demonstrated self-oscillations of twisted SWCNT films and obtained AC currents as high as 30 μA. Their films were 5–20 mm long, 0.1–3 mm wide, and 0.1 mm thick, the counter electrode was at a few mm distance, and the applied voltage was about 1 kV. The self-oscillation frequency is rather low, between 10 and several hundred Hz. The model they propose describe fairly well their experiments and is rather close to the one proposed in Refs. [11, 12] for SiC nanowires. It can be noticed that the same authors, apparently, obtained similar self-oscillations with field emission from MWCNT yarns [25] although in that case the mechanism is not well explained and the AC current was not measured.

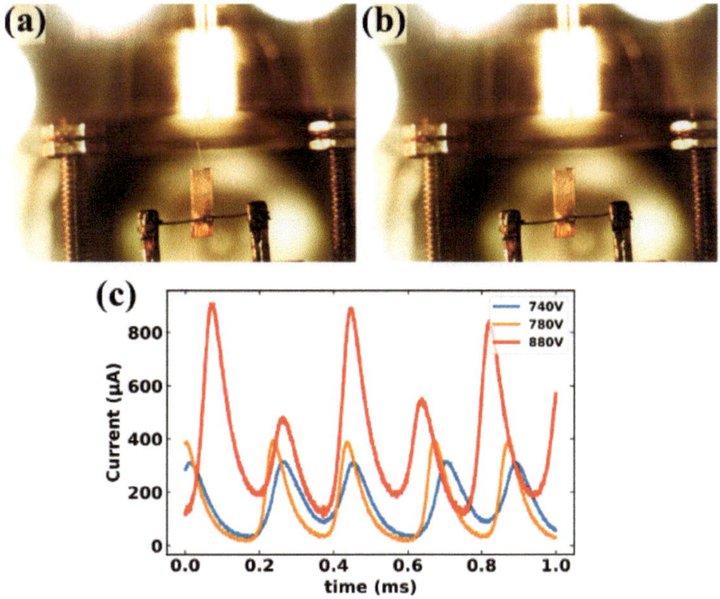

Figure 2.4 (a) Optical image of a carbon fiber at rest, (b) the same fiber in self-oscillation, and (c) field-emission current as a function of time for the same carbon fiber at different voltages (unpublished results).

2.4.2 Very High Current with Carbon Fiber Self-Oscillators

To go to higher currents, our team performed field-emission measurements on self-oscillating carbon fibers with a diameter of 10 μm, a length of 2 mm, and a counter electrode at a distance of 200 μm (see Fig. 2.4). Self-oscillations start at a voltage of 740 V and it is then possible to reach almost 1 mA of AC current but the oscillating frequency is in the kHz range. For the highest currents the signal clearly shows the presence of higher harmonic terms due to nonlinearities.

2.5 Conclusion

Electromechanical self-oscillators in field emission have been explored from nanometer to millimeter scales in a bottom-up and top-down approach. Although experimental details in the literature are often lacking in order to determine the origin of the self-oscillations, there is no doubt that (i) self-oscillation from pure field emission has been demonstrated and (ii) observing a field emitter with an electron beam can induce or help the onset of self-oscillations. Due to scaling laws, it appears that nanometer emitters can generate high frequency signals in the MHz range but at low currents. It seems possible that self-oscillations of SWCNT in the GHz range is within reach or may have been already realized but the weakness of the signal would make challenging to give a clear proof of such an achievement. Until now, the only way to reach high currents has been to increase the size of the emitter at the expense of the frequency of oscillation.

Acknowledgments

The authors acknowledge M. Choueib and T. Barois for their contributions to the original research. This work was carried out within the framework of the Plateforme Nanofils et Nanotubes Lyonnaise de l'Université Lyon 1.

References

1. Horowitz, P., & Hill, W. (1980). *The Art of Electronics.* Cambridge University Press, p. 162.
2. Andronov, A. A., Vitt, A. A. F., & Khaikin, S. E. (2013). *Theory of Oscillators: Adiwes International Series in Physics* (Vol. 4), Elsevier.
3. Miller, J. M. L., Ansari, A., Heinz, D. B., Chen, Y., Flader, I. B., Shin, D. D., Villanueva, L. G., & Kenny, T. W. (2018). Effective quality factor tuning mechanisms in micromechanical resonators. *Applied Physics Reviews,* **5**(4), 041307.
4. Mathieu, É. (1868). Mémoire sur le mouvement vibratoire d'une membrane de forme elliptique. *Journal de mathématiquespures et appliquées,* **13**, 137–203.
5. Jenkins, A. (2013). Self-oscillation. *Physics Reports,* **525**(2), 167–222.
6. Lamb, W. E. (1995). Anti-photon. *Applied Physics B,* **60**(2–3), 77–84.
7. Van der Pol, B. (1926). Over Relaxatietrillingen. *Physica,* **6**, 154–157.
8. Jensen, K., Weldon, J., Garcia, H., & Zettl, A. (2007). Nanotube radio. *Nano Letters,* **7**(11), 3508–3511.
9. Vincent, P., Poncharal, P., Barois, T., Perisanu, S., Gouttenoire, V., Frachon, H., Lazarus, A., de Langre, E., Minoux, E., Charles, M., Ziaei, A., Guillot, D., Choueib, M., Ayari, A., & Purcell, S. T. (2011). Performance of field-emitting resonating carbon nanotubes as radio-frequency demodulators. *Physical Review B,* **83**(15), 155446.
10. Jensen, K., Kim, K., & Zettl, A. (2008). An atomic-resolution nanomechanical mass sensor. *Nature Nanotechnology,* **3**(9), 533.
11. Ayari, A., Vincent, P., Perisanu, S., Choueib, M., Gouttenoire, V., Bechelany, M., Cornu, D., & Purcell, S. T. (2007). Self-oscillations in field emission nanowire mechanical resonators: A nanometric dc–ac conversion. *Nano Letters,* **7**(8), 2252–2257.
12. Barois, T., Perisanu, S., Vincent, P., Purcell, S. T., & Ayari, A. (2013). Role of fluctuations and nonlinearities on field emission nanomechanical self-oscillators. *Physical Review B,* **88**(19), 195428.
13. Yamashita, K., Sun, W., Kakushima, K., Fujita, H., & Toshiyoshi, H. (2006). RF microelectromechanical system device with a lateral field-emission detector. *Journal of Vacuum Science & Technology B: Microelectronics and Nanometer Structures Processing, Measurement, and Phenomena,* **24**(2), 927–931.
14. White, R., & Kivioja, J. (2011). *U.S. Patent No. 8,084,336.* Washington, DC: U.S. Patent and Trademark Office.

15. Aleman, B. J., & Zettl, A. (2016). *U.S. Patent No. 9,450,550*. Washington, DC: U.S. Patent and Trademark Office.
16. Tadokoro, Y., & Tanaka, H. (2019). *U.S. Patent No. 10,444,267*. Washington, DC: U.S. Patent and Trademark Office.
17. Steele, G. A., Hüttel, A. K., Witkamp, B., Poot, M., Meerwaldt, H. B., Kouwenhoven, L. P., & van der Zant, H. S. (2009). Strong coupling between single-electron tunneling and nanomechanical motion. *Science*, **325**(5944), 1103–1107.
18. Urgell, C., Yang, W., De Bonis, S. L. et al. (2020). Cooling and self-oscillation in a nanotube electromechanical resonator, *Nature Physics***16**, 32–37.
19. Barois, T., Ayari, A., Vincent, P., Perisanu, S., Poncharal, P., & Purcell, S. T. (2013). Ultra low power consumption for self-oscillating nanoelectromechanical systems constructed by contacting two nanowires. *Nano Letters*, **13**(4), 1451–1456.
20. Wang, Z. L., Gao, R. P., de Heer, W. A., & Poncharal, P. (2002). In situ imaging of field emission from individual carbon nanotubes and their structural damage. *Applied Physics Letters*, **80**(5), 856–858.
21. Saito, Y., Seko, K., & Kinoshita, J. I. (2005). Dynamic behavior of carbon nanotube field emitters observed by in situ transmission electron microscopy. *Diamond and Related Materials*, **14**(11–12), 1843–1847.
22. Vincent, P., Perisanu, S., Ayari, A., Choueib, M., Gouttenoire, V., Bechelany, M., Brioude, A., Cornu, D. & Purcell, S. T. (2007). Driving self-sustained vibrations of nanowires with a constant electron beam. *Physical Review B*, **76**(8), 085435.
23. Weldon, J. A., Aleman, B., Sussman, A., Gannett, W., & Zettl, A. K. (2010). Sustained mechanical self-oscillations in carbon nanotubes, *Nano Letters*, **10**,1728–1733.
24. Kleshch, V. I., Obraztsov, A. N., & Obraztsova, E. D. (2009). Self-oscillations in an electromechanical system with a field emitter. *JETP Letters*, **90**(6), 464.
25. Zakhidov, A. A., Nanjundaswamy, R., Obraztsov, A. N., Zhang, M., Fang, S., Klesch, V. I., Baughman, R. H., & Zakhidov, A. A. (2007). Field emission of electrons by carbon nanotube twist-yarns. *Applied Physics A*, **88**(4), 593–600.

Chapter 3

Performance of Point-Typed Carbon Nanotube Field Emitters

Cheol Jin Lee
School of Electrical Engineering, Korea University,
145 Anam-Ro, Seongbuk-Gu, Seoul 02841, South Korea
cjlee@korea.ac.kr

In this chapter, fabrication methods and field-emission properties of point-typed carbon nanotube (CNT) field emitters are demonstrated. The point-typed CNT field emitters indicate high performance, such as a low turn-on voltage, good emission stability, and a small electron beam spot compared to conventional CNT field emitters. Especially, the triangle-shaped carbon nanotube film (TCF) field emitter shows much excellent performances as electron emitters compared to other point-typed CNT field emitters, indicating its potentiality enlarge the realm of field-emission device applications in medical diagnosis as well as industrial non-destructive inspection systems.

3.1 Introduction

Carbon nanotubes (CNTs) have attracted much interest for a cold cathode field emitter due to their unique properties such as excellent

Nanostructured Carbon Electron Emitters and Their Applications
Edited by Yahachi Saito

ISBN 978-981-4877-62-6 (Hardcover), 978-1-003-14199-0 (eBook)
www.jennystanford.com

electrical and thermal conductivity, high mechanical strength, good chemical stability, and a high aspect ratio [1–4]. Since some reports of field-emission properties of CNTs [5–7], several groups have studied field-emission properties from diverse CNT materials and also tried diverse approaches to enhance field-emission properties of CNTs [8–10]. From 1990s, many research groups have studied to fabricate high performance CNT field emitters and also have investigated field-emission performance for various CNT field emitters. The CNT field emitters have much attention for various applications including flat panel displays, lamps, X-ray sources, and microwave amplifiers [11–15]. Several methods have been adopted to fabricate the CNT field emitters such as as-grown CNT emitter, CNT paste emitter, CNT yarn emitter, CNT spray emitter, CNT electrophoresis emitter, and CNT film emitter [9, 11, 16–22]. It is well known that for realization of CNT field emitter applications, both a high emission current and a stable electron emission are an important issue. In general, field emitters can be classified into two types according to the size of field emitters. The first one is a planar-typed field emitter, which has a large emission area, and can be suitably used for a field-emission display or lamps. The other is a point-typed field emitter, which has a small emission area, and can be used for X-ray sources, electron microscope, and so on. For the planar-typed field emitter, good emission uniformity over a large area is a key issue. However, for the point-typed field emitter, a high emission current and a small focal spot is a very important issue. Much effort has been made to fabricate the point-typed CNT emitters by mounting a single CNT or a CNT bundle on the tip [23–26], or by using a CNT yarn [20, 22], or by attaching a CNT film on the rod [27, 28], or by coating CNT paste on the rod [29], or by using a free-standing CNT film [21]. In fact, the point-typed field emitter is a better choice for X-ray source applications because the focusing of electron beam can be more easily achieved compared to the planar-typed field emitters. Moreover, the small size of the emitting area of CNT point-emitter is favorable to get the high brightness, which is one of the important parameter for the high-resolution electron sources. Especially, point-typed CNT field emitters, namely CNT emitters with a small emission area, are suitable for X-ray sources because they can provide a high emission current density with a small focal spot. Recently, several groups have reported cold cathode X-ray sources fabricated by CNT field emitter, which indicate high possibility to replace the thermionic X-ray sources in the future [30–40]. The achieved results indicate that point-typed CNT field emitter can be applicable to cold cathode X-ray sources with high performance. In this chapter, diverse methods to fabricate the point-typed CNT field emitters are discussed and their field-emission properties are investigated in detail.

3.2 Point-Typed CNT Field Emitter Made of a CNT Bundle

The point-typed CNT field emitter was fabricated by attaching a single-walled CNT (SWCNT) bundle or a double-walled CNT (DWCNT), which was produced by DC arc discharge, onto the tip of an atomic force microscope (AFM) probe by using an electrophoresis method [23]. Scanning electron microscope (SEM) images show that a point-typed SWCNT field emitter has a very sharp tip and the SWCNT bundle attached on the AFM probe has the length of 15 μm and the diameter of 0.7 μm, respectively (Fig. 3.1).

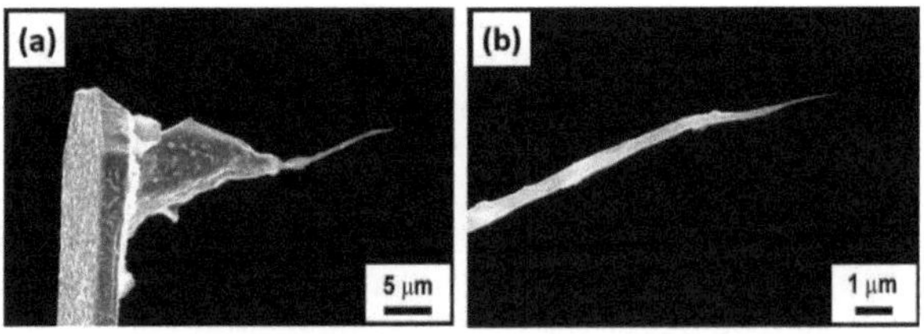

Figure 3.1 SEM images of (a) the point-typed SWCNT field emitter, and (b) the sharp end of an SWCNT bundle [23].

The field-emission current of the point-typed SWCNT field emitter was measured in the diode configuration in a vacuum chamber. The gap between the CNT bundle and the planar anode electrode was 400 μm, and the vacuum pressure of a measurement chamber was 2×10^{-7} Torr. Figure 3.2 shows the field-emission current of a point-typed CNT field emitter. The turn-on voltage (650 V) at an emission current density of 0.1 $\mu A/cm^2$ of the point-typed SWCNT field emitter is a little lower than that the DWCNT field emitter (750 V). The lower turn-on voltage is caused by a smaller diameter of SWCNT. The emission current of the SWCNT field emitter is about 5 μA, and the emission current of the point-typed DWCNT field emitter is about 1.8 μA. A higher emission current of the SWCNT point emitter is attributed to more emission sites within a single SWCNT bundle compared to a single DWCNT bundle.

Figure 3.3 shows the Fowler–Nordheim (F–N) plot of the point-typed SWCNT field emitter. Since the point-to-plane configuration was used in the measurement, the slope of F–N plot is inversely proportional to the field conversion factor defined in Eq. (3.1), where E and V are the electric field at the tip and the applied voltage, respectively:

$$E = \beta V \tag{3.1}$$

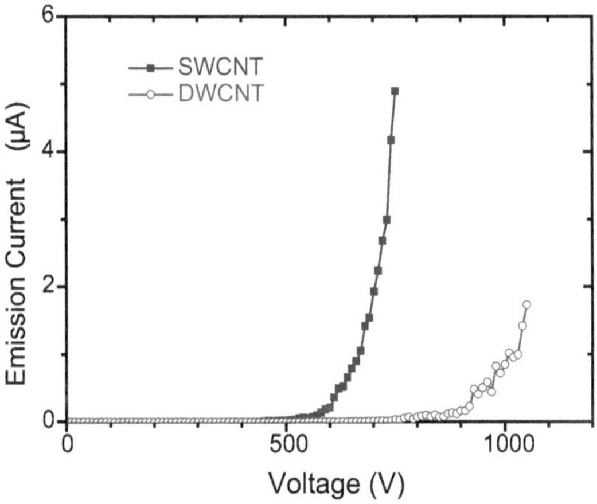

Figure 3.2 Field-emission current of the point-typed SWCNT and DWCNT field emitters. The SWCNT point emitter shows a lower turn-on voltage and a higher emission current than that of the DWCNT point emitter [23].

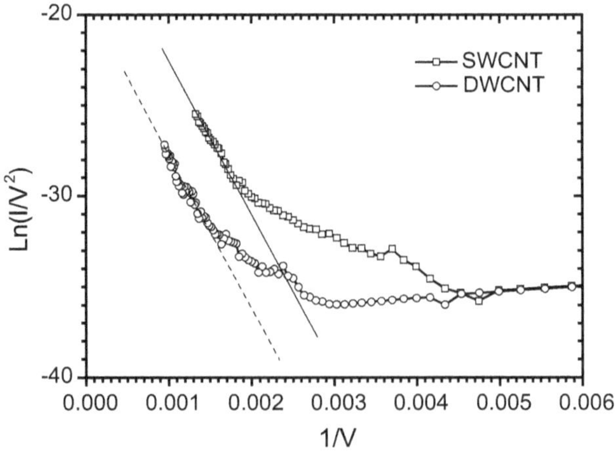

Figure 3.3 Fowler–Nordheim plots of the point-typed SWCNT and DWCNT field emitters [23].

The field conversion factor is given by

$$\beta = 1/(kr) \tag{3.2}$$

where k is a constant that depends on the geometry and r is the radius of curvature at the tip. Therefore, it can be seen that the field conversion factor is nearly independent of the length of CNT bundle but is dependent on the radius of CNT bundle. The slope of F–N plot of the point-typed SWCNT field emitter is almost same to that of the point-typed DWCNT field emitter in the high voltage region. By the way, the y-intercept value of the point-typed SWCNT field emitter (−14.02) is higher than that of the point-typed DWCNT field emitter (−19.27). It is considered that the difference in the y-intercept values is corresponding to the different numbers of an emission site in the CNT bundle [41]. The obtained result reveals that the SWCNT bundle has more emission sites compared to the DWCNT bundle. In general, the numbers of an individual CNT in the SWCNT bundle are larger than those of an individual CNT in the DWCNT bundle when the bundle has the same diameter.

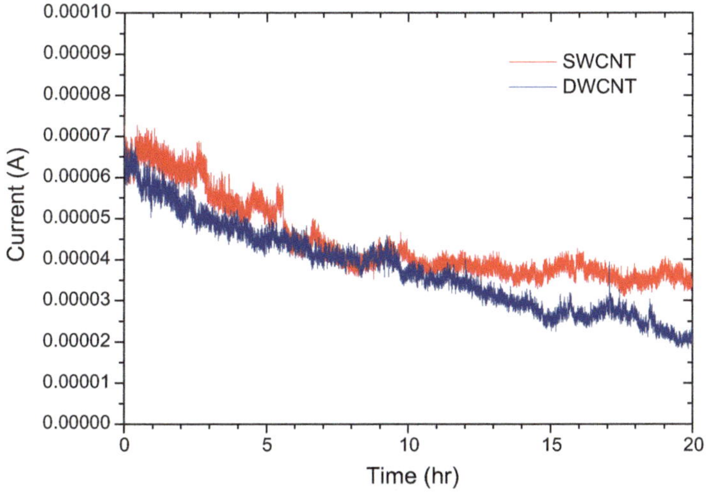

Figure 3.4 Long-term emission stability of the point-typed SWCNT and DWCNT field emitters [23].

The long-term emission stability of the point-typed CNT field emitter was investigated at DC constant voltage as shown in Fig. 3.4. The degradation rate of the emission current is about 46% for the

SWCNT point emitter and about 67% for the DWCNT point emitter, respectively. The point-typed CNT field emitters show a relatively low degradation of the emission current compared to other CNT paste field emitters or as-grown CNT field emitters [42–44] but they indicate a large fluctuation of the emission current during long-term emission stability test. It is attributed to the vibration of the CNT bundle at a high electric field during emission test [45, 46].

3.3 Point-Typed CNT Field Emitter Made of a CNT Yarn

The CNT point emitter was made out of a yarn of super aligned MWCNTs using a simple spinning technique, and field-emission properties of the CNT yarn have been investigated. The pristine MWCNT yarn has a diameter of about 50–60 μm. In order to obtain stiff and tight MWCNT yarn, the pristine MWCNT yarn was dipped in an ethylene glycol solution. Then, the MWCNT yarn was put in an oven for 10 min at 200 °C to completely evaporate the ethylene glycol. With twisting the MWCNT yarn in the ethylene glycol solution, an MWCNT point emitter with a sharp tip (S-point emitter) was fabricated [20]. After an ethylene glycol treatment, the diameter of the MWCNT yarn was decreased from 50–60 μm to about 40 μm. By cutting the pristine MWCNT yarn with a knife, the point-typed MWCNT field emitter with a blunt tip (B-point emitter) was prepared. The cathode electrode was a stainless steel pin and the anode electrode was also a stainless steel plate. An MWCNT yarn with a length of 5 mm was carefully attached to the pin by a silver paste. Since the segment of MWCNT yarn can be handled using a tweezer, a simple manipulation system is enough to attach the MWCNT yarn to the stainless steel pin. Field-emission properties were measured in a vacuum chamber at a pressure of less than 2×10^{-7} Torr. The gap between the anode and the emitter tip was adjusted to 1, 5, 10, and 15 mm separately by using glass spacers.

The tip morphology of MWCNT point emitters is shown in Fig. 3.5. The B-point CNT yarn emitter has a stem with a diameter of about 40 μm and a tip with a diameter of about 18 μm as shown in Fig. 3.5a. The magnified tip image of the B-point CNT yarn emitter is shown in the inset of Fig. 3.5a. On the other hand, the S-point

CNT yarn emitter has a stem with a diameter of about 40 μm and a very sharp tip like a needle less about 60 nm as shown in Fig 3.5b. Figure 3.5c shows the low and high magnification SEM images of the stem of the MWCNT S-point emitter. The SEM image indicates that the MWCNT yarn is composed of arrays of aligned MWCNTs along the axis. TEM image of the dispersed MWCNTs indicates the inside hollow of the MWCNTs with a high purity as shown in Fig. 3.5d. The individual MWCNTs generally have an outer diameter in the range of 10–20 nm and the graphene layers of the MWCNTs are very well resolved and lined up.

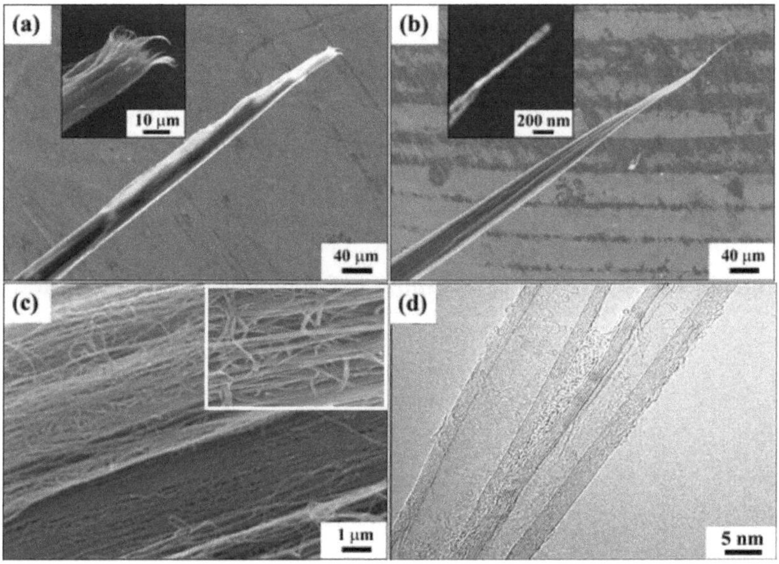

Figure 3.5 SEM image of the tip morphology of an MWCNT blunt tip (B-point emitter) (a), an MWCNT sharp tip (S-point emitter) (b), low magnification SEM image of the stem of the S-point emitter and the inset shows high magnification SEM image of the stem of the S-point emitter (c), TEM image of the dispersed MWCNTs with an inside hollow (d) [20].

The field-emission characteristics of MWCNT yarn point emitters are shown in Fig. 3.6. The gap between the emitter tip and the anode electrode was 1 mm and electron emission was started at about 250 V for the two point emitters. The CNT yarn emitter (B-point emitter) has a maximum emission current of 2.12 mA at an applied voltage of 750 V, which is correspondent to the emission current density about

833 A/cm^2. However, The CNT yarn emitter (S-point emitter) has the maximum emission current of 3.01 mA at an applied voltage of 720 V, which is correspondent to the emission current density about 1.1×10^8 A/cm^2. The higher emission current of the S-point emitter is attributed to the small diameter of the emitter tip compared to the B-point emitter. The small diameter of the tip is suitable for increasing the enhancement factor due to the concentration of electric field at the apex of the tip [22, 47]. The tip of the S-point emitter consists of small numbers of MWCNT bundles compared to the B-point emitter, resulting in a higher electric field at the individual MWCNT bundle. The inset of Fig. 3.6 shows the emission current stability of the S-point emitter, which has a very stable emission current at the current level of about 0.7 mA for over 20 h with a little current degradation. The diameter of the MWCNT yarn is reduced after an ethylene glycol treatment. It means that the network of CNTs within the yarn becomes tight and strong. As a result, the CNT yarn can endure the high local electric field during field-emission. Therefore, it is understood that good long-term emission stability may be caused by the tight binding of neighboring CNTs within the CNT yarn.

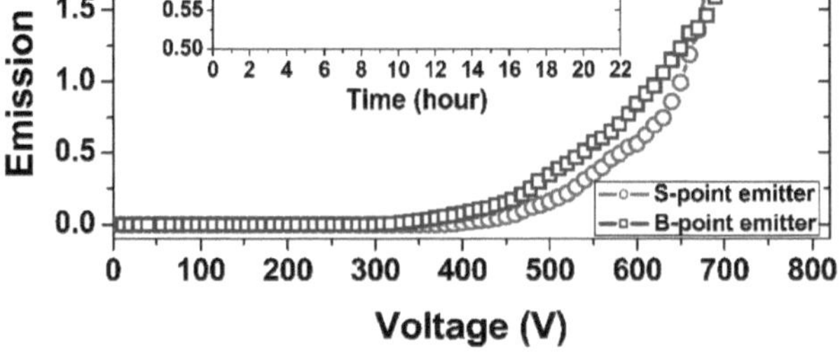

Figure 3.6 The field-emission characteristics of two types of MWCNT point emitters [20].

The field-emission properties of the point-typed MWCNT yarn field emitter (S-point emitter) were investigated according to the gap distance between the point emitter and the anode electrode. The gap of the S-point emitter was adjusted to 1, 5, 10, and 15 mm, separately. Figure 3.7 shows the field-emission current of the S-point emitter dependent of the gap. The turn-on field is decreased by decreasing the gap but the emission current is rapidly increased by decreasing the gap. The inset of Fig. 3.7 shows F–N plots that indicate the quantum mechanical tunneling behavior of the field electron emission. The field enhancement factor can be calculated using the simplified F–N equation:

$$I = (A\alpha F^2/\varphi)\exp(-B\varphi^{3/2}/F) = (A\alpha\beta^2 V^2/d^2\varphi) \times \exp[-B\varphi^{3/2}d/(\beta V)] \quad (3.3)$$

where I is the emission current, $A = 1.54 \times 10^{-6}$ A V^{-2} eV, $B = 6.83 \times 10^9$ eV$^{-3/2}$ V m^{-1}, α is the effective emission area, β is the field enhancement factor, φ is the work function, d is the gap between the anode and the point emitter tip, V is the applied voltage, and $F = \beta E = \beta V/d$ is the local field at the tip, where E is the macroscopic field. The calculated field enhancement factors from the F–N plots is 7.55×10^3, 3.93×10^4, 6.60×10^4, and 8.40×10^4 for the gap of 1, 5, 10, and 15 mm, respectively. The S-point emitter shows a very large field enhancement factor due to the sharp tip. As a result, it has the very low turn-on field at a considerably low applied voltage. In addition, it is well understood that the field enhancement factor strongly increases with increasing the gap distance.

The field-emission pattern of the CNT emitter was obtained to investigate the electron emission status. The experiment was performed by replacing the anode electrode from the stainless steel plate to a phosphor-coated indium tin oxide (ITO) glass substrate. A DC voltage was applied between the anode and the cathode to obtain an emission current of 0.2 mA, which protects the phosphor material from electron collision without burning out. The gap between the CNT emitter tip and the anode electrode was 1 mm. Figure 3.8 shows the dark image of the emission pattern of the point-typed CNT field emitter (S-point emitter). The emission pattern indicates a small electron beam spot (about 2 mm), revealing that the S-point emitter is suitable to get a small focal spot that can promise high-resolution X-ray sources.

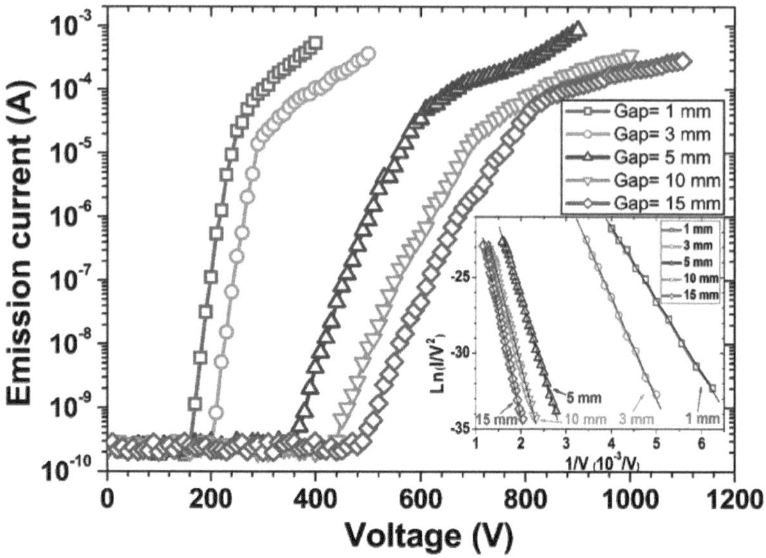

Figure 3.7 Field-emission current of the S-point emitter dependent of the gap [20].

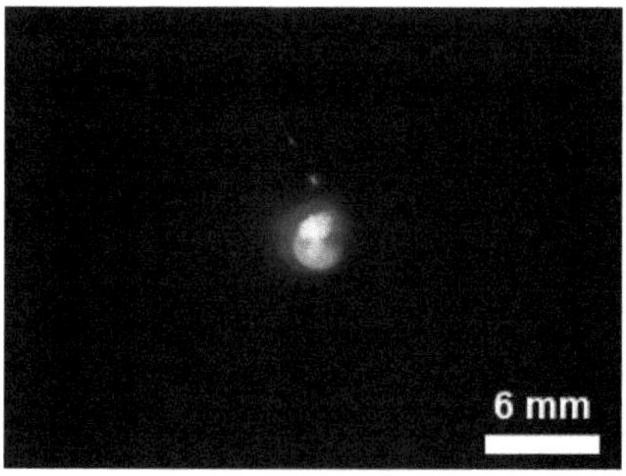

Figure 3.8 Emission pattern of the S-point emitter.

3.4 Point-Typed CNT Field Emitter Made of a CNT Film

In order to increase the emission current of point-type CNT field emitters, a lot of emission sites at the surface of CNT filed emitter are inevitably necessary. The new method to fabricate a point-typed CNT emitter with a high emission current was suggested. The point-typed CNT emitter was fabricated by attaching a CNT film onto a graphite rod with a small diameter using a graphite adhesive material [27, 28]. The graphite rod was made by graphite adhesive materials or it was simply prepared by a sharp pencil lead. The graphite adhesive material was prepared by mixing graphite powder (200 mg, COTRONICS Cor., 931 graphite powder) and organic binder (200 mg, COTRONICS Cor., 931 graphite binder) for 20 min using a paste mixer (Thinky, AR-250). Then, a silica capillary tube with an inner diameter of 1 mm was dipped into the graphite adhesive material and was filled with the graphite adhesive automatically owing to the capillary phenomenon. The silica tube was placed in an oven and cured in air ambient for 4 h at 100 °C, following which it was cured for 16 h at 130 °C. Finally, the silica capillary tube was dissolved in hydrofluoric acid, and a graphite rod of diameter 1 mm was produced. By adjusting the inner diameter of the silica capillary tube, the outer diameter of graphite rod is easily controlled. By using this method, the emission area and the length of the CNT emitter can be easily controlled. To fabricate the CNT film, thin multiwalled CNTs with diameters in the range of 5–8 nm were inserted into the ethanol and a CNT solution was prepared by tip sonication process for 20 min. The CNT solution was poured on a Teflon filter membrane and after vacuum filtration process, a uniform CNT film was deposited on the membrane. The residual solvent existing on the surface of the CNT film was disappeared after a thermal baking at 70 °C in air ambient. To fabricate the CNT emitter, the CNT film was attached to the graphite rod using the graphite adhesive material. After the first baking of the CNT emitter in an oven for 4 h at 100 °C in air ambient, the second baking was followed for 16 h at 130 °C in air ambient. Consequently, thermal annealing was conducted for 1 h at 400 °C in air ambient to remove remaining organic materials. The mechanical treatment using a scotch tape was employed to remove organic residues and/or the weakly binding CNTs on the surface of the CNT film on the graphite rod. The edge of the point-typed CNT film emitter was ground to avoid edge electron emission. The CNT emitter was thermally annealed at 800–900 °C in vacuum ambient. The stainless steel plate was used as the anode and the cathode electrodes, and the alumina plate was used as the spacer. The graphite rod was fixed using a copper holder on the cathode electrode. The gap between the emitter tip and anode was set to 550 μm. The field-emission measurements were performed using a diode configuration in a vacuum chamber under a pressure of 2×10^{-7} Torr. To obtain emission patterns from the CNT emitters, a phosphor deposited the indium tin oxide (ITO)-coated glass substrate was used as the anode electrode instead of the stainless steel plate.

Figure 3.9a shows a SEM image of the point-typed CNT field emitter, which is fabricated by attaching the CNT film on the surface of the graphite rod. The CNT film is strongly attached onto the surface of the graphite rod of diameter 1 mm. Figure 3.9b presents a SEM image that the edge of the point-typed CNT field emitter is well polished. The surface area which can contribute electron emission is reduced after edge polishing. Figure 3.9c shows the low magnification SEM image of the surface of the CNT field emitter. Many CNTs are quasi-vertically aligned on the surface even though the surface flatness is poor and CNTs show a rugged morphology due to the graphite adhesive paste. Figure 3.9d presents the high magnification SEM image of some randomly and vertically aligned CNTs on the surface of the point-typed CNT emitter. Several protruding CNTs on the surface after mechanical taping treatment, CNTs can contribute to field electron emission.

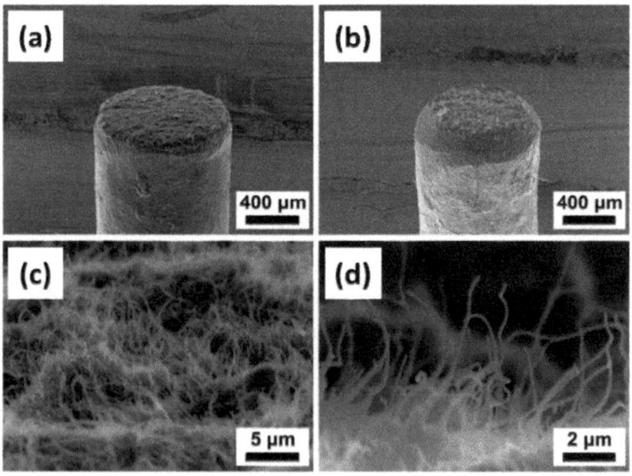

Figure 3.9 SEM image of the point-typed CNT field emitter, which is fabricated by attaching the CNT film on the surface of the graphite rod (a), SEM image of the point-typed CNT emitter after polishing the edge area of the graphite rod (b), low magnification SEM image of the surface of the point-typed CNT field emitter (c), high magnification SEM image of the protruding CNTs on the surface of the point-typed CNT field emitter (d) [27].

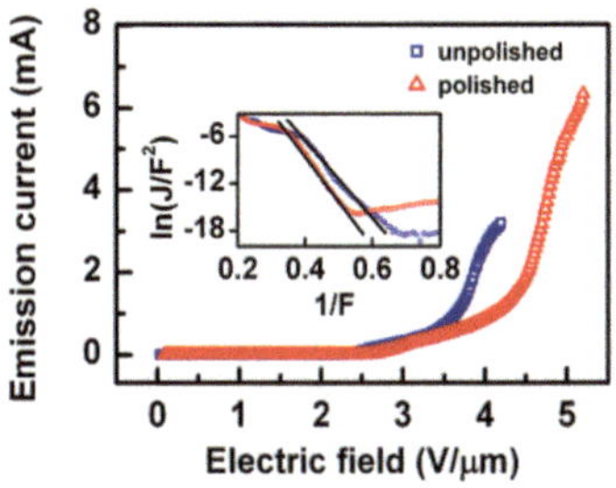

Figure 3.10 Field-emission properties of the point-typed CNT emitters which are fabricated by attaching the CNT film on the surface of the graphite rod [27].

Figure 3.10 shows the field-emission properties of the point-typed CNT emitters. The turn-on electric field of the CNT emitters is 1.56 V/μm at an emission current density of 0.1 μA/cm^2, and the threshold electric field is 2.24 V/μm, corresponding to an emission current density of 1.0 mA/cm^2. The turn-on electric field of the point-typed CNT field emitters is little increased from 1.56 to 1.78 V/μm after polishing the edge area of CNT field emitter. The threshold electric field is also increased from 2.24 to 2.44 V/μm. The maximum emission current of the point-typed CNT emitter is 3.20 mA (corresponding to an emission current density of 0.4 A/cm^2) at an applied electric field of 4.2 V/μm. However, after polishing an edge area of CNT emitter, it increases up to 6.34 mA (corresponding to an emission current density of 1.6 A/cm^2) at an applied electric field of 5.3 V/μm even though the emission area of the CNT emitter is decreased from 0.00785 to 0.00385 cm^2 after polishing the edge. In fact, the emission current of the CNT field emitter increases to be about 2 times higher than that of the pristine CNT emitter. It means that the emission current density of the CNT field emitter increases by 4 times after edge polishing. It is noteworthy that this obtained result reveals better emission performances compared to other point-typed CNT emitters [20, 48, 49–53]. The high field-emission current of this CNT field emitter is attributed to the high density of emission sites at the surface of the emitter by adopting CNT film on the graphite rod. In addition, the edge polishing treatment changes the shape of the edge of the point-typed CNT emitter from the sharp edge shape to the round edge shape, resulting in effective suppression of edge electron emission of the point-typed CNT emitters during the field emission. The inset of Fig. 3.10 shows F–N plot of the point-typed CNT emitter, which indicates quantum tunneling behavior.

The lifetime measurement was performed under the DC constant condition at the accelerated condition to evaluate the long-term emission stability of the point-typed CNT emitters. A constant voltage was applied between the anode and cathode to keep an initial emission current of 0.1 mA, corresponding to 26 mA/cm^2. Figure 3.11 shows that the degradation rate of the emission current of the CNT field emitter without edge polishing process is about 50% for 20 h. On the other hand, the emission current of the point-typed CNT emitter with edge polishing process indicates only a little degradation rate of less than 5% in 20 h. It reveals that edge polishing process can protect edge emission of the CNT field emitter, providing very strong long-term emission stability [54, 55].

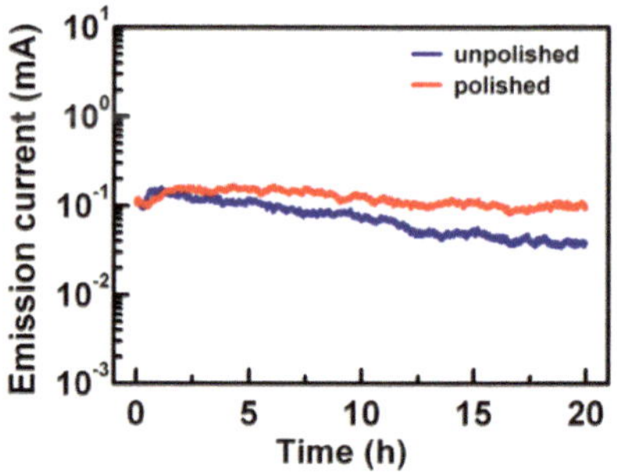

Figure 3.11 Long-term emission stability of the point-typed CNT emitters [27].

To analyze the electric field at the tip of the CNT emitters, a 3D model of the electric field distribution at the emitter tip was constructed using a commercial tool—CST EM Studio. The point-typed CNT emitter with a length of 5 mm and a diameter of 1 mm was placed in a diode configuration, and the gap between the emitter tip and the anode was set to 550 μm. Assuming that the condition of the CNT emitter on the two graphite rods is same, and a voltage of 2 kV was applied between the cathode and anode. Figure 3.12 shows the simulation result of the electric field at the emitter tip according to the edge polishing. Before polishing the edge area of the point-typed CNT emitter, the electric field at the edge of emitter tip is rapidly increased, even though it is nearly uniform at the central area of the emitter tip. On the other hand, the electric field at the edge of the point-typed CNT emitter after the edge polishing is smoothly increased and it is also nearly uniform at the central area of the emitter tip. This result is absolutely caused by the suppression of the electric field at the polished edge. With the edge polishing process, the high emission current at the edge area can be effectively reduced, resulting in good emission stability. This simulation result agrees well with the field-emission characteristics and long-term emission stability, as shown in Fig. 3.11.

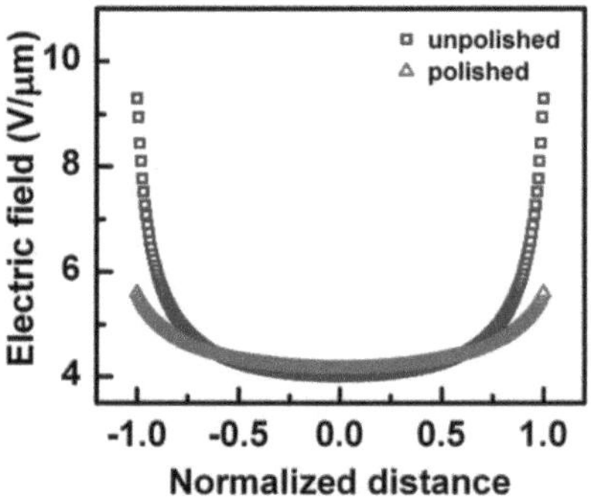

Figure 3.12 Simulation result of the potential distribution of the electric field at the emitter tips with and without the edge polishing process [27].

The field-emission pattern of the CNT emitter was obtained to investigate the electron emission status. A DC voltage was applied between the anode and the cathode to obtain an emission current of 0.2 mA. The experiment was performed by replacing the anode electrode from the stainless steel plate to a phosphor-coated ITO glass substrate. The gap between the emitter tip and the anode electrode was set to 550 μm. Figure 3.13a shows a dark emission pattern of the CNT emitter without edge polishing, which indicates a diameter of approximately 3 mm.

The emission pattern in Fig. 3.13b shows that the point-typed CNT emitter has a smaller spot of approximately 2 mm diameter and higher density of electron emission sites after edge polishing process. This result reveals that the number of emission sites of the CNT emitter can be increased after edge polishing process due to effective suppression of the edge emission.

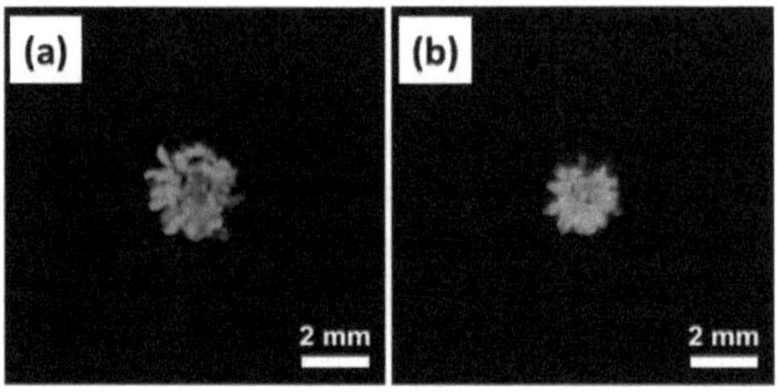

Figure 3.13 Emission patterns of the CNT emitters: (a) without edge polishing process and (b) with edge polishing process [27].

Field-emission properties of the CNT field emitters were investigated after the high temperature annealing at 900 °C in vacuum ambient for 1 h. Figure 3.14 shows that the turn-on electric field (corresponding to the emission current density of 0.1 μA/cm^2) is decreased after the high temperature thermal annealing and the maximum emission current is increased up to 5.9 mA (corresponding to the emission current density of 8.5 A/cm^2) at the electric field of 4.6 V/μm. This result means that both the improved crystallinity and reduced defects of the CNTs can enhance field-emission performance through the high temperature annealing [50, 51].

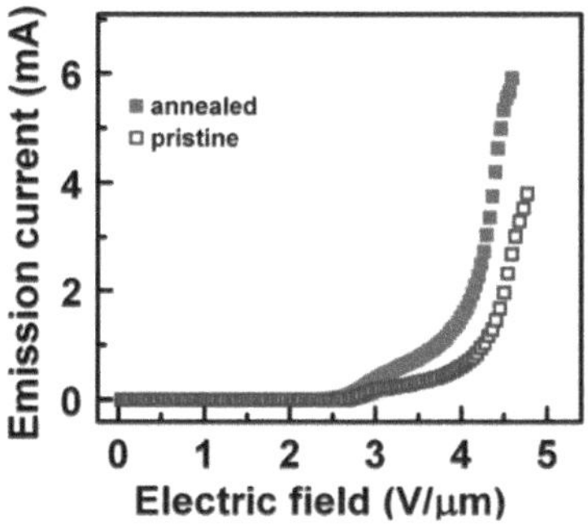

Figure 3.14 Maximum emission currents of the CNT field emitters after the high temperature thermal annealing at 900 °C in vacuum ambient. The CNT field emitters have a diameter of 0.3 mm and a length of 10 mm [28].

Figure 3.15 shows the long-term emission stability of the CNT emitters according to the high temperature vacuum annealing. The initial emission current was 2 mA (corresponding to the emission current density of 2.8 A/cm^2). The emission current of the pristine

CNT field emitter is decreased from 2 to 1.2 mA, indicating a degradation rate of 40% for 100 min. On the other hand, the emission stability of the CNT field emitter is much improved after the high temperature vacuum annealing. It presents nearly no degradation rate of the emission current and the little fluctuation of the emission current about 1.4%. These results reveal that the point-typed CNT field emitter after high temperature annealing can effectively endure severe emission condition during field-emission operation due to improved crystallinity and reduced defects in the CNTs.

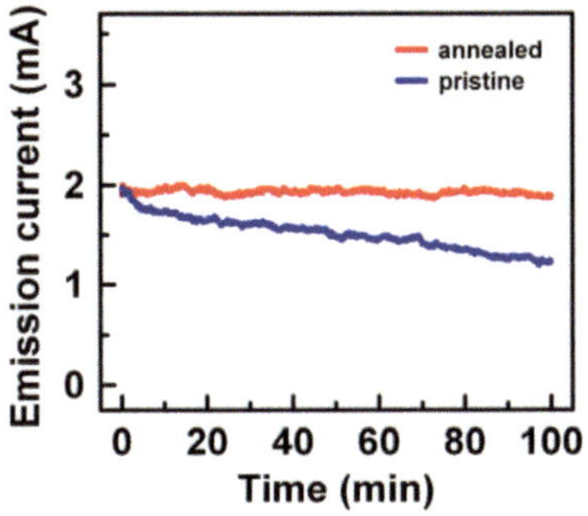

Figure 3.15 Long-term emission stability of the CNT field emitters after the high temperature thermal annealing at 900 °C in vacuum ambient [28].

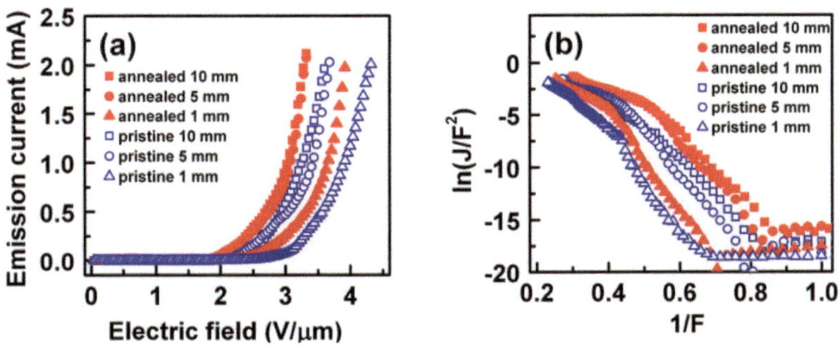

Figure 3.16 (a) Field-emission properties of the CNT field emitters according to the high temperature annealing and the length of the graphite rod. (b) The corresponding F–N plot [28].

Figure 3.16a shows field-emission properties of the point-typed CNT field emitter after high temperature annealing according to the length of the graphite rod. By increasing the length of the graphite rod from 1 to 10 mm, the turn-on electric field is decreased and the electric field at the emission current of 2 mA is also decreased. The enhanced field-emission performance by increasing the aspect ratio is attributed to the multistage effect [52, 56]. Figure 3.16b shows the corresponding F–N plot of the CNT field emitters according to the high temperature annealing. The field enhancement factor and the intercept at *y*-axis (*y*-intercept) were calculated from the straight lines in the F–N plot.

The field enhancement factor of the CNT field emitter is little increased after the high temperature annealing. It is considered that the increased field enhancement factor after the high temperature annealing is caused by both improved crystallinity and reduced defects of the CNTs. In addition, the field enhancement factor is also increased by increasing the length of the graphite rod. It means that the multistage effect is an important parameter to control the field-emission performance. It is understood that the above-mentioned two factors can strongly affect the field-emission performance of the CNT field emitter. From the above result, it is understood that the CNT field emitter (10 mm length) after the high temperature thermal annealing shows the lowest turn-on field of 1.15 V/μm, the lowest threshold field of 1.62 V/μm, and the maximum field enhance factor of 2233.

3.5 Point-Typed CNT Field Emitter Fabricated by CNT Paste

The point-typed CNT field emitter was fabricated by depositing CNT paste on the graphite rod. The CNT paste was made by blending of thin multiwalled CNTs (MWCNTs), graphite nanoparticles, and a graphite adhesive binder [29]. Different from previous methods, a graphite adhesive binder and graphite nanoparticles are used for fabrication of CNT paste. In general, ethyl cellulose (EC) binder and some fillers like SiC or Ni or Al_2O_3, or Ti nanoparticles are used for CNT paste [56–61]. For fabrication of CNT paste, the used MWCNTs have an entangled morphology with an average diameter of 20 nm and used graphite nanoparticles have a ball-shape with a diameter about 200 nm. First, the thin MWCNTs and the graphite nanoparticles (COTRONICS Co.) were dispersed in a sodium dodecyl sulfate (SDS) solution (1%) and the CNT solution was obtained by using ultra sonication process. The CNT film with the thickness of 5–10 μm was formed after the vacuum filtration of the CNT solution and it was baked at 70 °C in an air ambient to remove some residual water. Then, the CNT film was mixed with the graphite adhesive binder (COTRONICS Co.). The CNT paste was fabricated by mixing the CNT film, graphite nanoparticles, and the graphite binder via ball milling process for 1 h by using a paste mixer (Thinky, AR-250). By adjusting the amount of the graphite adhesive binder, the viscosity of the CNT paste can be controlled. The CNT paste was deposited on the tip of the graphite rod with a diameter of approximately 0.7 mm. The CNT paste on the graphite rod was thermally annealed to be solidified and attached in an oven at 100 °C in the air ambient for 4 h, followed by 130 °C for 16 h, and finally annealed at 350 °C for 1 h to further remove residual organic materials. The mechanical treatment using sandpaper was done to smoothen the surface of the CNT paste emitter and also to remove protruding CNTs on the surface of CNT paste emitters. The field-emission measurement was carried out using a diode configuration in a vacuum chamber under a pressure of 3×10^{-7} Torr, and the gap between the emitter tip and the anode electrode was set to 500 μm. A field-emission pattern was obtained at an emission current of 200 μA, and for the anode electrode, a phosphor-coated ITO glass was used instead of a stainless steel plate.

Figure 3.17a shows the SEM image of a point-typed CNT paste emitter fabricated on the tip of the graphite rod. The CNT paste is strongly attached on the graphite rod after the thermal annealing. Figure 3.17b–d shows the surface morphology of the CNT paste emitters fabricated by a different weight ratio of CNTs to graphite nanoparticles. SEM images show that by decreasing the amount of graphite nanoparticles, more CNTs are exited at the surface of the CNT paste emitter as shown in Fig. 3.17b–d. When the weight ratio of the CNTs to graphite nanoparticles is 10:500, only a few CNTs exist at the surface of the CNT paste emitter as shown in Fig. 3.17d. The role of graphite nanoparticle is to disperse CNTs within the paste and also provide good electrical route in the CNT paste.

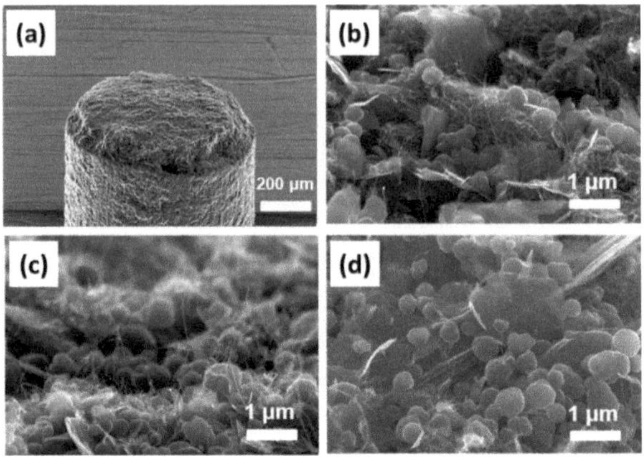

Figure 3.17 SEM images of the CNT paste emitter: (a) The tip of the point-typed CNT paste emitter fabricated on the graphite rod. The surface morphology of the CNT paste emitter according to the weight ratio of CNTs to graphite nanoparticles; (b) CNTs:graphite nanoparticles = 10:100, (c) CNTs:graphite nanoparticles = 10:300, and (d) CNTs:graphite nanoparticles = 10:500 [29].

Figure 3.18a shows the field-emission current of the point-typed CNT paste emitters according to the weight ratio of CNTs to graphite nanoparticles. The CNT paste emitter with the weight ratio of 10:300 shows the best field-emission properties, which indicates the low turn-on electric field of 2.8 V/μm and the maximum emission current of 4.1 mA at the electric field of 9.5 V/μm. It is well known that both the emission sites and the screening effect are considered as important factors to decide emission performance [62, 63]. Emission sites at the CNT paste emitter might be reduced with decreasing the amount of graphite nanoparticles, while the screening effect might be enhanced. In this work, the CNT paste emitter with the weight ratio of 10:300 is the optimized condition to enhance emission current. The F–N plot shows that the electron emission from the CNT paste emitter follows the quantum mechanical tunneling phenomenon as shown in Fig. 3.18b. The field enhancement factor calculated from the F–N equation indicates little different values about 768, 910, and 810 for the weight ratios of 10:100, 10:300, and 10:500, respectively. The y-intercept value of the CNT paste emitters for the weight ratios of 10:100, 10:300, and 10:500 is 4.9, 5.5, and 4.7, respectively. Among the three conditions, the CNT paste emitter with the weight ratio of 10:300 shows the highest value of y-intercept, which reveals more emission sites exist at the surface of the CNT paste emitter.

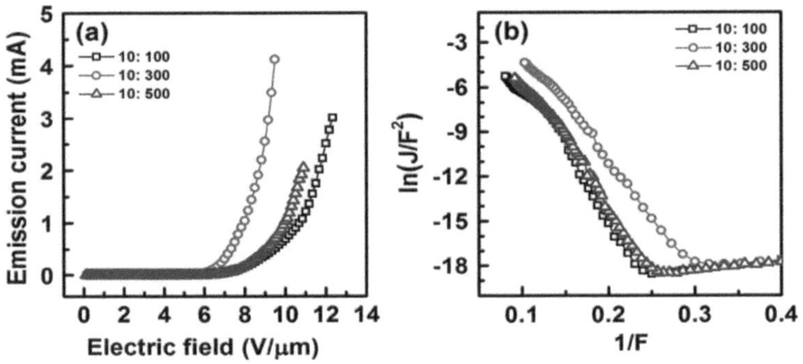

Figure 3.18 Field-emission properties of the CNT paste emitters with different weight ratios of the CNTs to the graphite nanoparticles. (a) Emission current characteristics of the CNT paste emitters and (b) F–N plots derived from the corresponding I–F curves [29].

The mechanical treatment by using a sandpaper was performed to obtain a further increase of the emission current for the point-typed CNT paste emitter. In this work, sandpaper treatment with a grit of 2000 (~10 μm) was used to increase the flatness of the surface of CNT paste emitter and also to remove protruding CNTs on the surface of the CNT paste emitter. Figure 3.19 shows the emission properties of the CNT paste emitters with the weight ratio of 10:300 according to the surface treatment by using sandpaper and a rubber roller. Figure 3.19a shows that the maximum emission current is largely increased from 4.6 to 8.4 mA (corresponding to an increased current density from 1.2 to 2.2 A/cm^2) after the sandpaper treatment, and the turn-on field is also increased from 2.8 to 3.2 V/μm. The enhanced maximum emission current is caused by the increased numbers of emission sites after the sandpaper treatment. By effectively removing the protruding CNTs after the sandpaper treatment, many CNTs can contribute electron emission, resulting in the increased emission current. By the way, the increased turn-on field is attributed to the reduced CNT length after the sandpaper treatment. Compared to the other contemporary CNT paste emitters, the point-typed CNT paste emitter shows a much higher emission current density at the same turn-on electric field [59, 64–69]. Figure 3.19b shows the F–N plot of the CNT paste emitters according to the surface treatment. The field enhancement factor calculated from the F–N plot is decreased from 956 to 713 after the sandpaper treatment. The reduced field enhancement factor is caused by shortening CNT lengths. The *y*-intercept value calculated from the F–N plot is increased from 5.4 to 8.0 after the sandpaper treatment, indicating that the number of emission sites at the surface of the CNT paste emitter is increased.

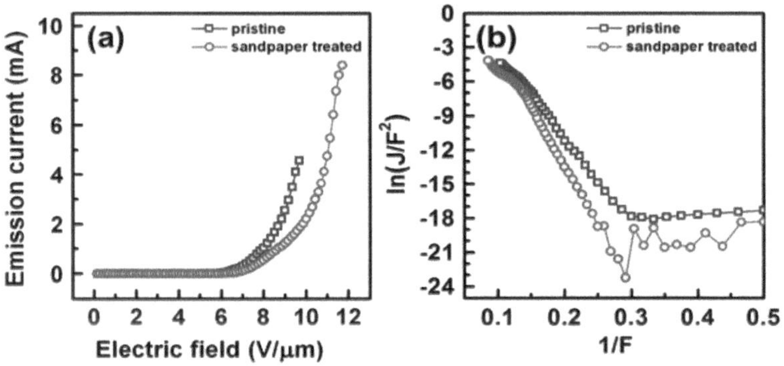

Figure 3.19 Field-emission properties of the CNT paste emitters according to the surface treatment. The weight ratio of the CNTs to the graphite nanoparticles is 10:300. (a) The emission current of the CNT paste emitters and (b) the corresponding F–N plots [29].

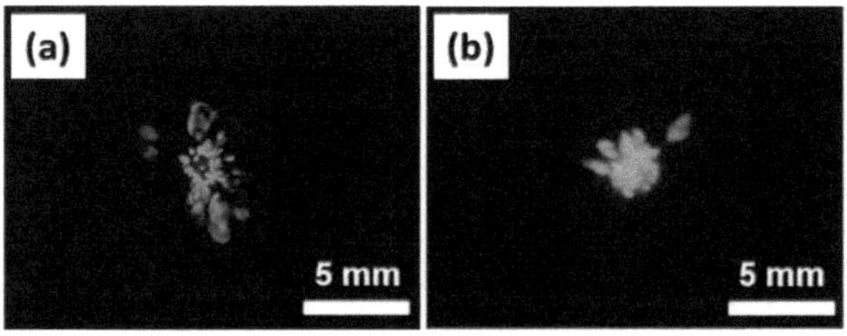

Figure 3.20 Emission patterns of the CNT paste emitter (a) without surface treatment and (b) with surface treatment [29].

The emission pattern was obtained by replacing the anode electrode from the stainless steel plate to a phosphor-coated ITO glass substrate. The gap between the emitter tip and the anode electrode was set to 500 μm. The electron emission patterns of the CNT paste emitters are shown in Fig. 3.20. The obtained emission patterns present that the number of emission sites are much improved after the sandpaper treatment and also support that the improved maximum emission current is largely caused by more emission sites after the sandpaper treatment.

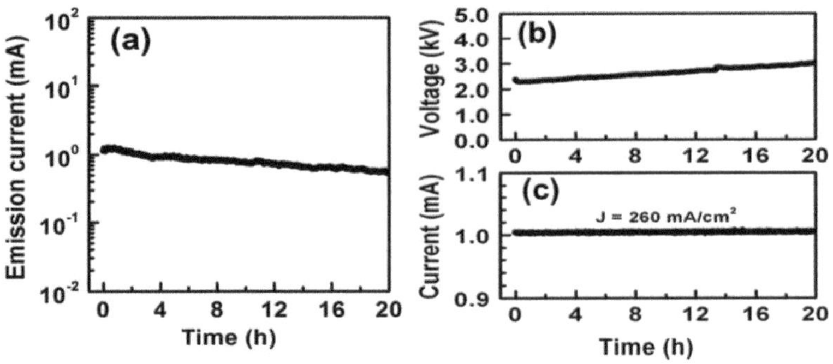

Figure 3.21 Lifetime measurement of the CNT paste emitters. (a) Emission current degradation at a constant voltage mode under DC power for 20 h. (b) The change of the anode voltage at a constant emission current mode under DC power for 20 h. (c) The recorded emission current at 1 mA at a constant emission current mode under DC power for 20 h [29].

The long-term emission stability was firstly measured by applying a constant anode voltage at the starting emission current of 1 mA (corresponding to a current density of 260 mA/cm^2). Figure 3.21a shows the lifetime measurement of the CNT paste emitter at a constant anode voltage. The emission current is degraded from 1 to 0.6 mA over 20 h, indicating the degradation rate about 40%. It reveals that the CNT paste emitter has good long-term emission stability enough to apply it to real field-emission devices. Sometimes the lifetime measurement at a constant current mode is crucial for field-emission applications such as X-ray systems. A voltage-changeable power supply was applied to the anode electrode to maintain a constant emission current of 1 mA, and the anode voltage and the emission current were recorded for 20 h. Figure 3.21b shows that the applied anode voltage is increased from 2.37 to 3.01 kV at the emission current of 1 mA (corresponding to a current density of 260 mA/cm^2) for 20 h, indicating a slight increased anode voltage about 0.64 kV (corresponding to an increased electric field of 1.8 V/μm). The increased anode voltage is caused by the degradation of the CNTs due to the heat generation by the high emission current density at the CNT tip. It is noteworthy that the heat is generated continually at the high current density when a DC power supply is used [70]. On the other hand, the heat generation at the tip is less occurred when a pulsed power supply is used because the CNT paste emitter is cool down during the turn off. Therefore, it is considered that the performance of the CNT paste emitters could be further improved if a pulsed power supplier is used instead of a DC power supplier. The emission stability result under a constant current mode also indicates that the point-typed CNT paste emitter has good long-term emission stability which can be used for real applications [70, 71].

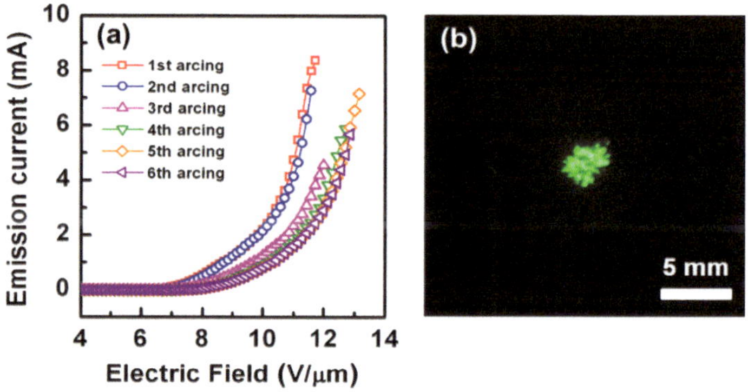

Figure 3.22 (a) Field-emission properties of the CNT paste emitter after several electrical arcing event. (b) The electron emission pattern of the CNT paste emitter after several electrical arcing events [29].

The electrical arcing endurance of the point-typed CNT paste emitters was also investigated to confirm the robustness of the CNT paste emitters. In general, the electrical arcing can severely destroy the CNT tips, which induces destruction of CNT field emitters. The anode voltage was increased to a sufficiently high value to induce the electrical arcing. After the electrical arcing event, the measurement was stopped and the next measurement was repeated. Figure 3.22a shows the field-emission properties of the CNT paste emitter according to the arcing event. Even though the electrical arcing occurs 6 times, the CNT paste emitter was still well operated and indicated a high emission current over 5 mA (higher than 1 A/cm^2). Typically, CNT emitters have shown much degraded emission properties after electrical arcing. However, the point-typed CNT paste emitters still showed good emission properties even though several arcing events exist. This interesting result explains that CNTs have very strong mechanical adhesion at the CNT paste material. It means that the graphite adhesive binder and graphite nanoparticles have good mechanical binding with CNTs and the graphite rod. It is a special advantage for CNT paste emitter process, which is different from EC binder and other filler materials such as ITO, Ag, Ni, SiO_2, and SiC nanoparticles. The emission pattern was observed to check the damage of emission sites after electrical arcing events. Figure 3.22b shows the emission pattern after electrical arcing events occurred 6 times. It reveals that a lot of active emission sites are still existed. In general, a single CNT or many CNTs on a sharp metal tip or CNT yarns or conventional CNT paste emitters would be heavily damaged after electrical arcing events, resulting in the disabling of the efficient electron emission capability. However, the obtained result supports a strong evidence that the proposed point-typed CNT paste emitter is robust and reliable, thereby enabling its use for real field-emission devices in harsh conditions.

In order to enhance field-emission performance, tip sonication process was inserted to fabrication process of the CNT paste. The CNT solution was prepared by blending CNT powder and graphite nanoparticles, SDS, and DI water. Before ball milling process, tip sonication process was done for 1 h and the CNT film was fabricated by vacuum filtration for 2 h. The produced CNT film was baked at 85 °C in an oven for 30 min. And then, the CNT film was mixed with the graphite binder. By adopting ball milling process for 1 h, the CNT paste was fabricated. Baking process for the CNT paste was the same as the previous method. The point-typed CNT paste emitter by using tip sonication process shows much high emission current as shown in Fig. 3.23. The emission current of the CNT paste emitter is increased from 6 to 13.5 mA after including tip sonication process even though the turn-on electric field is increased. This result is caused by a uniform length of CNTs caused by the cutting effect of CNTs through tip sonication process. The uniform length of CNT at the CNT paste induces that more CNTs can take part in electron emission. The increased turn-on electric field is attributed to the reduced length of CNTs during tip sonication process.

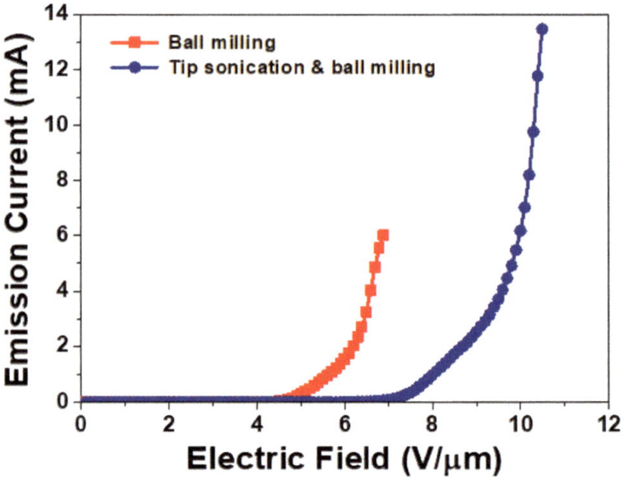

Figure 3.23 Field-emission properties of the CNT paste emitter by adopting tip sonication process.

The point-typed CNT paste emitter shows the enhanced long-term emission stability and the reduced emission current fluctuation by adopting tip sonication process as shown in Fig. 3.24. The degradation rate of the emission current is reduced from 33.6 to 28.7% by adopting tip sonication process as shown in Fig. 3.24a. The emission current fluctuation is much enhanced by adopting tip sonication process as shown in Fig. 3.24b. These results are mainly caused by an increased uniformity of the CNT length after tip sonication process during fabrication of the CNT paste.

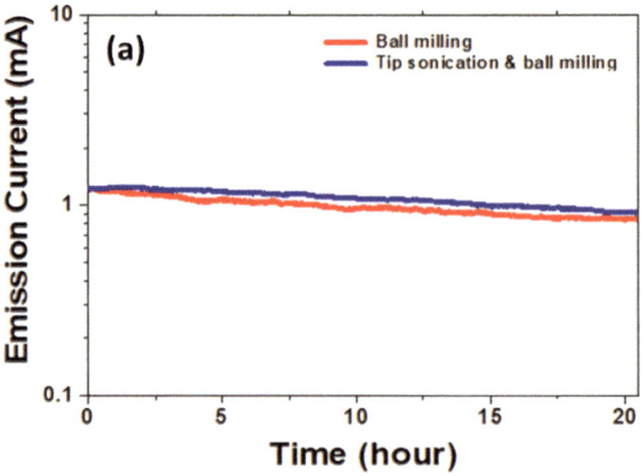

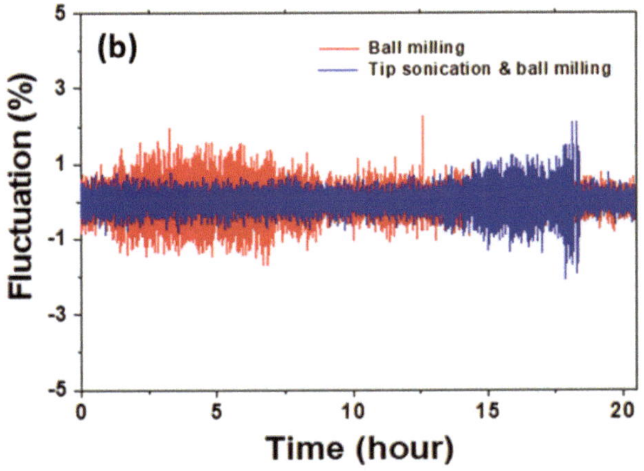

Figure 3.24 Long-term emission stability (a) and emission current fluctuation (b) of the CNT field emitters by adopting tip sonication process.

3.6 Point-Typed CNT Field Emitter Made of a Free-standing CNT Film

Field-emission properties of the point-typed CNT field emitter made by a free-standing CNT film have been studied [21]. The point-typed CNT field emitter was fabricated by cutting SWCNT film into a triangular shape using a razor as illustrated in Fig. 3.25a. The triangular-shaped CNT film (TCF) emitter has a tip angle of 10° (T10), 30° (T30), 90° (T90), and 150° (T150), respectively. SEM image of the T30-CNT tip fabricated by using a free-standing CNT film is shown in Fig. 3.25b. To fabricate the point-typed CNT field emitter, the triangular-shaped CNT film was tightly fixed by clamping a free-standing CNT film between two flat stainless steel plates. A magnified SEM image shows that the TCF emitter has a vertically aligned sharp tip which consists of multiple CNT bundles as shown in Fig. 3.25c. Figure 3.25d shows a SEM image of the side surface of the free-standing CNT film, in which the SWCNTs are tightly packed due to van der Waals forces, resulting in very mechanically stable film formation. The CNT film with the thickness of 5 μm presents good electrical conductivity and good mechanical strength and flexibility [21]. The fabrication process of the point-typed TCF emitters is very simple and reliable. Moreover, the process does not need any organic binder material which is fatal to vacuum devices.

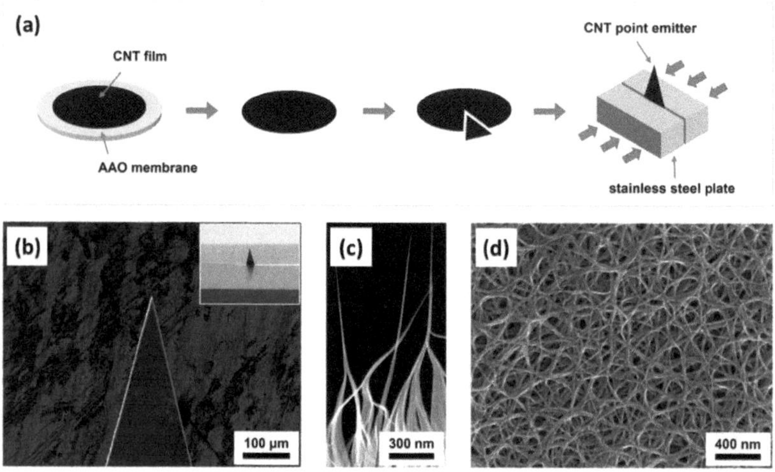

Figure 3.25 (a) A schematic of fabrication process of the point-typed CNT field emitter. (b) SEM image of the T30 CNT tip fabricated by using a free-standing CNT film. The inset represents a photograph of the TCF emitter clamped by stainless steel plates. (c) A magnified SEM image of the tip of the TCF emitter. (d) A SEM image of the side surface of the CNT film [21].

Figure 3.26a shows the field-emission properties of the point-typed TCF emitters according to tip angles of the CNT film. The current was measured continuously until the field emitter electrically broke down. Interestingly, both the turn-on electric field and the maximum emission current are increased as the tip angle increases. The turn-on electric fields of the TCF emitter with T10, T30, T90, and T150 angle are 0.24, 0.34, 0.46, and 0.54 V/μm, respectively. The maximum emission current of the TCF emitters is dramatically increased from 2.3 up to 22.4 mA as the tip angle increases from T10 to T150. It is noteworthy that the emission current of 22.4 mA (corresponding to the emission current density of ~10^5 A/cm^2) is a much higher level compared to those of previous point-typed CNT emitters [20–29, 72–78]. It is understood that the tensile stresses of the 2D free-standing CNT film is much lower than that of 1D materials and it decreases as the tip angle increases. F–N plots of the emission current are shown in Fig. 3.26b. The field enhancement factor for T10, T30, T90, and T150 TCF emitters, which is calculated from the F–N plots, was estimated to be 10418, 7569, 6052, and 5004, respectively.

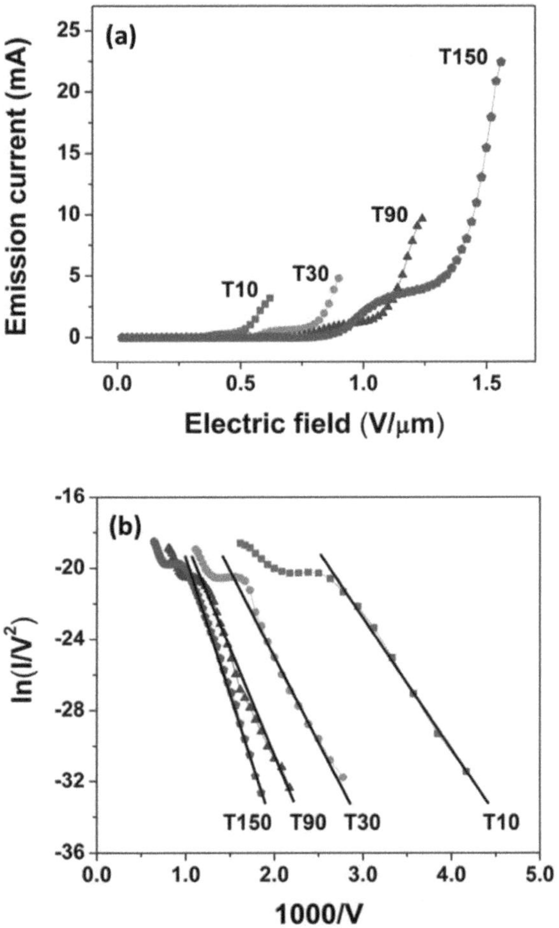

Figure 3.26 Field-emission characteristics of the point-typed TCF emitters according to tip angles (i.e., 10°, 30°, 90°, and 150°). (a) Field-emission current characteristics according to tip angles. (b) F–N plots (plotted points: experimental data) [21].

Figure 3.27a shows the long-term emission stability of the point-typed TCF emitter. The measurement was performed by applying a constant voltage to maintain an initial emission current of 1 mA for 20 h. The emission current of T10, T30, T90, and T150 TCF emitters

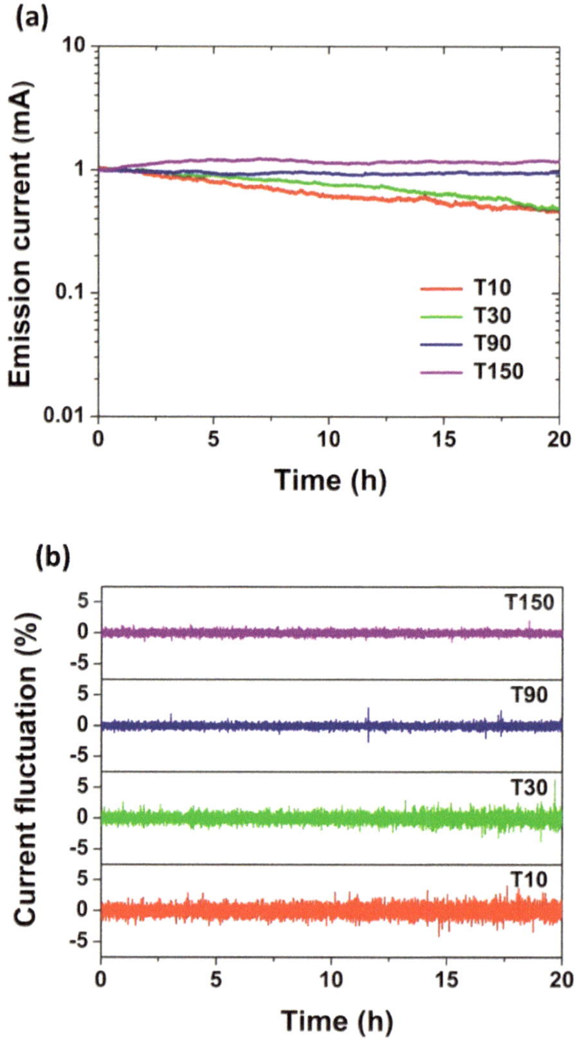

Figure 3.27 Long-term emission stability of the point-typed TCF emitter for 20 h (a) and fluctuation of the emission current of the TCF emitters (b) [21].

changes from 1 mA to 0.46, 0.49, 0.96, and 1.24 mA, respectively, during the field-emission measurement. It is understood that the long-term emission stability of the TCF emitter is largely dependent on the tip angle. The obtained degradation rate of the TCF emitters is high enough to be applicable for field-emission devices.

Interestingly, the emission current of the T150 TCF emitter indicates no degradation for 20 h. It is noteworthy that the TCF emitters show excellent emission stability even at the relatively large emission current of 1 mA compared to the results of previous CNT point emitters which are mostly measured at less than 0.1 mA [20–29, 72–80]. Figure 3.27b shows the fluctuation of the emission current of the TCF emitters. The fluctuation ratio of the emission current of the TCF emitters was calculated by $\Delta I/I \times 100$ (%) and was found to be −4.1–4.1%, −3.7–3.0%, −2.7–3.0%, and −1.5–2.1%, respectively. It can be clearly seen that the emission current is more stable for a larger tip angle.

Figure 3.28 shows the dark image of the emission pattern of the point-typed TCF field emitter. The gap between the CNT emitter tip and the anode electrode was 1 mm. The emission pattern indicates a small electron beam spot of 1.5 mm, revealing a small focal spot that can be used for high-resolution X-ray sources. The emission pattern presents much smaller and more focused electron beam spot compared to previous point-typed CNT field emitters, in which the TCF field emitter is appropriate for high resolution X-ray sources [20, 23, 27–29].

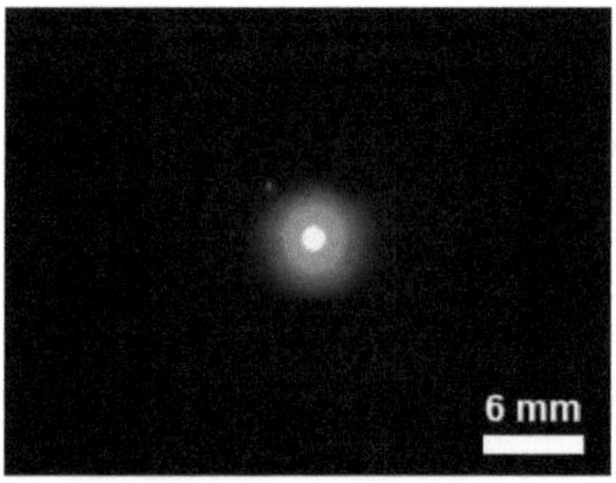

Figure 3.28 Emission patterns of the TCF field emitter [21].

Figure 3.29 shows a comparison of the maximum emission current from the point-typed CNT emitters achieved from various emission areas. The point-typed TCF field emitter shows the highest emission performance compared to other point-typed CNT field emitters. Most of the experimental results indicate that the achieved emission current from point-typed CNT emitters is less than 10 mA [20, 23, 33, 72–83]. However, the point-typed TCF field emitter indicates a very high emission current of 22.4 mA corresponding to a high emission current density of 10^5 A/cm^2 at the low electric field of 1.5 V/μm. (See red star shown in Fig. 3.29.)

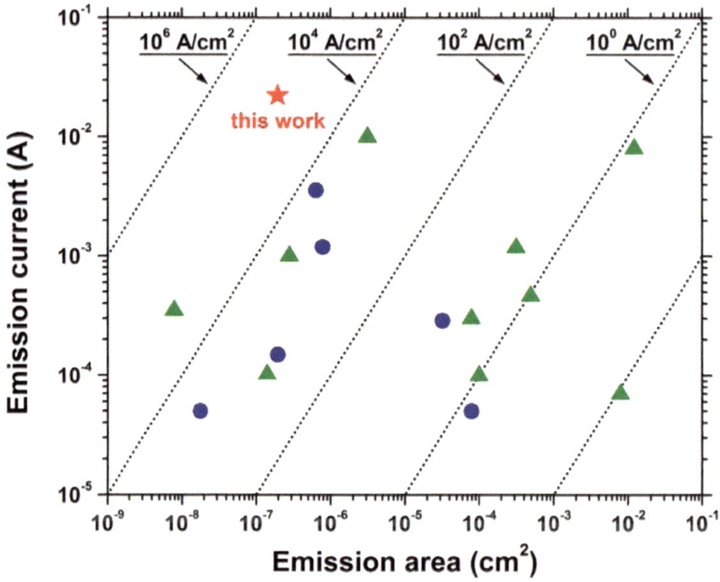

Figure 3.29 Comparison of the maximum emission current versus the emission area obtained from point-typed CNT field emitters. The point-typed CNT field emitters were fabricated by a CNT yarn (blue circles), by multiple CNTs on a metal tip (green triangles), and by TCF emitter (red star, this work), respectively [20, 21, 23, 33, 72–83].

3.7 Summary

Fabrication methods and field-emission properties of point-typed CNT field emitters were investigated. The point-typed CNT field emitters show a higher emission current density and a smaller electron beam spot compared to planar-typed CNT field emitters. In general, it is very important to get not only a high emission current but also a high emission current density with a small focal spot to realize high performance field-emission devices. Performance of field-emission properties of the point-typed CNT field emitters is summarized in Table 3.1. The point-typed CNT field emitters indicate not only a high emission current but also a very high emission current

Table 3.1 Comparison of field-emission performance of the point-typed CNT field emitters

	CNT bundle emitter	**CNT yarn emitter**	**CNT film emitter**	**CNT paste emitter**	**Free standing CKT film emitter**
Max emission current	4.9 μA	3.2 mA	6.3 mA	13.5 mA	22.4 mA
Emission current density	$\sim 1.2 \times 10^3$ A/cm^2	$\sim 10^3$ A/cm^2	~ 1.6 A/cm^2	~ 5 A/cm^2	$\sim 10^5$ A/cm^2
Turn-on voltage (0.1 μA/cm^2)	650 V	200 V	980 V	900 V	700 V
Threshold voltage (1.0 mA/cm^2)	–	500 V	1340 V	1600 V	1400 V
Emission stability	Fair	Excellent	Excellent	Excellent	Excellent
Electron beam spot	Excellent	Good	Good	Excellent	Excellent

density, which are important for realistic applications for diverse field electron devices. In addition, they show a low turn-on voltage, good emission stability, and a small electron beam spot compared to conventional CNT field emitters. It means that the point-typed CNT field emitters can be widely used for various field-emission devices such as X-ray sources, electron microscope, and so on in the future. Interestingly, the point-typed TCF field emitter indicates the excellent field-emission properties compared to ever reported CNT field emitters. Thus, it can enlarge the realm of field-emission device application not only in medical diagnosis systems but also in diverse industrial nondestructive inspection systems.

References

1. Treacy, M., Ebbesen, T. W. and Gibson, J. M. (1996). Exceptionally high Young's modulus observed for individual carbon nanotubes, *Nature*, **381**, pp. 678–680.
2. White, C. T. and Todorov, T, M. (1998). Carbon nanotubes as long ballistic conductors, *Nature*, **393**, pp. 240–242.
3. Ajayan, P. M. (1999). Nanotubes from carbon, *Chem. Rev.*, **99**, pp. 1787–1799.
4. Kim, P., Shi, L., Majumdar, A. and McEuen, P. (2001). Thermal transport measurements of individual multiwalled nanotubes, *Phys. Rev. Lett.*, **87**, 215502.
5. Rinzler, A. G., Hafner, J. H., Nikolaev, P., Lou, L., Kim, S. G., Tománek, D., Nordlander, P., Colbert, D. T. and Smalley, R. E. (1995). Unraveling nanotubes: Field emission from an atomic wire, *Science*, **269**, pp. 1550–1553.
6. De Heer, W. A., Châtelain, A. and Ugarte, D. (1995). A carbon nanotube field-emission electron source, *Science*, **270**, pp. 1179–1180.
7. Chernozatonskii, L. A., Gulyaev, Y. V., Kosakovskaja, Z. J., Sinitsyn, N. I., Torgashov, G. V., Zakharchenko, Y. F., Fedorov, E. A. and Val'chuk, V. P. (1995). Electron field emission from nanofilament carbon films, *Chem. Phys. Lett.*, **233**, pp. 63–68.
8. Saito, Y., Hamaguchi, K., Hata, K., Uchida, K., Tasaka, Y., Ikazaki, F., Yumura, M., Kasuya, A. and Nishina, Y. (1997). Conical beams from open nanotubes, *Nature*, **389**, pp. 554–555.
9. Lee, C. J., Kim, D. W., Lee, T. J., Choi, Y. C., Park, Y. S., Lee, Y. H., Choi, W. B., Lee, N. S., Park, G. S. and Kim, J. M. (1999). Synthesis of aligned carbon

nanotubes using thermal chemical vapor deposition, *Chem. Phys. Lett*, **312**, pp. 461–468.

10. Saito, Y. and Uemura, S. (2000). Field emission from carbon nanotubes and its application to electron sources, *Carbon*, **38**, pp. 169–182.
11. Lee, N. S., Chung, D. S., Kang, J. H., Kim, H. Y., Park, S. H., Jin, Y. W., Choi, Y. S., Han, I. T., Park, N. S. and Yun, M. J. (1999). Carbon nanotube-based field-emission displays for large-area and full-color applications, *Appl. Phys. Lett.*, **39**, pp. 7154–7158.
12. De Jonge, N., Lamy, Y., Schoots, K. and Oosterkamp, T. H. (2002). High brightness electron beam from a multi-walled carbon nanotube, *Nature*, **420**, pp. 393–395.
13. Teo, K. B. K., Minoux, E., Hudanski, L., Peauger, F., Schnell, J. P., Gangloff, L., Legagneux, P., Dieumegard, D., Amaratunga, G. A. J. and Milne, W. I. (2005). Carbon nanotubes as cold cathodes, *Nature*, **437**, 968.
14. Cheng, Y., Zhang, J., Lee, Y. Z., Gao, B., Dike, S., Lin, W., Lu, J. P. and Zhou, O. (2004). Dynamic radiography using a carbon-nanotube-based field-emission x-ray source, *Rev. Sci. Instrum.*, **75**, 3264.
15. De Jonge, N. and Bonard, J. M. (2004). Carbon nanotube electron sources and applications, *Phil. Trans. R. Soc. A*, **362**, 2239.
16. Chen, G., Shin, D. H., Iwasaki, T., Kawarada, H. and Lee, C. J. (2008). Enhanced field emission properties of vertically aligned double-walled carbon nanotube arrays, *Nanotechnology*, **19**, 415703.
17. Jung, S. I., Jo, S. H., Moon, H. S., Kim, J. M., Zang, D. S., Lee, C. J. (2007). Improved crystallinity of double-walled carbon nanotubes after a high-temperature thermal annealing and their enhanced field emission properties, *J. Phys. Chem. C*, **111**, pp. 4175–4179.
18. Ha, B., Shin, D. H., Park, J., Lee, C. J. (2008). Electronic structure and field emission properties of double-walled carbon nanotubes synthesized by hydrogen arc discharge, *J. Phys. Chem. C*, **112**, pp. 430–435.
19. Gao, B., Yue, G. Z., Qiu, Q., Cheng, T., Shimoda, H., Fleming, L., Zhou, O. (2001). Fabrication and electron field emission properties of carbon nanotube films by electrophoretic deposition, *Adv. Mater.*, **13**, 1770.
20. Chen, G., Shin, D. H., Roth, S., Lee, C. J. (2009). Field emission characteristics of point emitters fabricated by a multiwalled carbon nanotube yarn, *Nanotechnology*, **20**, 315201.
21. Shin, D. H., Yun, K. N., Jeon, S. -G., Kim, J. I., Saito, Y., Milne, W. I. and Lee, C. J. (2015). High performance field emission of carbon nanotube film emitters with a triangular shape, *Carbon*, **89**, pp. 404–410.

22. Wei, Y., Weng, D., Yang, Y., Zhang, X., Jiang, K., Liu, L. and Fan, S. (2006). Efficient fabrication of field electron emitters from the multiwalled carbon nanotube yarns, *Appl. Phys. Lett.*, **89**, 063101.

23. Jung, S. I., Choi, J. S., Shim, H. C., Kim, S., Jo, S. H. and Lee, C. J. (2006). Fabrication of probe-typed carbon nanotube point emitters, *Appl. Phys. Lett.*, **89**, 233108.

24. Yabushita, R. and Hata, K. (2008). Development of high spatial resolution x-ray radiography system equipped with multiwalled carbon nanotube field emission cathode, *J. Vac. Sci. Technol. B*, **26**, pp. 702–705.

25. Asaka, K., Nakahara, H. and Saito, Y. (2008). Nanowelding of a multiwalled carbon nanotube to metal surface and its electron field emission properties, *Appl. Phys. Lett.*, **92**, 023114.

26. Sakai, Y., Tone, D., Nagatsu, S., Endo, T., Kita, S. and Okuyama, F. (2009). Characterization of field emission from carbon nanofibers on a metal tip, *Appl. Phys. Lett.*, **95**, 073104.

27. Sun, Y., Song, Y., Shin, D. H., Yun, K. N., Jeon, S.-G., Kim, J.-I., Saito, Y. and Lee, C. J. (2014). Fabrication of carbon nanotube emitters on the graphite rod and their high field emission performance, *Appl. Phys. Lett.*, **104**, 043104.

28. Sun, Y., Shin, D. H., Yun, K. N., Hwang, Y. M., Song, Y., Leti, G., Jeon, S. -G., Kim, J. -I., Saito, Y. and Lee, C. J (2014). Field emission behavior of carbon nanotube field emitters after high temperature thermal annealing, *AIP Adv.*, **4**, 077110.

29. Sun, Y., Yun, K. N., Leti, G., Lee, S. H., Song, Y. H. and Lee, C. J. (2017). High-performance field emission of carbon nanotube paste emitters fabricated using graphite nanopowder filler, *Nanotechnology*, **28**, 065201.

30. Sugie, H., Tanemura, M., Filip, V., Iwata, K., Takahashi, K. and Okuyama, F. (2001). Carbon nanotubes as electron source in an x-ray tube, *Appl. Phys. Lett.*, **78**, pp. 2578–2580.

31. Yue, G. Z., Qui, Q., Gao, B., Cheng, Y., Zhang, J., Shimoda, H., Chang, S., LU, J. P. and Zhou, O. (2002). Generation of continuous and pulsed diagnostic imaging x-ray radiation using a carbon-nanotube-based field-emission cathode, *Appl. Phys. Lett.*, **81**, pp. 355–357.

32. Zhang, J., Yang, G., Cheng, Y., Gao, B., Qui, Q., Lee, Y. Z., LU, J. P. and Zhou, O. (2005). Stationary scanning x-ray source based on carbon nanotube field emitters, *Appl. Phys. Lett.*, **86**, 184104.

33. Kawakita, K., Hata, K., Sato, H. and Saito, Y. (2006). Development of microfocused x-ray source by using carbon nanotube field emitter, *J. Vac. Sci. Technol. B*, **24**, pp. 950–952.
34. Heo, S. H., Ihsan, A. and Cho, S. O. (2007). Transmission-type microfocus x-ray tube using carbon nanotube field emitters, *Appl. Phys. Lett.*, **90**, 183109.
35. Heo, S. H., Kim, H. J., Ha, J. M. and Cho, S. O. (2012). A vacuum-sealed miniature X-ray tube based on carbon nanotube field emitters, *Nanoscale Res. Lett.*, **7**, 258.
36. Jeong, J. W., Kang, J. T., Choi, S. Y., Kim, J. W., Ahn, S. J. and Song, Y. H. (2013). A digital miniature x-ray tube with a high-density triode carbon nanotube field emitter, *Appl. Phys. Lett.*, **102**, 023504.
37. Kang, J. T., Lee, H. R., Jeong, J. W., Kim, J. W., Park, S., Shin, M. S., Yeon, J. H., Jeon, H. J., Kim, S. H., Choi, S. Y. and Song, Y. H. (2015). Fast and stable operation of carbon nanotube field-emission X-ray tubes achieved using an advanced active-current control, *IEEE Electron Device Lett.*, **36**, pp. 1209–1211.
38. Lee, H. R., Lee, S. W., Shikili, C., Kang, J. S., Lee, J. and Park, K. C. (2016). Enhanced electron emission of paste CNT emitters with nickel buffer layer and its X-ray application, *J. Nanosci. Nanotechnol.*, **16**, pp. 12053–12058.
39. Lee, S. H., Han, J. S., Go, H. B. and Lee, C. J. (2018). High performance x-ray source using point-typed CNT field emitter, *IVNC*, **31**, 86.
40. Lee, S. H., Han, J. S., Go, H. B. and Lee, C. J. (2019). Fabrication and high performance of CNT cold cathode x-ray tube, *IVNC*, **32**, 88.
41. Forbes, R. G. (2001). Low-macroscopic-field electron emission from carbon films and other electrically nanostructured heterogeneous materials: Hypotheses about emission mechanism, *Solid-State Electron*, **45**, pp. 779–808.
42. Bonard, J.-M., Klinke, C., Dean, K. A. and Coll, B. F. (2003). Degradation and failure of carbon nanotube field emitters, *Phys. Rev. B*, **67**, 115406.
43. Kim, D. H., Yang, H. S., Kang, H. D. and Lee, H. R. (2003). Novel emission degradation behavior of patterned carbon nanotubes by field emission, *Chem. Phys. Lett.*, **368**, pp. 439–444.
44. Lee, J. H., Lee, S. H., Kim, W. S., Lee, H. J., Heo, J. N., Jeong, T. W., Baik, C. W., Park, S. H. and Yu, S. G. (2006). Current degradation mechanism of single wall carbon nanotube emitters during field emission, *Appl. Phys. Lett.*, **89**, 253115.

45. Vincent, P., Poncharal, P., Barois, T., Perisanu, S., Gouttenoire, V., Frachon, H., Lazarus, A., de Langre, E., Minoux, E., Charles, M., Ziaei, A., Guillot, D., Choueib, M., Ayari, A. and Purcell, S. T. (2011). Performance of field-emitting resonating carbon nanotubes as radio-frequency demodulators, *Phys. Rev. B*, **83**, 155446.

46. Kleshch, V. I., Obraztsov, A. N. and Obraztsova, E. D. (2010). Electromechanical self-oscillations of carbon nanotube field emitter, *Carbon*, **48**, pp. 3895–3900.

47. Liu, P., Jiang, K., Wei, Y., Liu, K., Liu, L. and Fan, S. (2010). Field emission behavior study of multiwalled carbon nanotube yarn under the influence of adsorbents, *J. Vac. Sci. Technol. B*, **28**, 736.

48. Asaka, K., Nakahara, H. and Saito, Y. (2008). Nanowelding of a multiwalled carbon nanotube to metal surface and its electron field emission properties, *Appl. Phys. Lett.*, **92**, 023114.

49. Heo, S. H., Ihsan, A., Yoo, S. H., Ali, G. and Cho, S. O. (2010). Stable field emitters for a miniature X-ray tube using carbon nanotube drop drying on a flat metal tip, *Nanoscale Res. Lett.*, **5**, 720.

50. Minoux, E., Groening, O., Teo, K. B. K., Dalal, S. H., Gangloff, L., Schnell, J.-P., Hudanski, L., Bu, I. Y. Y., Vincent, P., Legagneux, P., Amaratunga, G. A. J. and Milne, W. I. (2005). Achieving high-current carbon nanotube emitters, *Nano Lett.*, **5**, pp. 2135–2138.

51. Jung, S. I., Jo, S. H., Moon, H. S., Kim, J. M., Zang, D.-S. and Lee, C. J. (2007). Improved crystallinity of double-walled carbon nanotubes after a high-temperature thermal annealing and their enhanced field emission properties, *J. Phys. Chem. C*, **111**, pp. 4175–4179.

52. Huang, J. Y., Kempa, K., Jo, S. H., Chen, S. and Ren, Z. F. (2005). Giant field enhancement at carbon nanotube tips induced by multistage effect, *Appl. Phys. Lett.*, **87**, 053110.

53. Seelaboyina, R., Boddepalli, S., Noh, K. N., Jeon, M. and Choi, W. (2008). Enhanced field emission from aligned multistage carbon nanotube emitter arrays, *Nanotechnology*, **19**, 065605.

54. Smith, R. C. and Silva, S. R. P. (2009). Maximizing the electron field emission performance of carbon nanotube arrays, *Appl. Phys. Lett.*, **94**, 133104.

55. Song, Y., Shin, D. H., Jeon, S.-G., Kim, J.-I. and Lee, C. J. (2013). Field emission properties of carbon nanotube emitters dependent on electrode geometry, *J. Vac. Sci. Technol. B*, **31**, 052203.

56. Choi, W. B., Jin, Y. W., Kim, H. Y., Lee, S. J., Yun, M. J., Kang, J. H., Choi, Y. S., Park, N. S., Lee, N. S. and Kim, J. M. (2001). Electrophoresis deposition

of carbon nanotubes for triode-type field emission display, *Appl. Phys. Lett.*, **78**, 1547.

57. Lee, H. J., Lee, Y. D., Cho, W. S., Ju, B. K., Lee, Y. -H., Han, J. H. and Kim, J. K. (2006). Field-emission enhancement from change of printed carbon nanotube morphology by an elastomer, *Appl. Phys. Lett.*, **88**, 093115.

58. Kim, Y. C., Hur, J. N., Kim, I. H., Park, S. H., Jung, T. W., Kim, D. Y., Kim, H. J., Cha, S. N., Han, I. T., Kim, J. M. and Kim, Y. H. (2011). Activation-free printed carbon nanotube field emitters, *Nanotechnology*, **22**, 435601.

59. Jeong, J. -W., Kim, J. -W., Kang, J. -T., Choi, S., Ahn, S. and Song, Y. -H. (2013). A vacuum-sealed compact x-ray tube based on focused carbon nanotube field-emission electrons, *Nanotechnology*, **24**, 085201.

60. Kim, J. -W., Jeong, J. -W., Kang, J. -T., Choi, S., Park, S., Shin, M. -S., Ahn, S. and Song, Y. -H. (2015). Great improvement in adhesion and uniformity of carbon nanotube field emitters through reactive nanometer-scale SiC fillers, *Carbon*, **82**, pp. 245–253.

61. Floweri, O., Jo, H. J., Seo, Y. H. and Lee, N. S. (2018). Improving the thermal stability of carbon nanotubes and their field emission characteristics by adding boron and phosphorus compounds, *Carbon*, **139**, pp. 404–414.

62. Nilsson, L., Groening, O., Emmenegger, C., Kuettel, O., Schaller, E., Schlapbach, L., Kind, H., Bonard, J.-M. and Kern, K. (2000). Scanning field emission from patterned carbon nanotube films, *Appl. Phys. Lett.*, **76**, 2071.

63. Chen, P. -L., Chang, J. -K., Kuo, C. -T. and Pan, F. -M. (2005). Field emission of carbon nanotubes on anodic aluminum oxide template with controlled tube density, *Appl. Phys. Lett.*, **86**, 123111.

64. Jeong, J. W., Kang, J. T., Choi, S., Kim, J. W., Ahn, S. and Song, Y. H. (2013). A digital miniature x-ray tube with a high-density triode carbon nanotube field emitter, *Appl. Phys. Lett.*, **102**, 023504.

65. Kim, J. W., Jeong, J. W., Kang, J. T., Choi, S., Park, S., Shin, M. S., Ahn, S. and Song, Y. H. (2015). Great improvement in adhesion and uniformity of carbon nanotube field emitters through reactive nanometer-scale sic fillers, *Carbon*, **82**, pp. 245–253.

66. Yonathan, P., Kim, H. T. and Yoon, D. H. (2008). Field emission from multi-walled carbon nanotubes with various fillers, *Mater. Lett.*, **62**, pp. 2795–2798.

67. Kim, J. H., Lee, H. S., Goak, J. C., Seo, Y. H., Kim, K. B., Park, K. S. and Lee, N. S. (2010). Effect of micro and nanoparticle inorganic fillers on the field

emission characteristics of photosensitive carbon nanotube pastes, *Appl. Surf. Sci.*, **256**, pp. 2636–2642.

68. Vink, T. J., Gillies, M., Kriege, J. C. and van de Laar, H. W. J. J. (2003). Enhanced field emission from printed carbon nanotubes by mechanical surface modification, *Appl. Phys. Lett.*, **83**, pp. 3552–3554.
69. Kim, Y. C., Sohn, K. H., Cho, Y. M. and Yoo, E. H. (2004). Vertical alignment of printed carbon nanotubes by multiple field emission cycles, *Appl. Phys. Lett.*, **84**, pp. 5350–5352.
70. Calderon-Colon, X., Geng, H. Z., Gao, B., An, L., Cao, G. H. and Zhou, O. (2009). A carbon nanotube field emission cathode with high current density and long-term stability, *Nanotechnology*, **20**, 325707.
71. Liao, A., Alizadegan, R., Ong, Z. Y., Dutta, S., Xiong, F., Hsia, K. J. and Pop, E. (2010). Thermal dissipation and variability in electrical breakdown of carbon nanotube devices, *Phys. Rev. B*, **82**, 205406.
72. Zhang, J., Tang, J., Yang, G., Qiu, Q., Qin, L. C. and Zhou, O. (2004). Efficient fabrication of carbon nanotube point electron sources by dielectrophoresis, *Adv. Mater.*, **16**, pp. 1219–1222.
73. Ferrer, D., Tanii, T., Matsuya, I., Zhong, G., Okamoto, S., Kawarada, H., Shinada, T. and Ohdomari, I. (2006). Enhancement of field emission characteristics of tungsten emitters by single-walled carbon nanotube modification, *Appl. Phys. Lett.*, **88**, 033116
74. Ci, L., Punbusayakul, N., Wei, J. Q., Vajtai, R., Talapatra, S. and Ajayan, P. M. (2007). Multifunctional macroarchitectures of double-walled carbon nanotube fibers, *Adv. Mater.*, **19**, pp. 1719–1723.
75. Wei, Y., Jiang, K. L., Liu, L., Chen, Z. and Fan, S. S. (2007). Vacuum-breakdown-induced needle-shaped ends of multiwalled carbon nanotube yarns and their field emission applications, *Nano Lett.*, **7**, pp. 3792–3797.
76. Wei, Y., Liu, L., Liu, P., Xiao, L., Jiang, K. L. and Fan, S. S. (2008). Scaled fabrication of single-nanotube-tipped ends from carbon nanotube micro-yarns and their field emission applications, *Nanotechnology*, **19**, 475707
77. Kim, W. J., Lee, J. S., Lee, S. M., Song, K. Y., Chu, C. N. and Kim, Y. H. (2011). Better than 10 mA field emission from an isolated structure emitter of a metal oxide/CNT composite, *ACS Nano*, **5**, pp. 429–435.
78. Kim, J. P., Chang, H. B., Kim, B. J. and Park, J. S. (2012). Emission stability enhancement of a tip-type carbon-nanotube-based field emitter via hafnium interlayer deposition and thermal treatment, *Appl. Phys. Lett.*, **100**, 123103.

79. Lee, H., Goak, J., Choi, J., Kong, B., Lee, C. H., Kim, K. B., Park, J. Y., Seo, Y., Choi, Y. C., Song, Y. H. and Lee, N. (2012). High-current field emission of point-type carbon nanotube emitters on Ni-coated metal wires, *Carbon*, **50**, pp. 2126–2133.

80. Behabtu, N., Young, C. C., Tsentalovich, D. E., Kleinerman, O., Wang, X., Ma, A. W. K., Bengio, E. A., ter Waarbeek, R. F., de Jong, J. J., Hoogerwerf, R. E., Fairchild, S. B., Ferguson, J. B., Maruyama, B., Kono, J., Talmon, Y., Cohen, Y., Otto, M. J. and Pasquali, M. (2013). Strong, light, multifunctional fibers of carbon nanotubes with ultrahigh conductivity, *Science*, **339**, pp. 182–186.

81. Zhong, D. Y., Zhang, G. Y., Liu, S., Sakurai, T. and Wang, E. G. (2002). Universal field-emission model for carbon nanotubes on a metal tip, *Appl. Phys. Lett.*, **80**, pp. 506–508.

82. Liang, B., Ogino, A. and Nagatsu, M. (2010). Discharge characteristics of a nano-sized electrode with aligned carbon nanotubes grown on a tungsten whisker tip under various gas conditions, *J. Phys. D Appl. Phys.*, **43**, 275202.

83. Zakhidov, A. A., Nanjundaswamy, R., Obraztsov, A. N., Zhang, M., Fang, S. and Klesch, V. I. (2007). Field emission of electrons by carbon nanotube twist-yarns, *Appl. Phys. A*, **88**, pp. 593–600.

Chapter 4

Theoretical Field-Emission Patterns from Carbon Nanotubes

Toshiharu Higuchi,[a] Yoichi Yamada,[a] and Masahiro Sasaki[a,b]
[a]*Faculty of Pure and Applied Sciences, University of Tsukuba, 1-1-1 Tennoudai, Tsukuba, Ibaraki 305-8573, Japan*
[b]*Tsukuba Research Center for Energy Materials Science (TREMS), University of Tsukuba, 1-1-1 Tennoudai, Tsukuba, Ibaraki 305-8573, Japan*
toshihigu@r7.dion.ne.jp

In this chapter, field-emission patterns from carbon nanotubes (CNTs) are simulated by using the time-dependent density-functional theory (TD-DFT) and DFT. The TD-DFT and DFT calculations at the near field showed various emission patterns, e.g., five-, nine-fold symmetries, or five-, six-membered rings. These patterns reflect the positions of atoms in the CNT tips. The beam focusing characteristics were predicted by performing TD-DFT calculation in the far field.

4.1 Introduction

Carbon nanotube (CNT) field-emission patterns have been studied from both the experimental and theoretical perspectives. Saito et al. [1] acquired experimental images of capped multiwall carbon

Nanostructured Carbon Electron Emitters and Their Applications
Edited by Yahachi Saito

ISBN 978-981-4877-62-6 (Hardcover), 978-1-003-14199-0 (eBook)
www.jennystanford.com

nanotubes (MWCNTs) and described six bright pentagons, each with a dark region at its center. This is thought to be the first direct observation of pentagon structures on a MWCNT tip. The single-walled CNTs in field-emission microscope images published by Dean and Chalama [2] had five- and six-fold symmetries.

Kim et al. [3] performed a theoretical investigation of the electronic structure of a capped single-walled CNT under the applied electric field. They find that the charge densities at the top of the armchair cap show two- or five-lobed patterns depending on the field strength. Lee et al. [4] calculated the emission patterns of open-ended armchair CNTs on gold substrates by performing first-principles dynamic simulations. Hu et al. [5] used ab initio time-dependent density-functional theory (TD-DFT) to calculate the emission patterns of pristine CNTs and H_2-adsorbed CNTs. The works of Lee et al. and Hu et al. also contain examples of calculations performed using the TD-DFT method. The TD-DFT method makes it possible to predict the dependence of the electron density emitted into a vacuum on time. The emission current is determined from the relationship between electron density and the elapsed time. Therefore, it is possible to calculate the emission current almost directly using this method.

Calculations can also be performed using the density-functional theory (DFT) approach. Khazaei et al. [6] adapted the Penn–Plummer model [7] to develop a method for calculating field-emission patterns. They first performed DFT calculations to determine the potential, the electron density, and the local density of states (LDOS). They then calculated the tunneling probability and the slowly varying function of energy (details are described in subsection 4.3.1). Finally, the emission current was calculated by inserting these values into Khazaei's equation. Their method was thus a combination of the DFT and Khazaei methods. It will be referred to here as a DFT method for the sake of convenience.

We are also interested in emission patterns, and we discuss the simulated emission patterns of various CNTs obtained using the TD-DFT and DFT methods in this chapter [8, 9].

4.2 Field-Emission Patterns from CNTs Calculated Using TD-DFT

4.2.1 TD-DFT

The TD-DFT method is based on the time-dependent density-functional theory, solving the time-dependent Kohn-Sham (TD-KS) equation in real-time and real-space [10, 11]. With this method, the emission current can be calculated almost directly. It can be used to calculate near field– and far field–emission patterns, and transitions in electron energy levels due to electric field effects, although this is not covered in this chapter. The TD-DFT method can also be used to calculate wave function patterns. However, it cannot be used to calculate transmittance and the LDOS.

Equation (4.1) is the TD-KS equation used to describe the evolution of electron density (ρ) over time via a one-electron wave function Ψ_n,

$$i\frac{\partial \Psi_n(t)}{\partial t} = \left(-\frac{\Delta}{2} + \hat{V}[\rho,t]\right)\Psi_n(t)$$

$$\rho(t) = \sum_{n=1}^{N} |\Psi_n(t)|^2. \tag{4.1}$$

Here, $-\Delta/2$ is a kinetic operator, and $\hat{V}[\rho, t]$ is a potential operator. Atomic units are used hereafter. The solution to Eq. (4.1) has the exact expression:

$$\Psi_n(t) = T\exp\left[-i\int_0^t dt'\left(-\frac{\Delta}{2} + \hat{V}[\rho,t']\right)\right]\Psi_n(0), \tag{4.2}$$

where T exp is the time-ordered exponential. There are many numerical methods for computing Eq. (4.2). Although it does not afford an exact solution, the simplest method discretizes the elapsed time (t) into small slices (Δt) to approximate Eq. (4.2). Equation (4.3) represents the simplified form:

$$\Psi_n(t+\Delta t) \approx \exp\left[-i\,\Delta t\left(-\frac{\Delta}{2} + \hat{V}[\rho,t]\right)\right]\Psi_n(t). \tag{4.3}$$

There are many methods for computing a time evolution operator as a time-ordered exponential. Using an operator splitting technique, the solution can be approximated using Eq. (4.4):

$$\Psi_n(t+\Delta t) \sim \exp\left[\frac{i\,\Delta t}{2}\frac{\Delta}{2}\right] \times \exp\left[\frac{\Delta t}{i}\hat{V}[\rho,t]\right] \times \exp\left[\frac{i\,\Delta t}{2}\frac{\Delta}{2}\right] \cdot \Psi_n(t). \tag{4.4}$$

4.2.2 Method and Computational Details

Field-emission patterns under an applied external field were calculated using the Octopus code. Octopus is based on TD-DFT in the

Kohn-Sham scheme [12]. It is a pseudopotential real-space package designed for simulating the electron-ion dynamics in one-, two-, and three-dimensional finite systems subjected to time-dependent electromagnetic fields.

Our model structures included an open-ended (5,5) armchair hydrogen-terminated carbon nanotube (H-CNT); an open-ended (5,5) armchair CNT; an open-ended (6,6) armchair H-CNT; an open-ended (9,0) zigzag H-CNT; an open-ended (9,0) zigzag CNT; a capped (5,5) armchair CNT; and a capped (9,0) zigzag CNT, as shown on the top panel of Fig. 4. 1. The CNTs and H-CNTs had diameters of 6.8–7.2 Å and lengths of 9–12 Å. We relaxed the tube geometry in a zero electric field. The simulation box was a cylinder 20 Å in diameter and 60 Å in length, and the grid cutoff energy was set to 711 eV. To study the emission patterns, we analyzed the charge densities and wave functions under an external electric field of 7.5×10^7 V/cm. Our calculations were nonperiodic, and the wave functions and densities were zero at the simulation box boundary. The external electric field strength was varied with time along a trapezoidal shape that was constructed to have a ramp time of 2.63 fs and constant time of 1.32 fs.

We next describe a method used to obtain the emission current. Figure 4.2 shows the relationship between the electron density emitted into a vacuum and time. In the figure, 2.63 fs indicates the time at which the electric field reaches 7.5×10^7 V/cm. Thereafter, the electric field is held constant until t = 3.95 fs. The emission current, I, can be obtained from the gradient by $I = k\, \mathrm{d}Q/\mathrm{d}t$.

4.2.3 Results and Discussion

We obtained the emission patterns at varying distances from the tips of the CNTs. The calculated emission patterns at a distance of −3 Å from the tips are shown in Fig. 4.1. We found that an open-ended armchair CNT(b) afforded five-fold symmetry, while an open-ended zigzag CNT(e) generated nine-fold symmetry. A capped armchair CNT(f) generated a five-membered ring pattern, and a capped zigzag CNT(g) afforded a six-membered ring pattern. The emission patterns of the CNTs corresponded well to the positions of carbon atoms in the CNT tips. We also calculated the emission patterns of the open-ended armchair H-CNTs(a, c) and an open-ended zigzag H-CNT(d). The patterns in Fig. 4.1a,c,d show five-fold, six-fold, and nine-fold symmetries, respectively. The emission patterns of the H-CNTs corresponded to the positions of hydrogen atoms in the tips of the H-CNTs.

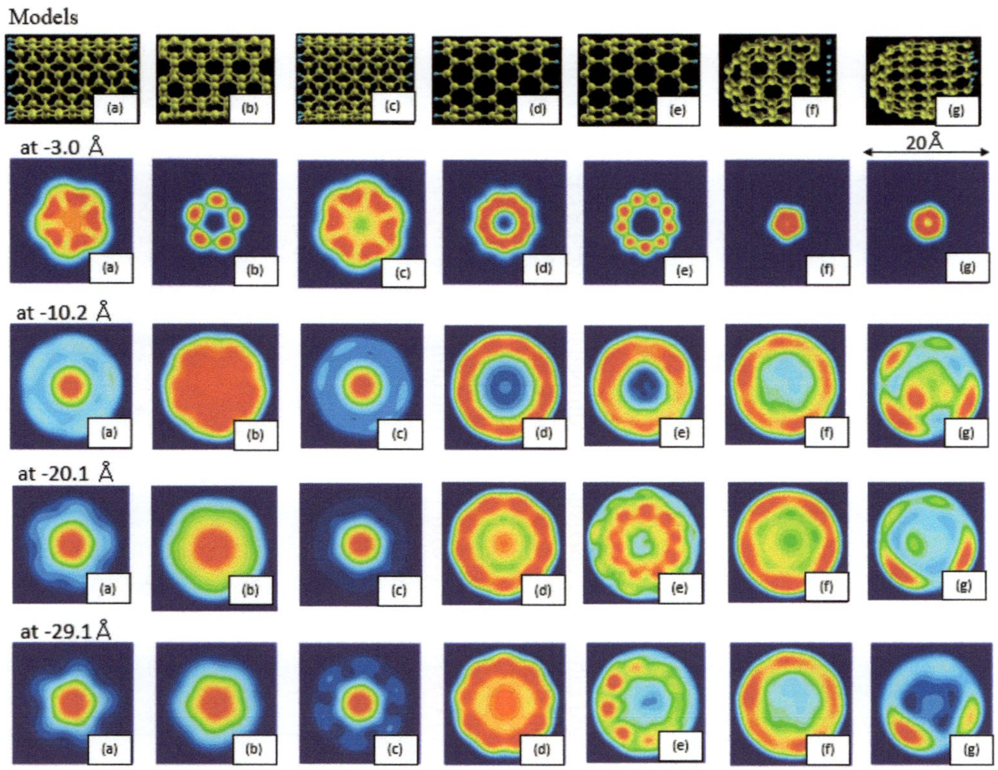

Figure 4.1 Atomic models and calculated field-emission patterns at distances of −3, −10.2, −20.1, and −29.1 Å from the tips of (a) an open-ended (5,5) armchair H-CNT, (b) an open-ended (5,5) armchair CNT, (c) an open-ended (6,6) armchair H-CNT, (d) an open-ended (9,0) zigzag H-CNT, (e) an open-ended (9,0) zigzag CNT, (f) a capped (5,5) armchair CNT, and (g) a capped (9,0) zigzag CNT. The electron beams in the models were emitted from the left sides.

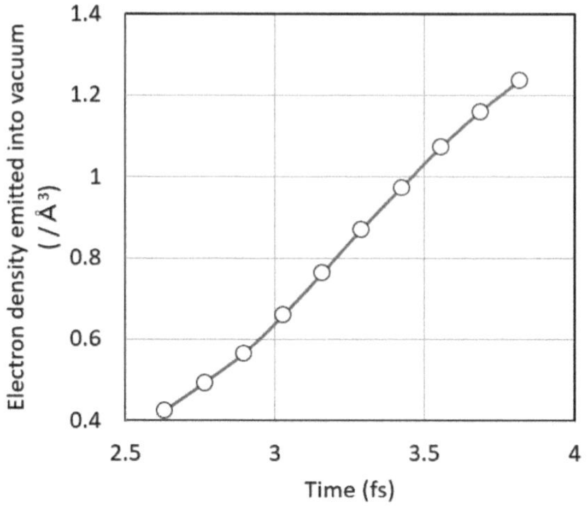

Figure 4.2 Evolution of the electron density emitted into a vacuum over time.

The beam-focusing characteristics of the CNT models fell into one of two categories, focused and unfocused (Fig. 4.3). Beams from the open-ended (5,5) and (6,6) armchair H-CNTs and the open-ended (5,5) armchair CNT were focused, but the beams from the open-ended zigzag (9,0) CNT and H-CNT were not. We checked the potential distributions and wave functions in the direction parallel to the tube axis to investigate the origin of these focusing differences. However, we were not able to identify the cause.

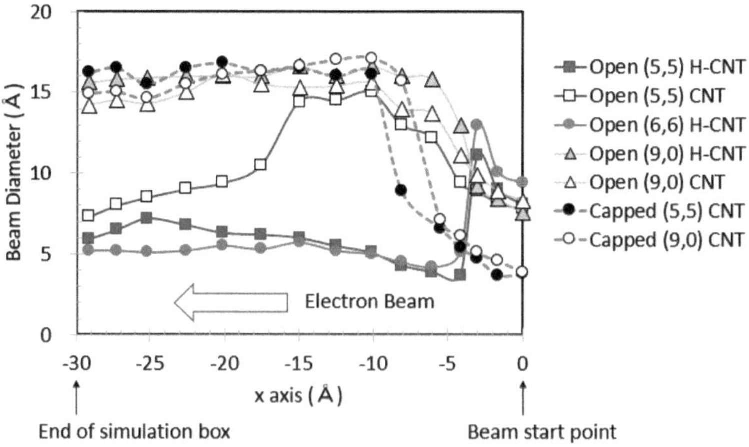

Figure 4.3 Beam-focusing characteristics of various CNTs.

Lee et al. predicted the beam-focusing characteristics of an armchair (10,10) H-CNT by performing dynamic first-principles simulations [4]. They described a focused beam and reported that the emitted electron density of s-like π states had a ring shape with a spot in the center.

We also calculated the wave functions at the HOMO levels of four open-ended CNTs. The calculated wave functions of (5,5) H-CNT, (5,5) CNT, (9,0) H-CNT, and (9,0) CNT are shown in Fig. 4.4. The surfaces of the calculated wave functions were located at the CNT tips, and the surfaces were perpendicular to the tube axes. We found that the number of wave functions calculated for the (5,5) H-CNT and (5,5) CNT were equal to the number of atoms at the tops of their tips. However, this was not the case for the (9,0) H-CNT and (9,0) CNT. The calculated patterns of the (5,5) H-CNT and CNT were in the π* state, which differed from the results reported by Lee et al. [4]. The origin of this difference is a topic for future discussion.

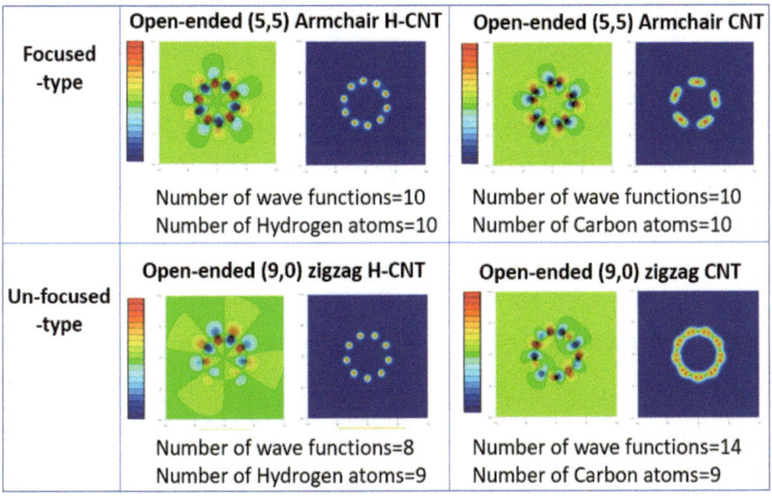

Figure 4.4 Calculated contour plots of the wave functions at the HOMO levels of four open-ended CNTs: (5,5) armchair H-CNT, (5,5) armchair CNT, (9,0) zigzag H-CNT, and (9,0) zigzag CNT. The wave functions are shown on the left sides of the images and the electron densities are shown on the right. The color codes on the left side of the wave functions denote the value of the wave function. The red and blue colors mean the maximum positive and the maximum negative values of the wave function, respectively.

The far field–emission patterns of CNTs—(Fig. 4.1e): an open-ended (9,0) zigzag CNT at −29.1 Å and (Fig. 4.1g): a capped (9,0) zigzag CNT at −10.2, −20.1, and −29.1 Å—were not symmetric. All of the atomic positions in the TD-DFT calculations were applied after structural optimization. Structural optimization generally breaks the symmetry of atomic positions. We thus assumed that structural optimization was the reason why the emission patterns were asymmetric. We then calculated the emission patterns using atomic positions without structural optimization. However, this did not improve the symmetry of the emission patterns. Further studies are needed to determine the reason for these findings.

4.3 Field-Emission Patterns from CNT Calculated Using DFT

In this section, we describe the DFT calculation method and the resulting emission patterns. The field-emission current could be calculated without the TD-DFT calculation code. The local density of states, transmittance, and the slowly varying function of energy could be calculated separately. This enabled us to clarify the influence of each factor that governed the emission patterns. Since we used the WKB approximation, these were quasi-classical approximations. The far field-emission patterns could not be calculated.

4.3.1 Theoretical Models for Calculating Field-Emission Patterns

We could calculate the field-emission current using the model developed by Khazaei based on the Penn–Plummer model [6, 7]. Equations (4.5), (4.6), and (4.7) are Khazaei's equations. The equations are only outlined here but described in detail in his report [6].

$$j_i(\varepsilon) = \frac{2e\hbar}{m_e} \cdot f(\varepsilon) \cdot S_i \cdot \lambda_i^{-2}(\varepsilon) \cdot D_i^2(\varepsilon) \cdot g_i(\varepsilon, x_{I,i}) \tag{4.5}$$

In Eq. (4.5), $j_i(\varepsilon)$ is the field-emission current along grid line i (Fig. 4.5). In our ab initio calculations, the supercell was generally discretized by introducing a fine grid. The total current, $j(\varepsilon)$, was thus calculated as the sum of the currents along grid line i by $j(\varepsilon) = \sum_i j_i(\varepsilon)$.

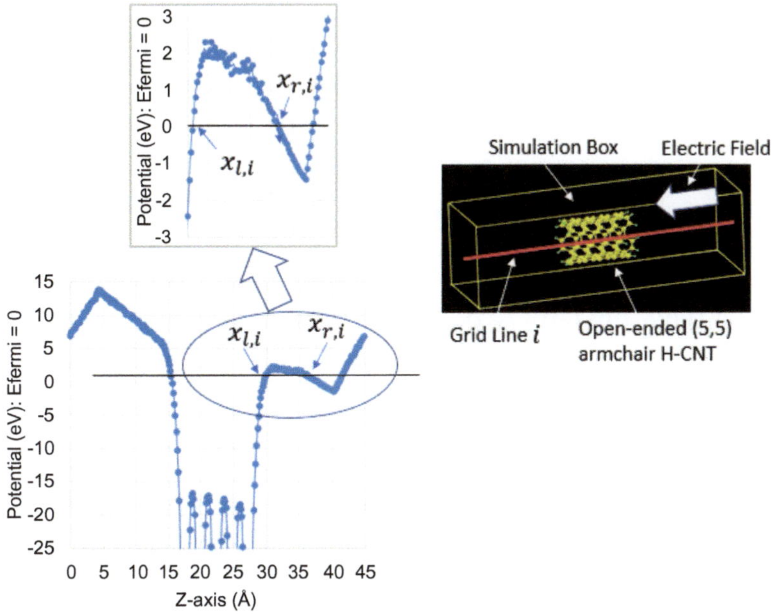

Figure 4.5 Potential distribution along grid line i on an open-ended (5,5) armchair H-CNT under an electric field. $x_{l,i}$ and $x_{r,i}$ are the left and right turning points, respectively, along grid line i.

In Eq. (4.5), m_e is the effective electron mass, $\hbar$ is the reduced Planck constant, ε is the electronic energy state, $f(\varepsilon)$ is the Fermi-Dirac distribution, and S_i is the area of surface element i, $\lambda_i(\varepsilon)$ is the slowly varying function of energy resulting from the asymptotic matching of the wave function of the emitting state at the left turning point $x_{l,i}$ with a WKB wave function within the barrier. The left and right turning points, $x_{l,i}$ and $x_{r,i}$, along grid line i (Fig. 4.5) are the points at which the energy of the emitting state is equal to the potential barrier energy, $u_i(x)$, where x is a coordinate in the emission direction. $D_i^2(\varepsilon)$ is the probability of an electron tunneling through the barrier between the nanostructure and vacuum, $u_i(x)$. The local density of states at the left turning point is represented by $g_i(\varepsilon, x_{l,i})$, and $\lambda_i(\varepsilon)$ is given by Eq. (4.6):

$$\lambda_i(\varepsilon)=(\pi/3)^{1/2}\cdot(c_i/3)^{-1/3}\left[\Gamma\left(\frac{2}{3}\right)\cos(\pi/6)\right]^{-1}, \tag{4.6}$$

where c_i is obtained by fitting the effective potential to $c_i^2(x-x_{l,i})$ at the left turning point and Γ is the gamma function. The tunneling probability, $D_i^2(\varepsilon)$, is calculated using Eq. (4.7):

$$D_i^2(\varepsilon)=\exp\left[-2\sqrt{\frac{2m_e}{\hbar^2}}\int_{x_{l,i}}^{x_{r,i}}\sqrt{u_i(x)-\varepsilon}\;dx\right]. \tag{4.7}$$

Using this approach, the information needed to calculate the emission current [$j_i(\varepsilon)$], the LDOS [$g_i(\varepsilon, x_{l,i})$], the tunneling probability [$D_i^2(\varepsilon)$], and the slowly varying function of energy [$\lambda_i(\varepsilon)$] was obtained from the Quantum ESPRESSO (QE) output data. This will be described in more detail later in the chapter.

For the actual calculations, we used a $2e\hbar/m_e$ value of 2.3153 × 10^{-4} C·eV·s/kg. m_e was the effective electron mass, but we assumed it to be the rest mass. We used an $f(\varepsilon)$ value of 0.5 at a temperature of 0 K. S_i was the grid mesh area on the face perpendicular to the emission direction in units of Å^2. We obtained c_i by fitting the potential distribution at the left turning point at the Fermi energy or HOMO (eV). The potential distribution was obtained from the output of the QE calculation. For this calculation, we defined ε as the Fermi energy or HOMO. We obtained $u_i(x)$ in eV using the potential distribution in the QE calculation output. We obtained $g_i(\varepsilon, x_{l,i})$, the local density of states at the Fermi energy or HOMO, in units of eV^{-1} Å^{-3} from the QE output data.

4.3.2 Method and Computing Details

The field-emission currents and the emission patterns of the carbon-based materials under applied external electric fields were calculated using Quantum ESPRESSO. QE was developed by the DEMOCRITOS National Simulation Center in Trieste, Italy [13]. This program is an integrated suite of computer codes used to perform electronic structure calculations and nanoscale modeling of materials. It is based on density-functional theory, plane waves, and pseudopotentials.

Calculations were performed for the capped (5,5) armchair CNT, the capped (9,0) zigzag CNT, the open-ended (5,5) armchair H-CNT, and the open-ended (9,0) zigzag H-CNT. The atomic structures shown in Fig. 4.6 are the same models used for the TD-DFT calculations (Fig. 4.1).

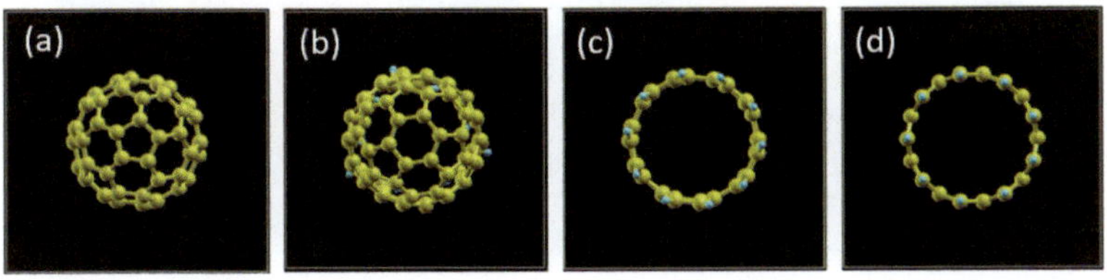

Figure 4.6 Atomic structures of (a) a capped (5,5) armchair CNT, (b) a capped (9,0) zigzag CNT, (c) an open-ended (5,5) armchair H-CNT, and (d) an open-ended (9,0) zigzag H-CNT.

For these calculations, we took structural optimization into account. The cutoff for wave function was 408 eV, and the cutoff for charge density was 1632 eV. The simulation box had the dimensions of 14 Å × 14 Å × 45 Å. We calculated their electronic properties under an external electric field of 7.5×10^7 V/cm. The external electric field had a sawtooth-like form.

4.3.3 Results and Discussions

The calculated distribution of the electron tunneling probability, $\lambda_i^{-2}(\varepsilon)$, the LDOS, and the field-emission pattern of the capped (5,5) armchair CNT are shown in Fig. 4.7. Each image in the figure represents an area of 14 Å × 14 Å. The current density was highest on top of the carbon atoms, and the emission pattern was a five-membered ring. In addition, the tunneling probability, $\lambda_i^{-2}(\varepsilon)$, and LDOS patterns corresponded to the emission pattern.

The emission patterns of four types of CNTs obtained from the DFT calculations are shown in Fig. 4.8. The capped (5,5) armchair CNT had a five-membered ring emission pattern, while the capped (9,0) zigzag CNT had a six-membered ring emission pattern. The pattern of the open-ended (5,5) armchair H-CNT had five-fold (ten-fold) symmetries, and the pattern of the open-ended (9,0) zigzag H-CNT exhibited nine-fold symmetry. For comparison, the calculated TD-DFT emission patterns at positions −1.62 Å from the tops of the CNTs are shown in the lower part of Fig. 4.8.

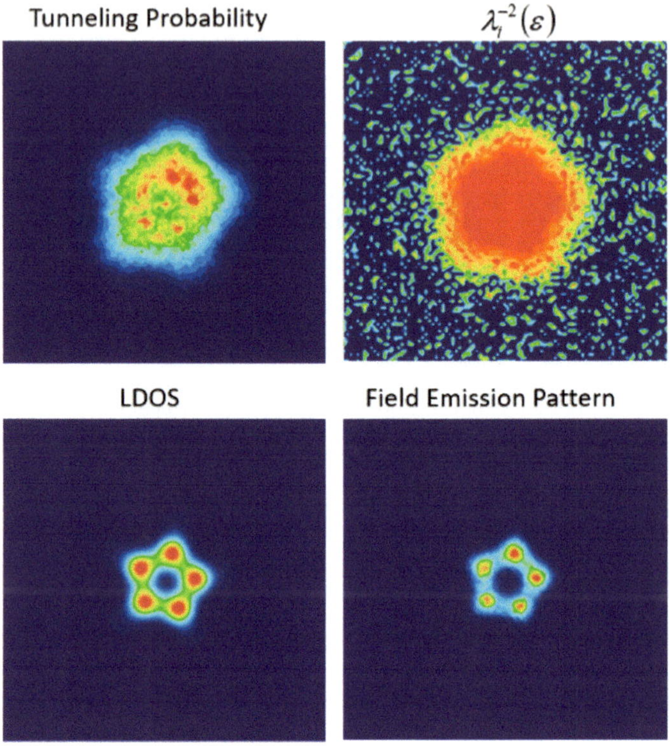

Figure 4.7 Calculated patterns of the probability of electron tunneling through a nanostructure–vacuum barrier, $\lambda_i^{-2}(\varepsilon)$, LDOS, and field-emission of the capped (5,5) armchair CNT.

These results indicated that the emission patterns calculated using the two methods were similar. However, unlike the TD-DFT method, the DFT method did not provide the focusing characteristics of the emitted beams. On the other hand, the advantage of using the DFT method was that it enabled the origin of emission to be determined.

The calculated tunneling probability, $\lambda_i^{-2}(\varepsilon)$, LDOS, and field-emission patterns of a hydrogen-terminated diamond (001) (1×1) are shown in Fig. 4.9. We found that more electrons were emitted through spaces between the hydrogen atoms. The tunneling probability pattern did not correspond to the LDOS and emission patterns, and the emission pattern was strongly dependent on the LDOS.

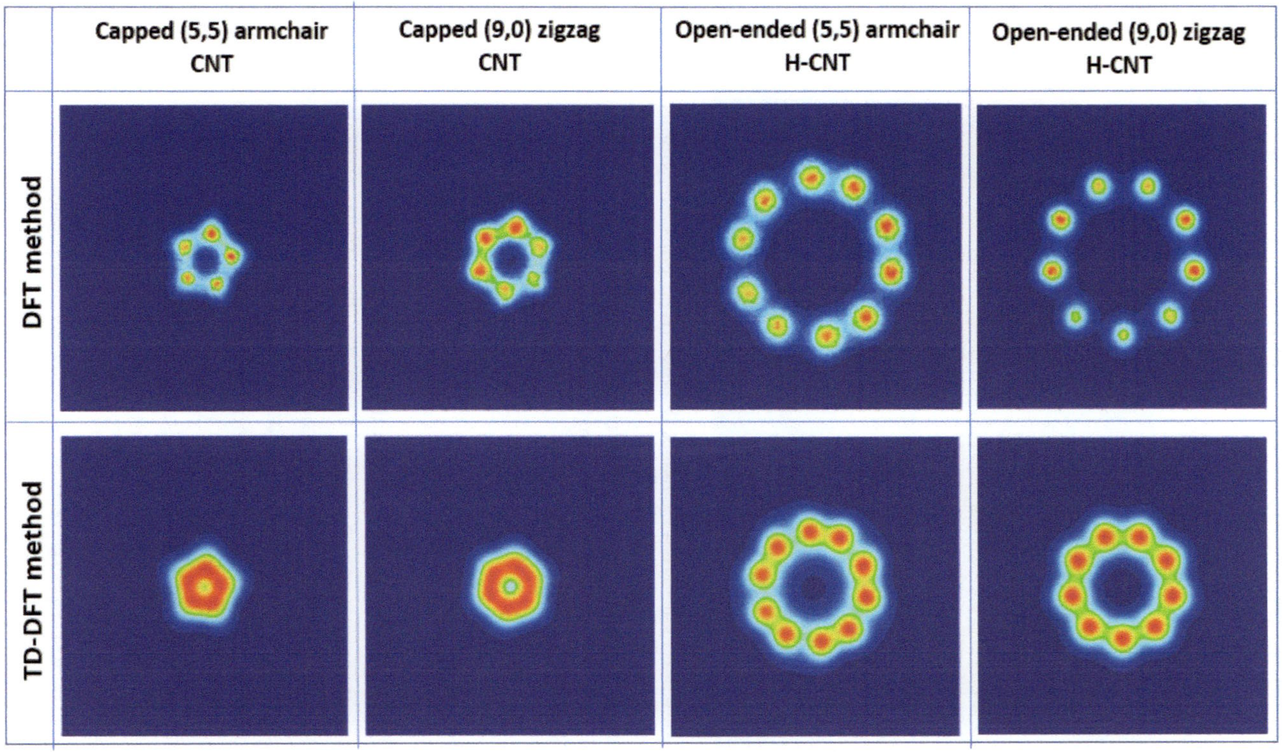

Figure 4.8 Emission patterns of various CNTs calculated using the DFT and TD-DFT methods.

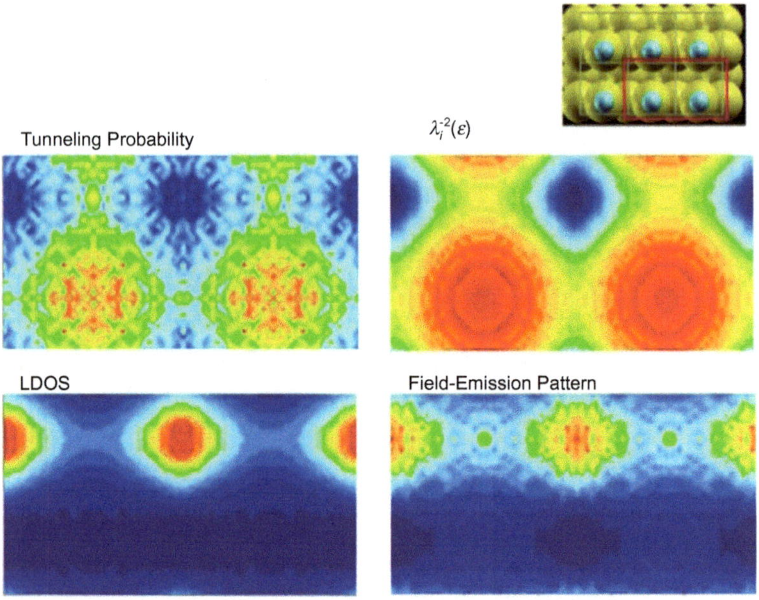

Figure 4.9 Calculated tunneling probability, $\lambda_i^{-2}(\varepsilon)$, LDOS, and field-emission patterns of a hydrogen-terminated diamond (001) (1 × 1). Inset: top-down view of an atomic structural model. The yellow and blue balls represent C and H atoms, respectively. The rectangle outlined in red corresponds to the areas shown in the patterns [8].

4.4 Conclusion

The TD-DFT and DFT methods are useful methods for calculating emission patterns of various CNTs. The TD-DFT method is assumed to be superior to the DFT method, because it seems to solve the time-dependent Schrödinger equation. However, this method does not solve the Schrödinger equation exactly. The TD-DFT method is based on the Kohn-Sham equation. Also, it is used for approximation, such as a split operator method. The DFT method is used for quasi-classic WKB approximations. The methods also differ in terms of their calculation codes. The Octopus code used for the TD-DFT calculations is suitable for nonperiodic structures, while the Quantum ESPRESSO code used for the DFT calculations is suitable for periodic structures. It is difficult to judge which method is better at this time. It is thus important to perform calculations with both methods and compare the results.

Acknowledgments

The authors thank the research group free Octopus code, which is composed by Miguel A. L. Marques, Alberto Castro, George F. Bertsch, and Angel Rubio, and thank the developers of Quantum ESPRESSO code, which is an integrated suite of open-source codes for electronic-structure calculations and materials modeling at the nanoscale. The authors wish to thank Dr. M. Khazaei from the RIKEN Advanced Institute for Computational Science (AICS) for their helpful discussions.

References

1. Saito, Y., Hata, K., and Murata, T. (2000). Field emission patterns originating from pentagons at the tip of a carbon nanotube, *Jpn. J. Appl. Phys.* **39**, L271.
2. Dean, K. A., and Chalama, B. R. (1999). Field emission microscopy of carbon nanotube caps, *J. Appl. Phys.* **85**, 3832.
3. Kim, C., Kim, B., Lee, S. M., Jo, C., and Lee, T. H. (2001). Effect of electric field on the electronic structures of carbon nanotubes, *Appl. Phys. Lett.* **79**, 1187.
4. Lee, S. B., Kim, S., and Ihm, J. (2007). First-principles dynamic simulations of field emission from carbon nanotubes on gold substrate, *Phys. Rev. B* **75**, 075408-1– 075408-5.
5. Hu, C., Mori, R., and Watanabe, K. (2014). Patterns of field electron emission from carbon nanotubes: Ab initio simulations by time-dependent density functional theory, *Proceedings of the 12th Asia Pacific Physics Conference, JPS Conference Proceedings* **1**, 012067-1–012067-4.
6. Khazaei, M., Farajian, A. A., and Kawazoe, Y. (2005). Field emission patterns from first-principle electronic structures: Applicayion to pristine and cesium-doped carbon nanotubes, *Phys. Rev. Lett.* **95**, 177602.
7. Penn, D. R., and Plummer, E. W. (1974). Field emission as a probe of the surface density of states, *Phys. Rev. B* **9**, 1216.

8. Higuchi, T., Sasaki, M., Horie, S., Yamada, Y., Matsumoto, S., and Fukuda, S. (2015). On the mechanism of improvement of field emission properties of carbon-coated field emitters, *J. Vac. Sci. Technol. B* **33**(3), 03C102-1–03C102-5, May/Jun.
9. Higuchi, T., Yamada, Y., and Sasaki, M. (2018). Field emission patters from carbon nanotubes calculated by time-dependent density functional theory, *Technical Digest of the 31st International Vacuum Nanoelectronics Conference (IVNC), Kyoto, Japan*, 9–13 July.
10. Castro, A., and Marques, M. A. L. (2006). Propagators for the time-dependent Kohn-Sham equations, *Lect. Notes Phys.* **706**, 197–210.
11. Watanabe, N., and Tsukada, M. (2002). Efficient method for simulating quantum electron dynamics under the time dependent Kohn-Sham equation. *Phys. Rev. E* **65**, 036705.
12. Marques, M. A. L., Castro, A., Bertsch, G. F., and Rubio, A. (2003). octopus: A first-principles tool for excited electron-ion dynamics, *Computer Phys. Commun.* **151**, 60–78.
13. Quantum ESPRESSO, http://www.quantum-espresso.org/

Chapter 5

Heat Localization and Thermionic Emission from Carbon Nanotubes

Alireza Nojeh
Department of Electrical and Computer Engineering, and Quantum Matter Institute, University of British Columbia, Vancouver BC, V6T 1Z4, Canada
alireza.nojeh@ubc.ca

Thermionic emission has been known for over a century and has played a key role in enabling the development of vacuum electronics. In recent decades, with progress in nanomaterials, there has been interest in studying thermionic emission from these materials. Carbon nanotubes, in particular, are excellent candidates for electron emission applications. This chapter starts with a brief overview of the status of research in thermionic emission from carbon nanotubes and then reviews in more detail a heat localization phenomenon that enables efficient heating and thermionic emission using a modest source of light. Thermal and photothermal effects are discussed along with the mechanism of heat localization and potential applications of this effect.

Nanostructured Carbon Electron Emitters and Their Applications
Edited by Yahachi Saito

ISBN 978-981-4877-62-6 (Hardcover), 978-1-003-14199-0 (eBook)
www.jennystanford.com

5.1 Introduction

Among electron emission mechanisms, field-electron (FE) emission has been studied extensively in carbon nanotubes (CNTs). This is perhaps primarily because of the obvious advantage of the high aspect ratio of a CNT in enhancing an applied electric field to induce quantum tunneling of electrons. Thermal electron emission (historically referred to as thermionic emission—TE) and photo-electron emission (PE) from nanotubes have also been investigated, but to a much smaller degree than FE. The topic of this chapter is TE from CNTs with a focus on a thermal confinement effect that may represent significant advantages for thermionic applications of this material and potentially other one-dimensional materials.

For a broad study of the phenomenon of electron emission, the book by Jensen [1] may be used. The works of Forbes and collaborators and Jensen and collaborators are referred to for ongoing developments in the theory of FE and combined thermal- and field- (and photo-) emission [2, 3]. The book edited by Saito offers the necessary background to the reader interested in FE, specifically in CNTs [4]. This chapter will not be concerned with the extensive topic of FE. The chapter will start with a brief overview of the status of the TE literature on CNTs before delving into the discussion of heat localization in this context.

5.2 Thermionic Emission from Carbon Nanotubes

The rich physics of CNTs, owing to their quasi-one-dimensional (1D) nature and related quantum confinement effects, greatly complicates the development of comprehensive theories of electron emission. Only a handful of attempts have been made so far at formulating TE from CNTs. These have included a comparative modeling of FE and TE based on a tight-binding band structure [5], a TE theory based on a linear band structure for Dirac electrons [6], and an estimate of the emission current by including the effect of thermal expansion on work function [7]. Related works, although not specifically concerned with TE in CNTs, have included the modeling of heating during FE [8–11], theoretical studies of heating and thermal-field (and photo-) emission from meso- and micro-scale structures [12], and development of TE theories for graphene [13].

Experimental studies of TE from intrinsic CNTs have dealt with individual CNTs [14–17] and collections of CNTs, using both resistive/substrate heating [18–25] and direct optical heating [26–32]. The Richardson–Laue–Dushman (RLD) theory of TE augmented with the Schottky effect to account for the reduction of the vacuum barrier due to an applied field has commonly been used to fit experimental TE data from CNT collections. For various samples consisting of individual CNTs or sheets, yarns, and arrays of vertically aligned CNTs (CNT forests), work functions in the range of 4.1–5.5 eV have been reported, with most data being around 4.5–5 eV. For CNT collections, in some cases a Richardson constant in agreement with the nominal value for bulk metals has been found to be suitable. Emission currents in these experiments have ranged from tens of nA typically for individual CNTs to hundreds of μA for CNT collections. A few works have investigated the temporal behavior [31–33] of TE from CNTs, reporting response times from hundreds of ns for individual CNTs to over 1 ms for collections, as well as the energy distribution of the emitted electrons [22, 34], seeing a broad spectrum consistent with a high-temperature thermal process.

A number of experimental studies have also claimed a reduction in the work function of CNTs through coating or intercalation with other materials down to as low as 1.5 eV [33–37], or enhancement in emission current, reportedly because of surface-plasmon resonance due to coating by silver nanoparticles [38]. CNTs have also been used as a backbone for other materials such as doped diamond crystals and h-boron nitride to form thermionic emitters [39, 40].

5.3 Heat Localization in Carbon Nanotube Forests Leading to Thermionic Emission

5.3.1 The Heat Trap Effect

In many of the above studies, sufficient power has been delivered in order to heat the entire CNT structure to TE temperatures. On the other hand, in a series of experiments by the author's research group, the emphasis has been on localized heating within multiwall CNT forests [41]. CNTs are generally considered to be good thermal conductors. It is therefore natural to expect that, regardless of where thermal energy is delivered, temperature would rise everywhere in the CNT structure. However, surprisingly, it has been observed that the heat can remain tightly confined to the point of delivery. In these experiments, a forest of multiwall CNTs with macroscopic dimensions (a height of 1–2 mm and side dimensions of 4–5 mm) has been used and a laser beam has been focused onto a spot typically around 50–100 μm in diameter, on the side surface of the forest. A photo of a CNT forest can be seen in Fig. 5.1 with a locally hot region exhibiting strong incandescent glow, while the rest of the forest, including a significant portion of the length of the CNTs whose middle section is hot, remains relatively cool [27, 42]. This implies low thermal dissipation along the CNTs, and temperature gradients greater than 10 K/μm have been observed [43]. Such heat localization in a thermally conductive material is counterintuitive and is referred to as the heat trap effect.

Figure 5.1 *Top right*: Photo of a multiwall carbon nanotube forest with a height of over a millimeter. *Bottom right*: Photo of the same CNT forest when the middle section of its side surface is illuminated by a focused infrared laser beam. A locally hot region is seen exhibiting bright incandescence. Adapted with permission from [42], Copyright 2018 by the American Physical Society. *Left*: Scanning electron micrograph of a small section of the side surface of a CNT forest showing the direction of the CNTs and their overall aligned nature.

The effective quenching of heat dissipation to the surroundings, combined with the excellent broadband optical absorption of CNT forests, creates a favorable condition whereby a low-input irradiance is sufficient to heat the illuminated spot to TE temperatures. To put this into perspective, consider that a typical metallic TE cathode requires a minimum irradiance of hundreds of kW/cm^2, necessitating the use of high-power or pulsed lasers. Heat trap experiments in CNT forests involve irradiance levels of three orders of magnitude lower and can thus be conducted using a handheld continuous-wave laser delivering only a few tens of mW or even a few mW [27, 28].

In these experiments, the temperature of the hot spot has been estimated using thermographic cameras or by fitting the incandescence spectrum to the Planck black body radiation law, as well as through modeling the system by considering conductive and radiative thermal dissipation. An example of TE current as a function of temperature is shown in Fig. 5.2(left), together with RLD-estimated behavior. Given the metallic nature of multiwall CNTs and the involvement of a relatively large number of nanotubes in emission, the overall agreement with RLD theory is not surprising, consistent with other experiments on CNT collections as discussed previously.

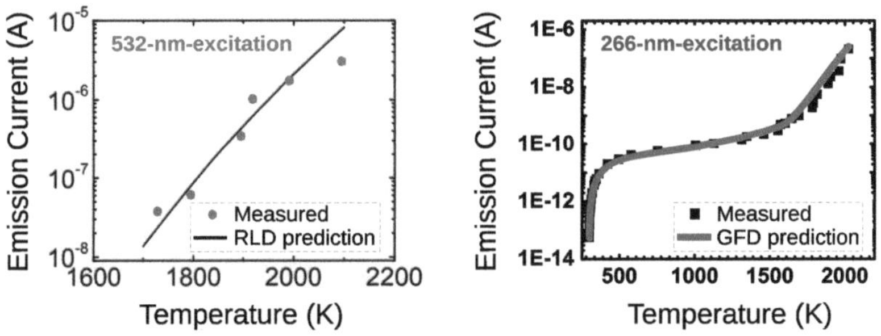

Figure 5.2 *Left*: Emission current from a CNT forest excited by a continuous-wave laser with a wavelength of 532 nm, showing overall agreement with the predictions of the Richardson–Laue–Dushman model. Adapted from [27], Copyright 2011, with permission from Elsevier. *Right*: Emission current from a CNT forest excited by a continuous-wave laser with a wavelength of 266 nm, showing overall agreement with the predictions of the generalized Fowler–DuBridge model. As the laser irradiance and thus the heat trap temperature are gradually increased, the emission behavior evolves from single-photon PE to two-photon PE to eventually being dominated by TE. Adapted with permission from [30], Copyright 2015 American Chemical Society.

5.3.2 The Effect of Photon Wavelength and Combined Multiphoton Thermal Photoemission

The TE current shows minimal dependence on excitation wavelength in the visible-to-near-infrared range [28]. This is consistent with the broadband absorptive nature of the CNT forest, making it an extremely dark material. However, the situation becomes more complex when higher photon energies are used. Photons with a wavelength of 266 nm (i.e., a photon energy of 4.66 eV) can induce single-photon PE in CNT forests [44]. This is a linear process and the emission current scales directly with the input optical power at very low irradiance levels. However, as the irradiance is increased to beyond a few tens of W/cm^2, a heat trap is created and PE and TE take place simultaneously [30], with a gradual transition from room-temperature PE to dominant TE (Fig. 5.2, right). Particularly interesting is the fact that single-photon PE is not sufficient to explain the full range of the measured emission current, and a two-photon PE term must also be invoked. This can be seen through fitting the current to the generalized Fowler–DuBridge (GFD) model of multiphoton thermal-PE [45], where the zeroth-order term of the expansion corresponds to TE, the first-order term to single-photon PE, and the second-order term to two-photon PE.

In bulk metals, two-photon PE is typically weak and requires a very high input irradiance, normally available from a pulsed laser. It is thus worth reiterating that the above experiments in the CNT forest have used a low-irradiance continuous-wave laser. A possible explanation for such efficient two-photon PE is based on the extremely porous nature of the CNT forest. Electrons photogenerated deep inside a bulk metal can lose much of their energy to electron–electron scattering before reaching the surface. The CNT forest has a unique structure, with an average inter-nanotube distance of a few tens of nanometers. This is much smaller than the photon wavelength in use and does not lead to particularly deep photon penetration—optical absorption is indeed very strong as discussed before. On the other hand, the photoexcited electrons in each CNT, no matter how deep within the forest, can immediately find themselves in vacuum. Although these electrons are still surrounded by other CNTs, they may escape the forest through the large inter-nanotube empty regions with relative ease [30].

5.3.3 The Effect of Polarization and Temporal Behavior

When the input laser is polarized such that its electric field vector is aligned with the axis of the CNTs, a much higher emission current is observed compared to the case where the field is perpendicular to the CNTs (Fig. 5.3, left). This is likely due to the fact that the former polarization corresponds to more efficient optical absorption and, therefore, shallower penetration of the light, and thus easier electron escape [28]. This is also consistent with the justification given in the previous section for the high two-photon PE efficiency of the CNT forest compared to bulk metals. The temporal behavior of the emission current, which strongly depends on the temporal behavior of the heating effect itself, is also important for applications. Sub-ms radiative heating and cooling of the CNT forest have been reported and explained based on the material's high emissivity and low specific heat [46]. Comparative experiments have shown heating and cooling speeds similar to that of multilayer graphene and much faster than that of graphitic scrolls (Fig. 5.3, right) or bulk metals.

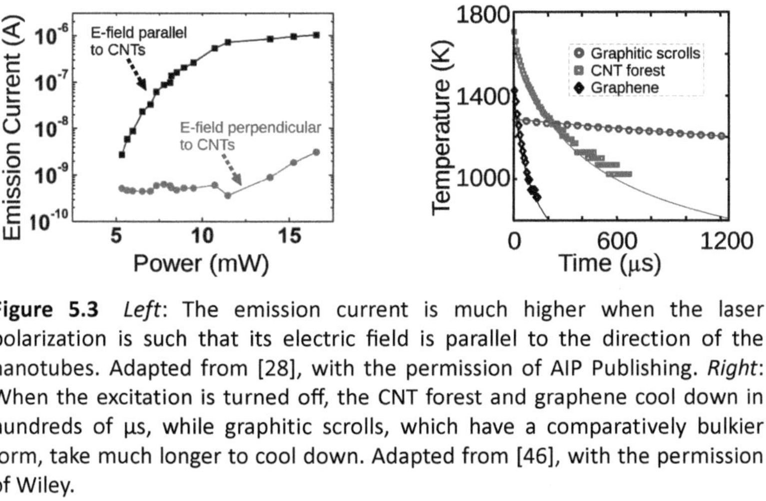

Figure 5.3 *Left*: The emission current is much higher when the laser polarization is such that its electric field is parallel to the direction of the nanotubes. Adapted from [28], with the permission of AIP Publishing. *Right*: When the excitation is turned off, the CNT forest and graphene cool down in hundreds of μs, while graphitic scrolls, which have a comparatively bulkier form, take much longer to cool down. Adapted from [46], with the permission of Wiley.

5.3.4 The Mechanism of Heat Localization

Given the thermally conductive nature of CNTs, what is the cause of the heat trap effect? In order to understand this phenomenon, the nature of thermal transport within the CNT forest needs to be considered. The study of the thermal conductivity of CNTs is an active area of research, but it is generally understood that CNTs are good conductors, with room temperature thermal conductivity as high as 3500 W/mK for individual pristine CNTs [47] and lower but still significant values such as 50 W/mK for CNT sheets and 15 W/mK for

CNT forests [48, 49] (although much lower values can be expected in highly disordered or defective CNT collections). However, regardless of the value of thermal conductivity, it is crucial to realize that heat transport within a CNT forest is highly anisotropic, with transverse conductivity being one to two orders of magnitude lower than conductivity in the direction of the CNTs [49, 50].

To explore the role of this anisotropy in thermal confinement, a bulk thermal transport model based on a thermal conductivity tensor has been developed, with results shown in Fig. 5.4 [42]. The dotted circle represents the illuminating laser spot. In the top row, the left panel corresponds to an isotropic material with a high thermal conductivity of 100 W/mK in all directions, the middle panel to a material with a thermal conductivity of 100 W/mK only along the two in-plane directions but two orders of magnitude lower along the depth (mimicking a layered material such as ordered graphite), and the right panel to a material with a thermal conductivity of 100 W/mK only along the vertical direction but two orders of magnitude lower along the other two directions (mimicking a material such as a CNT forest). It can be seen that the same input optical power leads to drastically different outcomes in the three cases, yielding much higher temperatures in the 1D case. While this outcome may appear surprising, it may be understood by considering the fact that heat flow between two given points takes all possible routes and, therefore, constraining it laterally leads to an overall restriction in heat flow, resulting in heat localization. The situation becomes even more drastic if thermal conductivity is taken not to be a constant, but to depend on temperature, which is in reality the case. Starting with a room-temperature value of 100 W/mK and assuming a $1/T$ dependence, as expected due to Umklapp phonon scattering, a stronger thermal confinement is observed in all three materials, with an even lower input power requirement for attaining the same maximum temperature (compare the second row of Fig. 5.4 to the first row). If thermal conductivity drops with temperature even more rapidly, for example, in the form of a $1/T^2$ dependence (which may be expected due to higher-order phonon scattering), the effect becomes even more drastic—see the bottom row of Fig. 5.4, where only 28 mW of input power is creating an extremely confined hot region with a peak temperature of 3000 K on the 1D material (despite the material's high room-temperature thermal conductivity value of 100 W/mK). This may be interpreted as a positive feedback mechanism: a local increase in temperature leads to a local reduction in thermal conductivity, which further increases the temperature locally and so on, and the effect is much more pronounced for the 1D material due to restricted lateral thermal transport.

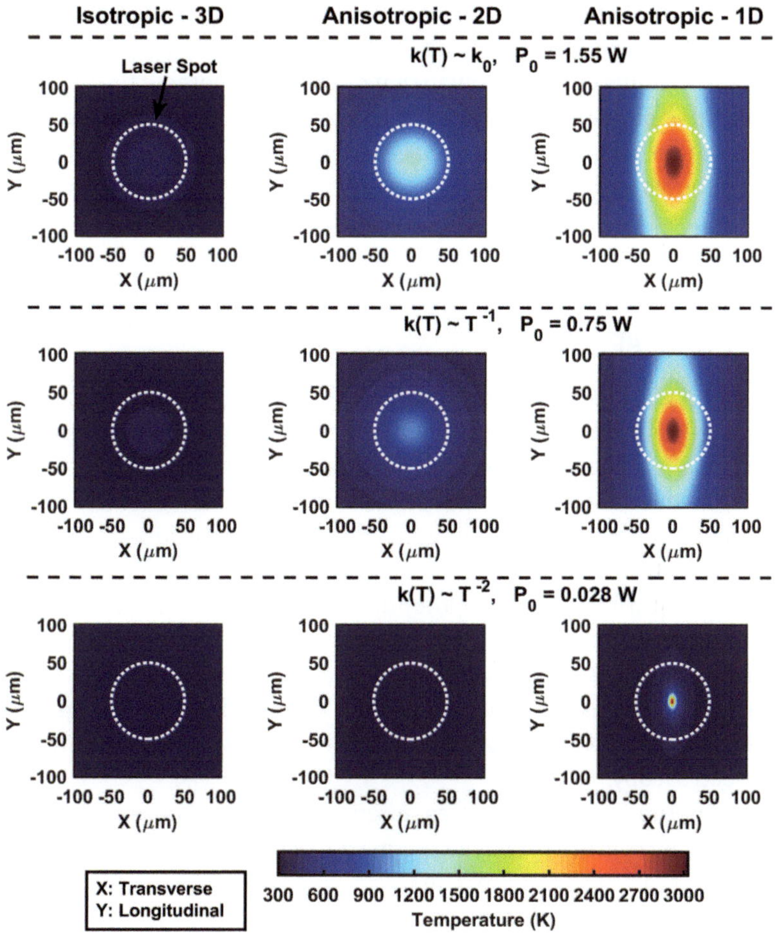

Figure 5.4 Temperature distribution on the surface of a material heated by focused light, showing heat localization. *Left to right*: increase in thermal anisotropy; *top to bottom*: increase in temperature dependency of thermal conductivity. Adapted with permission from [42], Copyright 2018, by the American Physical Society.

While the above provides a possible macroscopic explanation for the heat trap effect, it is important to emphasize that the underlying mesoscopic and nanoscale physics of thermal transport in the CNT forest are subjects of ongoing research. Within the CNT forest, each nanotube absorbs light, conducts heat along its axis, radiates thermal photons, and emits thermal electrons. Neighboring CNTs interact not only through occasional physical contact (mediated by the van der Waals force) but also through emission and reabsorption of photons and electrons. The optical interaction among the CNTs is itself a complex problem given that both the diameter of the CNTs and the spacing between adjacent CNTs are deeply sub-wavelength as far as thermal radiation is concerned. Indeed, recent modeling work has shown an increase in CNT temperature due to such interactions compared to the case of an isolated CNT [51].

5.3.5 Applications of Thermionic Emission Due to Heat Trap

Several features of the heat trap effect make it promising for a variety of applications. A CNT forest locally heated by a low-power laser can act as a compact and flexible electron source. The lack of heat spread to the surrounding structures makes the cathode assembly simple without a need for thermal management. There is also no need for power–electronic circuitry, and the cathode can conveniently be floated at a high electric potential if acceleration is required. The electron emission spot size and shape are effectively defined by the laser spot size and shape, which allows the creation of multiple electron beams or shaped electron beams. These are appealing features for electron beam systems for lithography, microscopy, and spectroscopy.

Another application area is energy conversion. A thermionic energy converter is a device where electrons leave a hot emitter, traverse a vacuum or plasma region, and reach a cold collector, creating a potential difference and a current that passes through an external load. This device concept has been known for over a century (see Ref. [52] for a review); significant developments in making thermionic energy converters took place starting in the middle of the twentieth century, with several waves of interest over the years. A particular area of focus has been solar thermionic energy conversion [53], notably in early works by NASA [54]. Considering the high irradiance required to heat a bulk cathode to thermionic emission temperatures as discussed before, it is not surprising that the developed devices needed large solar collectors and concentrators, together with elaborate thermal engineering in order to minimize the loss and detrimental spread of heat beyond the emitter. Those devices were thus large and complex and in some ways impractical. The heat trap effect immediately changes this landscape by eliminating the heat spread problem and enabling the attainment of TE temperatures using a simple handheld glass lens. This has led to the demonstration of small and compact solar thermionic energy converters based on this effect (Fig. 5.5) [29, 55, 56].

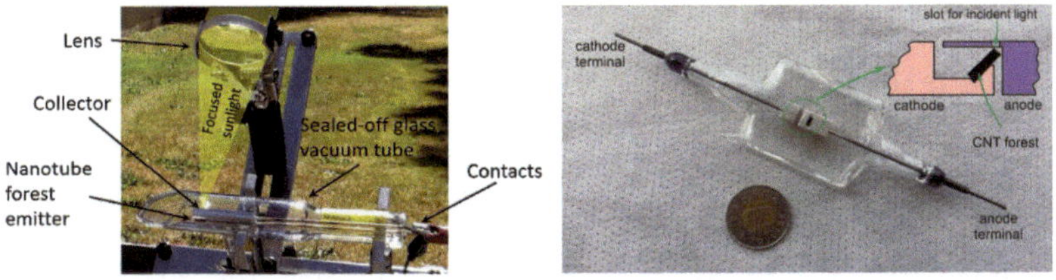

Figure 5.5 *Left*: A compact, sealed-off solar thermionic energy converter based on the heat trap effect in a CNT forest, under test [29]. Reprinted with permission from [56], Copyright 2019 IEEE. *Right*: Another embodiment of the device in a more compact form factor. A toonie (the Canadian two-dollar coin) is included to show the device size. Inset shows a schematic of the main part of the device. Reprinted with permission from [55], Copyright 2016 IEEE.

A close relative of the thermionic energy converter is the thermoelectric energy converter, in which the hot and cold sides are in direct contact. The primary challenge of thermoelectricity is to reduce heat flow between the two sides without hindering the flow of electricity. The heat trap effect, as a condition where heat flow is restricted and thus a high temperature gradient is maintained within a conductor, thus presents a new path for thermoelectricity. A basic thermoelectric converter has been demonstrated based on this effect [57]. It should be noted that both thermionic and thermoelectric conversion efficiencies based on pure CNTs are low due to high work function and low Seebeck coefficient. However, this is not a fundamental limitation as electronic properties may be engineered through the deposition or intercalation of other materials on/within CNTs. Also, there is no fundamental reason to assume that the heat trap effect is limited to carbon-based systems, and a particular opportunity is the exploration of this effect in other 1D materials with desirable electronic properties for thermionic and thermoelectric energy conversion.

5.4 Conclusion and Outlook: The Future of Thermionic Emission from Carbon Nanotubes

Much remains to be learned about not only heat localization but also, more generally, TE from CNTs and other one- and two-dimensional materials. A comprehensive theory is lacking; constructing such a theory will be difficult and will require the availability of comprehensive and quantitatively accurate experimental data. In experiments on large collections of CNTs, much of the nanoscale physics may be averaged out; systematic experiments on individual CNTs are needed. It will be crucial to raise the experimental sophistication by using ultra-high vacuum and avoiding leakage currents and other parasitic effects. Measurements of the angular distribution and energy distribution of the emitted electrons should become an integral part of the studies, and this will be challenging due to the difficult nature of TE electron spectroscopy partly because of space charge and Boersch effects [58, 59]. Accurate temperature measurements will be needed, taking into account the physics of nanoscale thermal transport [60] and the nature of thermal radiation from a CNT as a 1D object with a thickness much smaller than the thermal radiation wavelength [61, 62]. Therefore, optical spectroscopy covering the visible to long-wavelength infrared will be required, which is highly challenging, especially if temperature distributions with high spatial resolution are sought. The TE data will need to be closely correlated to the exact structure of the specific CNT under study. It will also be crucial to view TE from CNTs in the broader context of low-dimensional physics, through consideration of the dependence of thermal conductivity and other parameters of importance on temperature, or by taking into account the fact that the electrons and lattice can be at substantially different temperatures in strongly-out-of-equilibrium conditions, even in steady state. Studies of the various time scales, including ultrafast dynamics, in nanostructures and carbon systems are also of particular interest in this regard [63]. By advancing our experimental capabilities and strengthening those broader links, we will be able to usher in the era of nanothermionics and capitalize on the great degrees of freedom that nanostructures could offer in controlling and harnessing electron emission for a wide range of new applications.

5.5 Acknowledgments

The author thanks R. G. Forbes, K. L. Jensen, and M. M. Cahay for literature suggestions. Over the years, the author's research program on electron emission and related phenomena has been supported financially by the Natural Sciences and Engineering Research Council of Canada, the Canada Foundation for Innovation, the British Columbia Knowledge Development Fund, the BCFRST Foundation, the British Columbia Innovation Council, the Peter Wall Institute for Advanced Studies, the Canada First Research Excellence Fund, Quantum Materials and Future Technologies Program, CMC Microsystems, Compute Canada, and Modern Electron, all of whom the author gratefully acknowledges.

References

1. Kevin L. Jensen. *Introduction to the Physics of Electron Emission*. John Wiley & Sons, September 2017.
2. Richard G. Forbes. Renewing the mainstream theory of field and thermal electron emission. In: G. Gaertner, W. Knapp, R. G. Forbes (eds.) *Modern Developments in Vacuum Electron Sources. Topics in Applied Physics*, Vol. 135. Springer, Cham. October 2020.
3. Kevin L. Jensen. A thermal-field-photoemission model and its application. In: G. Gaertner, W. Knapp, R. G. Forbes (eds.) *Modern Developments in Vacuum Electron Sources. Topics in Applied Physics*, Vol. 135. Springer, Cham. October 2020.
4. Yahachi Saito. *Carbon Nanotube and Related Field Emitters: Fundamentals and Applications*. John Wiley & Sons, October 2010.

5. H. Tang, S.-D. Liang, S. Z. Deng, and N. S. Xu. Comparison of field and thermionic emissions from carbon nanotubes. *Journal of Physics D: Applied Physics*, 39(24):5280–5284, December 2006.
6. Shi-Dong Liang and Lu Chen. Theories of field and thermionic electron emissions from carbon nanotubes. *Journal of Vacuum Science & Technology B*, 28(2): C2A50–C2A57, March 2010.
7. Dilip K. De and Olukunle C. Olawole. A three-dimensional model for thermionic emission from graphene and carbon nanotube. *Journal of Physics Communications*, 3(1): 015004, January 2019.
8. S. T. Purcell, P. Vincent, C. Journet, and V. T. Binh. Hot nanotubes: stable heating of individual multiwall carbon nanotubes to 2000 K induced by the field-emission current. *Physical Review Letters*, 88(10):105502, February 2002.
9. P. Vincent, S. T. Purcell, C. Journet, and V. T. Binh. Modelization of resistive heating of carbon nanotubes during field emission. *Physical Review B*, 66(7): 075406, August 2002.
10. M. Cahay, W. Zhu, S. Fairchild, P. T. Murray, T. C. Back, and G. J. Gruen. Multiscale model of heat dissipation mechanisms during field emission from carbon nanotube fibers. *Applied Physics Letters*, 108(3): 033110, January 2016.
11. G. Tripathi, J. Ludwick, M. Cahay, and K. L. Jensen. Spatial dependence of the temperature profile along a carbon nanotube during thermal-field emission. *Journal of Applied Physics*, 128(2): 025107, July 2020.
12. Kevin L. Jensen, Michael McDonald, Oksana Chubenko, John R. Harris, Donald A. Shiffler, Nathan A. Moody, John J. Petillo, and Aaron J. Jensen. Thermal-field and photoemission from meso- and micro-scale features: effects of screening and roughness on characterization and simulation. *Journal of Applied Physics*, 125(23): 234303, June 2019.
13. S.-J. Liang and L.K. Ang. Electron thermionic emission from graphene and a thermionic energy converter. *Physical Review Applied*, 3(1): 014002, January 2015.
14. D. C. Cox, R. D. Forrest, P. R. Smith, and S. R. P. Silva. Thermionic emission from defective carbon nanotubes. *Applied Physics Letters*, 85(11):2065–2067, September 2004.
15. M. Sveningsson, K. Hansen, K. Svensson, E. Olsson, and E. E. B. Campbell. Quantifying temperature-enhanced electron field emission from individual carbon nanotubes. *Physical Review B*, 72(8): 085429, August 2005.

16. Xianlong Wei, Sheng Wang, Qing Chen, and Lianmao Peng. Breakdown of Richardson's law in electron emission from individual self-Joule-heated carbon nanotubes. *Scientific Reports*, 4: 5102, May 2014.
17. Yuwei Wang, Gongtao Wu, Li Xiang, Mengmeng Xiao, Zhiwei Li, Song Gao, Qing Chen, and Xianlong Wei. Single-walled carbon nanotube thermionic electron emitters with dense, efficient and reproducible electron emission. *Nanoscale*, 9(45): 17814–17820, 2017.
18. R. O. Loutfy, M. Samandi, A. Moravsky, and S. Strange. Carbon nanotubes as thermionic emitters. *AIP Conference Proceedings*, 699(1): 773–780, 2004.
19. R. P. Raffaelle, B. J. Landi, J. D. Harris, S. G. Bailey, and A. F. Hepp. Carbon nanotubes for power applications. *Materials Science and Engineering B*, 116(3): 233–243, February 2005.
20. Seong Chu Lim, Hee Jin Jeong, Keun Soo Kim, Im Bok Lee, Dong Jae Bae, and Young Hee Lee. Extracting independently the work function and field-enhancement factor from thermal-field emission of multi-walled carbon nanotube tips. *Carbon*, 43(13): 2801–2807, November 2005.
21. Franz A. M. Koeck, Yunyu Wang, and Robert J. Nemanich. Thermionic converters based on nanostructured carbon materials. *AIP Conference Proceedings*, 813(1): 607-613, 2006.
22. Peng Liu, Yang Wei, Kaili Jiang, Qin Sun, Xiaobo Zhang, Shoushan Fan, Shufeng Zhang, Chuangang Ning, and Jingkang Deng. Thermionic emission and work function of multiwalled carbon nanotube yarns. *Physical Review B*, 73(23): 235412, June 2006.
23. Yang Wei, Kaili Jiang, Xiaofeng Feng, Peng Liu, Liang Liu, and Shoushan Fan. Comparative studies of multiwalled carbon nanotube sheets before and after shrinking. *Physical Review B*, 76(4): 045423, July 2007.
24. S. K. Kolekar, S. P. Patole, P. S. Alegaonkar, J. B. Yoo, and C. V. Dharmadhikari. A comparative study of thermionic emission from vertically grown carbon nanotubes and tungsten cathodes. *Applied Surface Science*, 257(23):10306–10310, September 2011.
25. Andriy Sherehiy, Santoshrupa Dumpala, Abdelilah Safir, David Mudd, Ivan Arnold, Robert W. Cohn, Mahendra K. Sunkara, and Gamini U. Sumanasekera. Thermionic emission properties and the work function determination of arrays of conical carbon nanotubes. *Diamond and Related Materials*, 34: 1–8, April 2013.
26. Parham Yaghoobi, Mehran Vahdani Moghaddam, Mario Michan, and Alireza Nojeh. Visible-light induced electron emission from carbon

nanotube forests. *Journal of Vacuum Science & Technology B*, 29(2): 02B104, March 2011.

27. Parham Yaghoobi, Mehran Vahdani Moghaddam, and Alireza Nojeh. Heat trap: light-induced localized heating and thermionic electron emission from carbon nanotube arrays. *Solid State Communications*, 151(17): 1105–1108, September 2011.
28. Mehran Vahdani Moghaddam, Parham Yaghoobi, and Alireza Nojeh. Polarization-dependent light-induced thermionic electron emission from carbon nanotube arrays using a wide range of wavelengths. *Applied Physics Letters*, 101(25): 253110, December 2012.
29. Parham Yaghoobi, Mehran Vahdani Moghaddam, and Alireza Nojeh. Solar electron source and thermionic solar cell. *AIP Advances*, 2(4): 042139, December 2012.
30. Mehran Vahdani Moghaddam, Parham Yaghoobi, George A. Sawatzky, and Alireza Nojeh. Photon-impenetrable, electron-permeable: the carbon nanotube forest as a medium for multiphoton thermal-photoemission. *ACS Nano*, 9(4): 4064–4069, April 2015.
31. Zhenjun Li, Bing Bai, Chi Li, and Qing Dai. Efficient photo-thermionic emission from carbon nanotube arrays. *Carbon*, 96: 641– 646, January 2016.
32. Chi Li, Zhenjun Li, Ke Chen, Bing Bai, and Qing Dai. Edge effect enhanced photo-thermionic emission from a carbon nanotubes array. *Applied Physics Letters*, 110(9): 093105, February 2017.
33. Yuwei Wang, Li Fang, Li Xiang, Gongtao Wu, Yi Zeng, Qing Chen, and Xianlong Wei. On-chip thermionic electron emitter arrays based on horizontally aligned single-walled carbon nanotubes. *IEEE Transactions on Electron Devices*, 66(2): 1069–1074, February 2019.
34. Tyler L. Westover, Aaron D. Franklin, Baratunde A. Cola, Timothy S. Fisher, and Ronald G. Reifenberger. Photo- and thermionic emission from potassium-intercalated carbon nanotube arrays. *Journal of Vacuum Science & Technology B*, 28(2): 423–434, March 2010.
35. Feng Jin, Yan Liu, and Christopher M. Day. Thermionic emission from carbon nanotubes with a thin layer of low work function barium strontium oxide surface coating. *Applied Physics Letters*, 88(16): 163116, April 2006.
36. Lin Xiao, Peng Liu, Liang Liu, Kaili Jiang, Xiaofeng Feng, Yang Wei, Li Qian, Shoushan Fan, and Taihua Zhang. Barium-functionalized multiwalled carbon nanotube yarns as low-work-function thermionic cathodes. *Applied Physics Letters*, 92(15): 153108, April 2008.

37. Feng Jin and Dustin Carter. Thermionic cooling efficiency of functionalized carbon nanotube thin films. *Journal of Vacuum Science & Technology B*, 36(5): 051804, September 2018.

38. Mahta Monshipouri, Yaser Abdi, and Sara Darbari. Realization of low power-laser induced thermionic emission from Ag nanoparticle-decorated CNT forest: a consequence of surface plasmon resonance. *Applied Physics Letters*, 109(20): 203105, November 2016.

39. Andriy Sherehiy, Santoshrupa Dumpala, Mahendra K. Sunkara, Jacek B. Jasinski, Robert W. Cohn, and Gamini U. Sumanasekera. Thermionic emission from phosphorus (P) doped diamond nanocrystals supported by conical carbon nanotubes and ultraviolet photoelectron spectroscopy study of P-doped diamond films. *Diamond and Related Materials*, 50: 66–76, November 2014.

40. Xinhe Yang, Peng Liu, Duanliang Zhou, Feng Gao, Xinhe Wang, Shiwei Lv, Zi Yuan, Xiang Jin, Wei Zhao, Haoming Wei, Lina Zhang, Jiandong Gao, Qunqing Li, Shoushan Fan, and Kaili Jiang. High temperature performance of coaxial h-BN/CNT wires above 1,000 °C: thermionic electron emission and thermally activated conductivity. *Nano Research*, 12(8): 1855–1861, August 2019.

41. Alireza Nojeh. Carbon nanotube photothermionics: toward laser-pointer-driven cathodes for simple free-electron devices and systems. *MRS Bulletin*, 42(07): 500–504, July 2017.

42. Mike Chang, Harrison D. E. Fan, Mokter M. Chowdhury, George A. Sawatzky, and Alireza Nojeh. Heat localization through reduced dimensionality. *Physical Review B*, 98(15): 155422, October 2018.

43. Harrison D. E. Fan, George A Sawatzky, Jeff F. Young, and Alireza Nojeh. Hyperspectroscopy: a tool for high-spatial-resolution temperature mapping of electron emitters. In *2017 30th International Vacuum Nanoelectronics Conference (IVNC)*, pages 44–45, Regensburg, Germany, July 2017, IEEE.

44. Parham Yaghoobi, Mario Michan, and Alireza Nojeh. Middle-ultraviolet laser photoelectron emission from vertically aligned millimeter-long multiwalled carbon nanotubes. *Applied Physics Letters*, 97(15): 153119, October 2010.

45. J. H. Bechtel, W. Lee Smith, and N. Bloembergen. Two-photon photoemission from metals induced by picosecond laser pulses. *Physical Review B*, 15(10): 4557–4563, May 1977.

46. Faezeh Mohammadbeigi, Lorne Whitehead, Ranjith Divigalpitiya, and Alireza Nojeh. Observations of radiation-dominated rapid cooling

of structures based on carbon nanotubes and graphene. *Advanced Engineering Materials*, 22: 1901315, January 2020.

47. Eric Pop, David Mann, Qian Wang, Kenneth Goodson, and Hongjie Dai. Thermal conductance of an individual single-wall carbon nanotube above room temperature. *Nano Letters*, 6(1): 96–100, January 2006.

48. Ali E. Aliev, Marcio H. Lima, Edward M. Silverman, and Ray H. Baughman. Thermal conductivity of multi-walled carbon nanotube sheets: radiation losses and quenching of phonon modes. *Nanotechnology*, 21(3): 035709, January 2010.

49. Ilia Ivanov, Alexander Puretzky, Gyula Eres, Hsin Wang, Zhengwei Pan, Hongtao Cui, Rongying Jin, Jane Howe, and David B. Geohegan. Fast and highly anisotropic thermal transport through vertically aligned carbon nanotube arrays. *Applied Physics Letters*, 89(22): 223110, November 2006.

50. Michael B. Jakubinek, Mary Anne White, Ge Li, Chaminda Jayasinghe, Wondong Cho, Mark J. Schulz, and Vesselin Shanov. Thermal and electrical conductivity of tall, vertically aligned carbon nanotube arrays. *Carbon*, 48(13): 3947–3952, November 2010.

51. Kevin J. Voon, Mike Chang, George A. Sawatzky, and Alireza Nojeh. Modeling energy exchange and heating within nanorod arrays due to near-field radiative coupling. *Journal of Applied Physics*, 127(23): 234302, June 2020.

52. David B. Go, John R. Haase, Jeffrey George, Jochen Mannhart, Robin Wanke, Alireza Nojeh, and Robert Nemanich. Thermionic Energy conversion in the twenty-first century: advances and opportunities for space and terrestrial applications. *Frontiers in Mechanical Engineering*, 3: 13, November 2017.

53. Gang Xiao, Guanghua Zheng, Min Qiu, Qiang Li, Dongsheng Li, and Mingjiang Ni. Thermionic energy conversion for concentrating solar power. *Applied Energy*, 208: 1318–1342, December 2017.

54. Steven F. Adams. Solar thermionic space power technology testing: a historical perspective. *AIP Conference Proceedings*, 813(1): 590–597, 2006.

55. Kais Dridi, Amir Koshaman, Alireza Nojeh, and George Sawatzky. Towards compact solar thermionic converters based on carbon nanotubes forests. In *2016 29th International Vacuum Nanoelectronics Conference (IVNC)*, pages 1–2, Vancouver BC, Canada, July 2016, IEEE.

56. Alireza Nojeh. Thermionic energy conversion: fundamentals and recent progress enabled by nanotechnology. In *2019 19th International*

Conference on Micro and Nanotechnology for Power Generation and Energy Conversion Applications (PowerMEMS), pages 1–6, Krakow, Poland, December 2019, IEEE.

57. Harrison D. E. Fan. *Laser-induced thermoelectric energy generation using carbon nanotube forests.* PhD thesis, University of British Columbia, 2015.
58. E. H. Hirsch. Abnormal electron temperatures and electron reflection in the cylindrical thermionic diode. *Journal of Physics D: Applied* Physics, 35(21):2766–2771, November 2002.
59. M. S. Bronsgeest, J. E. Barth, G. A. Schwind, L. W. Swanson, and P. Kruit. Extracting the Boersch effect contribution from experimental energy spread measurements for Schottky electron emitters. *Journal of Vacuum Science & Technology B,* 25(6): 2049–2054, 2007.
60. David G. Cahill, Paul V. Braun, Gang Chen, David R. Clarke, Shanhui Fan, Kenneth E. Goodson, Pawel Keblinski, William P. King, Gerald D. Mahan, Arun Majumdar, Humphrey J. Maris, Simon R. Phillpot, Eric Pop, and Li Shi. Nanoscale thermal transport. II. 2003–2012. *Applied Physics Reviews,* 1(1): 011305, March 2014.
61. S. B. Singer, M. Mecklenburg, E. R. White, and B. C. Regan. Single-color pyrometry of individual incandescent multiwalled carbon nanotubes. *Physical Review B,* 84(19): 195468, November 2011.
62. Wade DeGottardi, Michael J. Gullans, Suraj Hegde, Smitha Vishveshwara, and Mohammad Hafezi. Thermal radiation as a probe of one-dimensional electron liquids. *Physical Review B,* 99(23): 235124, June 2019.
63. Shijing Tan, Adam Argondizzo, Cong Wang, Xuefeng Cui, and Hrvoje Petek. Ultrafast multiphoton thermionic photoemission from graphite. *Physical Review X,* 7(1): 011004, January 2017.

Chapter 6

Field Emission from the Edges of Single-Layer Graphene

S. T. Purcell, P. Vincent, S. Perisanu, A. Ayari, and P. Poncharal
Université Claude Bernard Lyon 1, CNRS, Institut Lumière Matière, F-69622, Villeurbanne, France
stephen.purcell@univ-lyon1.fr

There is now a body of theoretical and experimental work on field emission from the edges of single-layer graphene that we start by reviewing in some detail. Theory generally makes predictions on emission from idealized edges while experimental samples are mostly far from ideal. We present here experimental emission studies and analysis on well characterized, "closer to ideal," individual single-layer graphene edges in ultra high vacuum in an attempt to approach theoretical predictions and experimental realizations. Of note are the very narrow energy distributions in contrast to previous experimental work that can be an excellent guide for understanding the emission process.

Nanostructured Carbon Electron Emitters and Their Applications
Edited by Yahachi Saito

ISBN 978-981-4877-62-6 (Hardcover), 978-1-003-14199-0 (eBook)
www.jennystanford.com

6.1 Introduction

There has been an explosion of interest in graphene since 2003–2005, perhaps originally for the exotic nature of this one-atom-thick, high-quality, stable, two-dimensional (2d) material, but soon for the new paradigms it opened in fundamental physics related to 2d electron transport and in materials science, photonics, composites, batteries, etc. [1–3].

The edges of graphene were fairly quickly considered for field emission (FE) from a theoretical point of view [4] and somewhat later experimentally [5], firstly because its single atomic edge provides nature's smallest emitter radius and thus the promise of high field enhancement and low applied voltages. Other advantages of graphene as electron sources include low absorption, high melting temperatures, relatively high electrical and thermal conductivities, and the large experience obtained with carbon nanotube (CNT) emitters that raise similar questions on fabrication and applications. Graphene is not an obvious point source for electron microscopy and the planar geometry is a challenge for verticality, but this opens opportunities for in-plane nano vacuum devices where fabrication techniques can be transferred from the very active graphene device community [6].

Most experimental FE is on "multilayer" graphene layers (see review [7]) that has much overlap with work on "nanographite" films started earlier [8]. Here we address a much smaller body of work on single-layers or few layer graphene which has been defined as real graphene [1] and not thin graphite. The primary goal of such work is to advance in the fabrication, material characterization, and mechanisms of emission of more idealized samples that would then serve as guides for theory and applications.

First, a survey is made of pertinent theoretical work and then the experimental articles of interest for emission from graphene edges. The next section, and main part of the article, summarizes our recent article which attempts to carry out a classical FE characterization in ultrahigh vacuum (UHV) of an individual single-layer graphene flake attached to a W support tip. The chapter closes with a perspectives/summary section.

6.2 Survey of Theoretical Work

There is a considerable body of theoretical work on this subject. Here we present a nonexhaustive view of the ways that the problem has been approached to highlight some interesting predictions and to mark few examples where it appears useful to compare experiment and theory. The first scientific questions to answer are how to model the specific field enhancement and the tunneling I/V characteristics for very sharp edges and how is such a model related to the classic Fowler–Nordheim (F–N) equation.

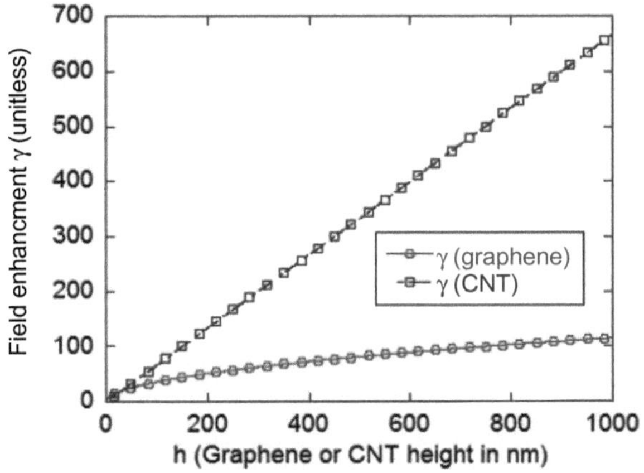

Figure 6.1 Relative γ for CNTs and graphene (see text).

Start by briefly examining the field enhancement factor γ (unitless) in the planar capacitor geometry where the emitters are on the negative inside surface. γ is defined by $F = \gamma V/d$, where F is the highest electric field at the emission zone, V is the applied voltage, d is the distance between the planar electrodes, and $F_0 = V/d$ is the applied macroscopic field. Electrostatic theory for γ at the center of a flat-topped wall of height h, thickness $2r$, infinite in the transverse direction, placed vertically on the inside cathode plane, $r << h << d$, gives $\gamma = \sqrt{(4h/\pi r)}$ [9, 10]. Our finite element calculations give empirically a logically somewhat higher value of $\gamma = 1.5\sqrt{(h/r)}$ when the wall is terminated by a hemispherical cap of radius r. These are very different and much weaker than for hemispherical capped cylinders that mimic nanotubes and nanowires and for which $\gamma \sim (2/3)h/r$ [11]. This means that a straight graphene edge loses much of its expected advantage of having a smaller radius because edges have smaller γ than cylinders. To make this more concrete, Fig. 6.1 compares γ_{CNT} of a (10,0) CNT, r = 0.5 nm, with γ_{Gr} of a graphene edge of van der Waals radius of r = 0.17 nm. Even at h = 1000 nm, γ_{Gr} = 115; a modest value when planar emitters covered with nanostructures generally attain 500–2000. This underlines the obvious fact that emission from surfaces covered by vertical, separated, nanolayer emitters comes principally from the sharp wall corners and jagged edges with high transverse curvature zones [8].

Electrostatic calculations on wall corners rounded to the wall thickness show typically 5 times increased γ[12]. Even more extreme amplification can be imagined with high sawtooth-shaped emitters. This alone means that great care must be taken when comparing calculations on idealized straight structures with all experimental results.

The high field on sharp objects and the local work function still combine to create a triangular tunneling barrier with width in the 1 nm range that plays a primordial role leading to exponential I/V characteristics. However, the 2d geometrical form of the emitter and the atomic scale nature of the electron density of states and their specific tunneling mechanism may lead to strong differences in all the FE characteristics including I/V's, field-emission microscopy (FEM) patterns, temperature dependence, high current behavior, and energy distributions.

At the next theoretical level, tunneling was reconsidered from a metallic band picture adapted to a 2d metal taking into account lateral confinement and the curvature of the potential barriers induced by small radius. According to Qin et al. [10] the "F–N like" equation for a straight edge morphs from $I(\text{flat}) = aF^2 \exp(-b/F)$ to $I(\text{edge}) = cF^3 \exp(-d/F^2)$ (a, b, c, d constants) when the radius decreases to well below the tunneling barrier width. Verifying such a law should be a primary goal for this research. It would have very large effects on the fit parameters of experimental data which are related to the atomic level processes of the supply function and the details of the tunneling barrier and the emission area. Note that plotted in an F–N plot this equation would curve downwards from a straight line.

Band structure calculations of field emission from graphene most often consider the electrons to come from the edge electronic structure, which is a very important theme in graphene research [13, 14] for its interest for electronic devices based on quantum confinement. Potentially this is where FE measurements could have a larger impact on the physics of graphene. Emission due to the bulk graphene band structure must also be considered and at present in the absence of pertinent experiments a clear preference for which feature of the density of states governs the supply function has not emerged. The physics is obscured by the large difficulty in producing well-characterized samples and a verification of the specific edge FE tunneling model.

Graphene edges have dangling bonds and π bonds and a special edge state based on π bonds on the zigzag edge with a high density of states at the Fermi level [15, 16]. The earliest calculations on FE from graphene appear to be on nanoribbons [4, 17] using time-dependent density-functional theory for the band structure and the tunneling currents. They brought out several recurrent themes: comparison of emission from zigzag and armchair edges, effect of edge adatoms, in this case hydrogen, and a treatment of the different tunneling probabilities for the different states. This work provided several predictions to which later authors could make comparisons: the dangling-bond states localized at edges of clean zigzag and clean armchair ribbons are the strongest sources of the FE current, the unique zigzag edge state does not emit despite its large density of states at the Fermi level (E_F) and H termination reduces the FE current because it fills the dangling bonds. They predict large total energy distributions (TEDs) [18] of width 1 eV and very surprisingly that emission of the armchair would be 2 eV below E_F. These cited papers do not discuss I/V characteristics and field enhancement.

Field emission from graphene was then calculated within the local-density approximation framework [19] which the authors claim is simpler, more widespread, can capture full atomic and electronic structures, allow explicit imposition of an external electric field, and a fuller examination of the complex and subtle supply function sources of the electron currents. Current is constant in time rather than time-dependent. Results on zigzag edges followed well the previous references while those for armchair edges revealed new emission channels near E_F. Again the calculation uses large electric fields in the 6 V/nm range but γ factors are not discussed; so it is rather difficult to place their I/V characteristics in context and test for an edge-type F–N behavior. Very narrow TEDs ~0.06 eV range are predicted in contradiction with the previous work. This obviously needs clarification. A useful discussion is given on the effect of edge dipole moments of clean and hydrogen-terminated zigzag and armchair edges on the different edge work functions.

The classical FE model for graphene [10] was extended to a more evolved and complete analytical treatment that includes the unique bulk band structure of graphene and field penetration which shifts the Dirac point and adds available occupied states at E_F [20].

Calculations are for armchair edges as the zigzag edges states are proposed to screen the field penetration. The relation between the effective vacuum barrier height and the applied field is obtained including consideration of the image charge. I/V characteristics are downward curved on classic F–N plots but were not compared to the new $cF^3 \exp(-d/F^2)$ model.

There appears as yet to be no calculations on graphene corners and jagged edges despite that experimentally most emission is likely to come from such topologies. Theory predicts interesting field-dependent band structure effects on jagged narrow ribbons [21].

The novel and still not completely explained FEM patterns for molecules [22] and nanoscale objects such as nanotubes [23] can show interesting interference phenomena. Two theoretical articles treat the FEM patterns from edges [24, 25] but they do not as yet resemble experimental results.

Quite a few later articles followed which we can only mention briefly. For example, the carbon edge and hydrogen passivating atoms were replaced with various other monoatomic and molecular species and corresponding edge work functions and FE currents estimated [26–28]. Four novel articles treat subjects of current interest to a wider graphene community including an intriguing study of Klein Tunneling assisted FE [29], which would radically change the I/Vs and lower the field necessary for emission, coherent line emission from the bulk band structure [25], spin-polarized FE from the zigzag edge state [27], and laser-driven femtosecond emission from graphene nanoribbons [30]. Ref. [31] provides a survey of some recent theoretical developments.

Summing this up, there is a wealth of predictions using different theoretical methods for the states at the origin of the supply function, estimating the surface barrier and calculating the specific state-dependent tunneling probabilities, energy distributions, and total currents. To permit such theoretical work to be tested, the priority is to produce samples of known edge structure and to establish a reliable semiclassical model as proposed in Ref. [10].

6.3 Survey of Experimental Work

Although considerable work has been done on FE from vertically dressed multilayer graphene (or nanographite) emitters [7, 8], usually covering a surface, there is a dearth of measurements on well-characterized, individual, single-layer graphene that can guide the community toward future developments. There are few clear proofs of single-layer edge emission and only two examples of UHV cleaning (one of which is on a multilayer edge), which is necessary for controlled FE studies.

A first work succeeded in obtaining FE from the edge of an exfoliated single-layer (SL) [32], whose nature was proven by Raman spectroscopy, using nanomanipulation tools in a scanning electron microscope (SEM). The deposited layer was scratched by a probe tip to dress an edge somewhat vertically. Stable currents and I/Vs were obtained which had novel upward curving F–N plots. They were obtained only with very large applied fields with the probe tip at the small distance of about 600 nm. The stability was obtained in the poor SEM vacuum by FE current conditioning. Current conditioning may or may not produce clean surfaces of intrinsic nature. By field-driven diffusion it automatically accumulates adsorbates and stabilizes nano-objects formed from the adsorbates, at the places of highest field, which then become the emitters. At the same time, local emission can physically transform the emission zone by local heating. Unexpected and still unexplained upward curving F–N plots were obtained. The authors did not discuss the possibility that the field could flex the edge of the tilted layer toward the probe which would increase the γ factor and thus create an upward curving F–N plot.

Mechanical manipulation was used in an optical microscope to place what was probably a graphene flake on a tip that was not visible with the optical resolution [33] though Raman spectroscopy supported that in the tip vicinity there was an object containing SL graphene. This was then mounted in a 10^{-7} Torr FEM chamber where again current conditioning was used to "clean" the sample. The team succeeded in obtaining FE patterns which showed elliptical shapes that could in principle correspond to emission from an edge emitter. Very unstable currents were obtained.

Multilayer, individual graphene flakes with gradual emission current conditioning were studied at the atomic level in a TEM [34] and in a UHV environment with sample heating [35], as in the work presented in the next section. Significant and original FEM patterns in the form of a large band composed of finer transverse elliptical forms were obtained which will be discussed in the chapter by Saito et al. The patterns were interpreted in terms of both the larger scale local knife edge form, which should be general to the edges of single- and multilayer samples (see below), and more speculatively that the elliptical forms are due to the local states theoretically predicted to be the source of electrons for emission.

Emission from chemical vapor deposition (CVD) graphene deposited on quartz, where the edges were nominally still in contact with the substrate, was studied at 10^{-6} Torr [36]. This allowed to examine field enhancement in the presence of a dielectric which can be an important consideration for planar devices. Two test geometries were used of which one is shown in Fig. 6.2a. Reproducible I/Vs were obtained (Fig. 6.2b), which resemble F–N behavior. The voltages were in the 200–300 V range for 500 μm distance to the anode. There was a highly original semiperiodic FE pattern that still needs explanation (inset in Fig. 6.2b). The technically important aspect of gradual degradation and peeling of the samples were followed in some detail.

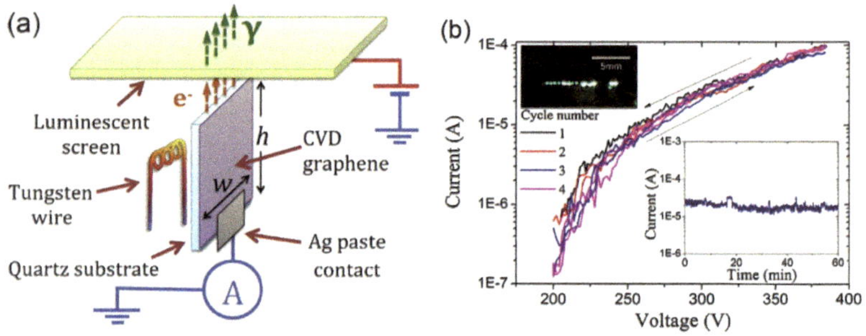

Figure 6.2 (a) Schematic of the setup [36] and (b) FEM and I/V characteristics.

Remarkable advances in sample preparation have been presented in a very recent article where graphene flakes were again mounted in a TEM and local evaporation at a displaceable counter tip probe was used to fashion the flakes to desired geometries [37]. The graphene used in this work was mostly few-layered (3–10) nanosheets scraped from highly oriented pyrolytic graphite (HOPG). Quite straight edges can be produced but with a certain jaggedness still present. In situ TEM FE from these multilayer edges were made and the I/V data fit reasonably well standard F–N theory.

6.4 UHV Studies of Free-Standing, Individual, Cleaned, Graphene Flakes

A study has been recently carried out on FE in UHV from the edges of individual, free-standing, cleaned, graphene flakes where their single graphene layer nature has been carefully characterized and the emission zones being from the edges clearly demonstrated [38]. The sample characterization started with the mounting of individual flakes in a very controlled way in a SEM (Fig. 6.3) followed by TEM imaging, Raman characterization, and high temperature and high field desorption treatments in UHV. Detailed FE characterization was carried out including I/V, FEM, and TEDs during subsequent stages of sample cleaning at room and liquid air temperatures.

Figure 6.3 Experimental procedure for graphene sample production (Taken from Ref. [38]). (a) Three-axis nanomanipulator installed in the SEM. (b)–(e) Pickup sequence of a graphene flake, starting from CVD-grown graphene on a TEM grid to finish with a graphene flake on a tip. The black circle shows the location of the TEM imaging analysis (see text). (f) The graphene sample studied, nicknamed "Italy."

The TEDs of the emitted electrons are of particular interest because they give access to the density of states near E_F and could perhaps make an easier connection to theory which has so far predicted very different energy widths of 1 and 0.06 eV and positions with respect to E_F [4, 19]. Peak widths for cold FE from metallic emitters are typically in the 0.25–0.4 eV range [18, 39]. Only one measurement [6] of the energy distributions of emitted electrons from what was reliably graphene (but uncleaned) has been reported. Peak widths of ~2 eV were shown which shift strongly to lower energy with increasing applied field. Obviously, this needs clarification and is a major point of this work.

The TED peak shape and position provides several types of information. For the simple metallic case, a broadening on the high-energy side is a signature of a temperature increase while a broadening on the low-energy side marks an increased electric field [18]. The peak maximum for a metal is slightly below E_F of the emission zone and thus generally at the voltage applied to the tip. Shifts occur if there are voltage drops leading up to the apex and also due to nanometric effects such as occur for quantum dots (QDs) formed from adsorbates whose formation is ubiquitous on poorly cleaned surfaces and in poor vacuum. Such objects act in FE exactly as deposited molecules with localized states separated from the main emitter Fermi sea by a thin tunnel barrier and then emit into vacuum through the vacuum tunnel barrier by resonant tunneling [40]. The TED scan consists of one or more peaks that shift by field penetration downwards in energy as the applied voltage is increased. When a state above E_F previously reaches the Fermi level, a new window for electron emission opens, which causes an upward kink in the I/V characteristics and a new peak in the TED. Peak widths are typically in the 0.2–0.3 eV range but peaks as narrow as 0.065 eV have been measured [41]. The complete spectra can widen to several volts and add numerous peaks as such nanoprotrusions increase in size. This occurs naturally with continued emission. A critical problem for this work is to distinguish this type of QD emission site from the true graphene edge and corner emitters, though we can "do our best" by careful in situ cleaning.

Graphene was produced by chemical vapor deposition with strict control of hydrogen etching to achieve a coverage of mostly SL graphene with some bilayer islands [42, 43] and transferred to TEM grids. Graphene flakes were torn from the layers and picked up at the apex of etched tungsten tips (Fig. 6.3). Complementary Raman spectroscopy and TEM diffraction measurements were performed on the very same grid sample. The Raman measurements showed that the sample was indeed graphene, i.e., small D-band and single sharp 2D-band [44]. The TEM diffraction [38] showed that the sample was, however, polycrystalline with an average single-crystal grain size of a few tens of nanometers. TEM diffraction on a single grain showed the hexagonal pattern. These tips were then mounted in a polyvalent field-emission system with a base pressure of 1×10^{-10} Torr, sample heating by a W loop, imaging by microchannel plate for FE microscopy, and an electron energy analyzer. The graphene tips are about 2 mm to intermediary extraction anodes. The flake withstood the fields necessary for field ion emission of oxygen, ~15 V/nm.

The sample was mounted in a UHV system which was then baked out for 24 h at about 120 °C. First, emission in the pA range occurred at about 300 V. The sample was then heated for 1-min cycles with the highest temperature of 770 °C. This is necessary to establish a minimum current stability and, in general, is what the team does with all metal emitters. One must heat a W tip to 1500 °C before it starts to show the characteristic single-crystal emission pattern. Although much better heat cleaned than most examples in the literature, this is termed as the "dirty" or "pristine" state of the sample. First, FEM images consisted of one spot (Fig. 6.4a) presumably from the dominant nanostructure formed by the first application of field and emission.

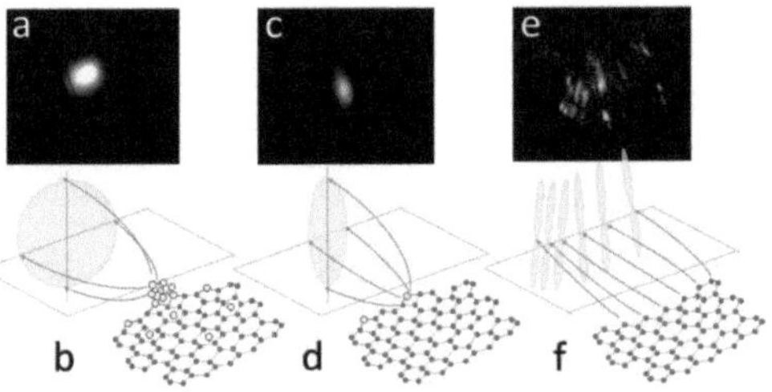

Figure 6.4 (a, b) FE pattern and schematic of emission from pristine graphene dominated by a single emitter (nanoprotrusion). (c, d) FE pattern and schematic of emission from thermally cleaned graphene. The local environment is no longer symmetric: the emitted electrons tend to focus along the graphene edge direction but less in the perpendicular direction. (e) Field ion microscopy (FIM) patterns and (f) schematic from the field desorbed graphene using oxygen. As field desorption eliminates local asperities until the field is more even, emission is no longer dominated by a single spot but multiple emitters are now visible [38].

A mechanical vibration test is performed while the sample is emitting. A small AC voltage is scanned in frequency until an electromechanical resonance occurs, which is observable by an elongation of the FE spot parallel to the vibration direction and perpendicular to the graphene flake. This test is also extremely useful to check that our sample is still present after outgassing or other treatments involving a change of the emission sites. The resonance frequencies were in the MHz range throughout this work and are discussed in another publication [45].

Heat cleaning consists of raising the sample to 1000 °C for about 4 min under a field sufficient for FE. After cleaning, mechanical resonance tests were performed to check that the sample was still undamaged. The FEM images still consisted of one spot but with a clear elliptical shape as expected for an edge emitter (Fig. 6.4c), which should electrostatically impose weaker field focusing perpendicular to the edge (see discussion in Refs. [34, 35]). The elliptical deformation was consistent with the flake direction inferred from the mechanical resonance experiment and the SEM imaging of the sample before mounting in the FEM system.

To better eliminate these nanostructures, field-induced desorption was used which is more local than thermal cleaning. An O_2 pressure of 10^{-4} Torr was introduced into the chamber and an increasing positive voltage was then applied to the tip until a field ion emission pattern was observed. Field desorption eliminates local asperities where the electric field is highest. As the first principal emitters were removed, new emission sites appeared at higher voltage (see Fig. 6.4e). The voltage was about 1200 V at the end of the process for reasonable ion imaging. Each emitter was elliptically shaped in both FIM and FEM and in the expected direction, although they were not perfectly aligned. This could have been a consequence of local crinkling of the graphene sheet which would tilt the average trajectory of each emitter. It is worth noting that each FIM zone exhibited two symmetric lobes with a darker central band that reminds one of the FEM forms [35] that was attributed to edge state symmetry wave functions.

The F–N and TED curves were recorded after each cleaning stage. The F–N plots (Fig. 6.5) can be useful for examining the existence of QD effects which induce higher order curving by resonant tunneling through nano-objects whose resonances shift through E_F on increasing field. Several changes in the slope can be seen as expected from FE from a "QD" (Fig. 6.5, upper curve). On the thermally cleaned sample the plot is straighter, and on the field-desorbed sample perfectly straight though the measurement was carried out over a reduced range. Note also the increase of extraction voltage to reach a given emission current is consistent with reducing high field zones by first elimination of adsorbates and edge QDs with thermal cleaning and then a small reduction of sample roughness and perhaps length by the field desorption.

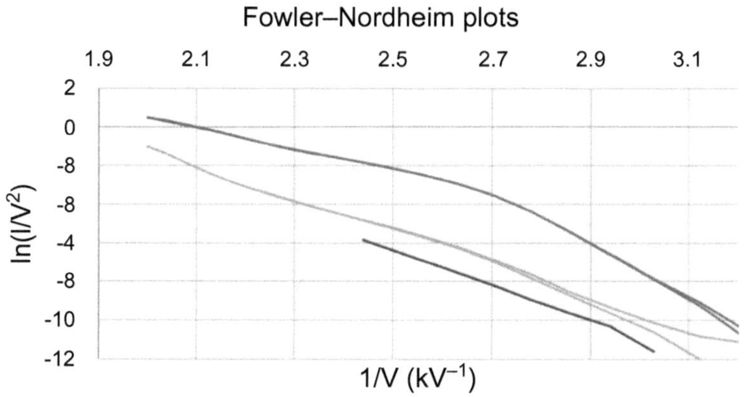

Figure 6.5 Fowler–Nordheim curves on pristine graphene (top), after high temperature curing (middle), and after field desorption (bottom). These plots cover ranges of 300–500 V for 1 pA–400 nA (5 orders of magnitude in current) [38].

The big uncertainty in proportionality between the pertinent field F for tunneling and the applied voltage V ($F = \beta V$) due to the mounting on a tip support and the jagged form of the flake at many length scales means that it is extremely hazardous to interpret the F–N slopes in terms of a predictable local field enhancement at the emitting zone. In addition, different sections of F–N plots for the pristine and heated sample have quite different slopes. This is the source of major confusion in the field and we did not wish to repeat that error. One should compare fits with other versions of the F–N theory for edges [10] only if the sample is well defined and stable during the I/V and the data covers many orders of magnitude. In this same sample-mounting geometry, FE from a small radius nanotube grown directly in situ on similar W tips was achieved at as low as 65 V. Although not very quantitative, this agrees with the prediction that edge emitters have quite lower γ factors than cylinders.

TED spectra of pristine, heat treated, and desorbed graphene are compared in Fig. 6.6. These three TEDs were recorded around 10 pA of emission current at room temperature. Two peaks are present for the pristine sample (adsorbate QD) of which one disappears after higher temperature curing. Actually, as we discuss below, the peak at higher energy is emission from near E_F and the QD peak is dominant but not enough so as to obscure all the Fermi level emission. This happens at intermediate stages of field-driven formation of QDs [46].

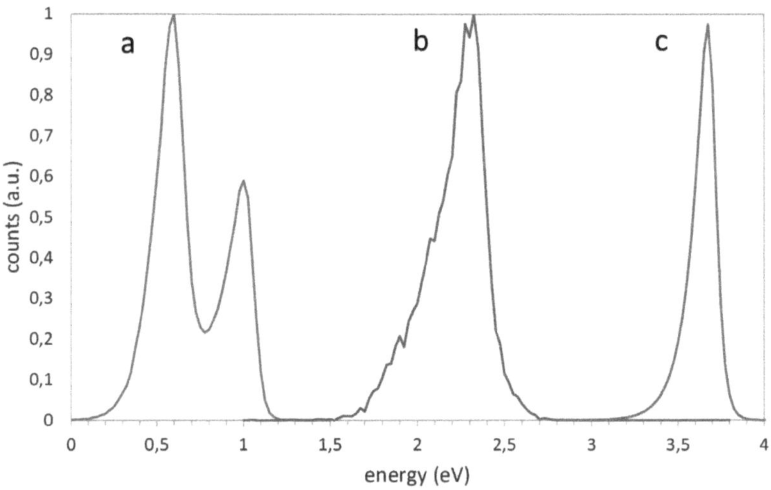

Figure 6.6 TEDs of (a) pristine, (b) heated, and (c) desorbed states for comparable emission currents (respectively 7.5, 12, and 8 pA). FWHMs are respectively 0.20 (highest peak), 0.26, and 0.15 eV. Curves have been arbitrarily shifted for clarity [38].

Consider now the shift behavior of the TED peaks with extraction voltage because strong linear shifting is a good indicator of the existence of QD-like behavior. In Fig. 6.7 we show TEDs for the desorbed state for different applied voltages and emitted currents. The desorbed state TEDs fit extremely well to the free electron TED [18] if one allows for a voltage-dependent shift. One notes there are two weaker peaks that appear at higher voltages with one "arriving" at E_F.

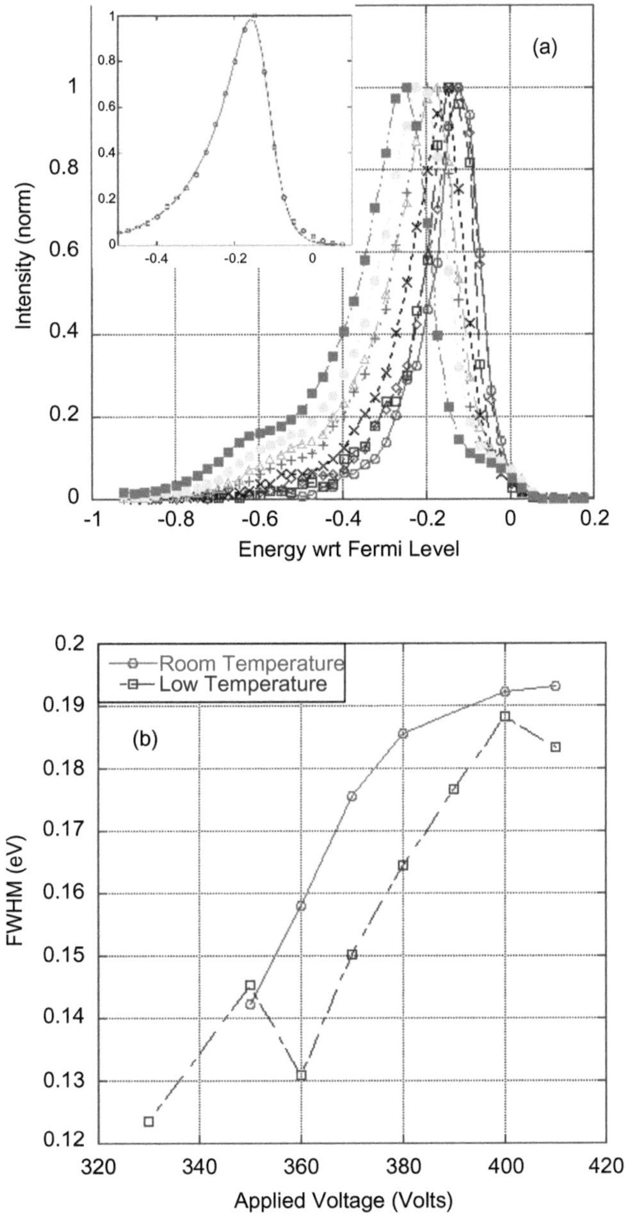

Figure 6.7 (a) TED spectra of the desorbed state for different applied voltages at low temperature. Inset: Spectrum fitted with TED formula [18]. (b) FWHMs for TEDs at 300 and 110 K. They widen roughly linearly in field.

The principal peak at lower energy of the pristine state shifts by an ~6 meV/applied volt; in the range we sometimes measure for adsorbate QDs, but the Fermi level peak and both our heat-treated and desorbed graphene peaks all have a considerably smaller shifts of ~2 meV/applied volt. Normally, the QD type emission is very easy to distinguish from its metal tip "substrate," because the metal TED has no shift at all. Here we are in a new situation of a weak linear shift even though the spectra are metallic-like.

Three interpretations can be proposed that will need theoretical support. First, it is possible that there is a Schottky barrier at the W-tip/graphene interface which can create a voltage-dependent shift. Thus, the emitted electrons would come directly from E_F of the graphene flake and, in particular, its edge. This is developed in more detail in Ref. [38]. Second, the actual edge emitter region even for the desorbed state may not yet have an idealized straight form and still has a topology that lends itself to a weak QD behavior. There may be atomic/nm scale disorder allowing some insulating sp3 coordination or robust adsorbate complexes. It was necessary to heat multiwall CNTs to 1300 °C to obtain metallic emission [47]. A third possibility is that the shifting is a manifestation of the theoretically predicted field penetration into an armchair edge [20], possible because of the atomic thickness and low electron density. In this case the peaks come from the pertinent density of states for emission of the edge or the bulk electron densities and they are pulled to lower energy with increasing field similar to a QD, but there is always emission present at E_F.

Finally, let us focus on the narrowness of the TEDs which are all sensibly narrower than TEDs from metallic emitters [39]. The narrowness of the adsorbate QD (0.2 eV) is consistent with a localized state which as mentioned have been measured as narrow as 0.065 eV [41], though the metallic-like shape is not elucidated. However, once adsorbate structures are thermally removed, the electrons should be emitted from a wider density of states through a triangular barrier. The thermally treated TEDs widen only slightly (0.26 eV). The FWHM (full width at half maximum) shrinks down to even lower values for the desorbed state (see Fig. 6.7), reaching as low as 0.12 eV and increasing roughly linearly with field as with a metallic emitter (Fig. 6.7b). The main effect is that the low energy side is especially narrow. We propose that the narrowness of the peak is a direct manifestation of the curving of the barrier potential due to the atomic scale of the graphene edge which to first order is controlled by the slope of the tunnel barrier at the second crossing point where the potential crosses E_F. The measured FWHM implies that the field at 1 nm from the edge is about one-half of the surface value, in line with an atomic scale tip. A similar argument for energy filtering has been used to explain the narrow TEDs for LaB_6 nanowires, down to 0.2 eV, where the low slope at the E_F crossing occurs, because lower fields are necessary to induce emission on the low work function material [48].

These spectra are narrower by ~2–3× than a metallic emission peak, ~15× than previous measurements on graphene [6], 6× than one of the theoretical values [17] but curiously ~3× larger than the other [19], which is surprisingly narrow at ~0.06 eV. All the peaks presented here have the asymmetric form characteristic of tunneling from a metal through a triangular barrier which is not the case of the theoretical predictions in Ref. [17] where the TEDs do not appear to have a high energy temperature edge.

6.5 Summary and Perspectives

The preparation of clean, or at least cleaner, graphene samples for FE studies has been explored. Sample characterization by TEM and Raman spectroscopy, cleaning procedures, ultrahigh vacuum environment, and mechanical vibration control strategy confirm that the measured object is indeed carefully desorbed SL graphene. The extraction voltages are in the 400 V range which is low but not extraordinarily low as one might naively have expected for graphene edges mounted on a tip, which agrees with simple considerations of the edge field enhancement factors.

Clean graphene shows a straight F–N behavior over the presented measurement range but a wider range is needed if viable comparisons with newer theories are to be drawn.

The TED peaks are very narrow (~0.15 eV) and this is proposed to be due to the curved potential barrier. It may be one of the first experimental manifestations of a distinctive tunneling behavior for graphene, in the absence of the confirmation of a new F–N-like law. It will be interesting to see if other authors confirm this, as well as the residual shifting.

The TEDs maintain their metallic form even when they are below E_F. This known and very general behavior for molecular size emitters, raised in early work [40], shows that the simple resonant tunneling model is not sufficient. This is still awaiting theoretical explanation. Help may come from recent work by the Scanning Tunneling Microscopy community which has now established inelastic tunneling models [49].

The existence of the principal and extra peaks in the desorbed state TEDs should now be considered in the light of theoretical band structure predictions and an accompanying k state-dependent emission model.

Although this is a progress in sample preparation, one is still far from the well-defined zigzag or armchair emitters of determined length and electrostatic geometry that mimic theoretical models. Theoretical predictions such as coherent tunneling, spin-polarized emission, and femtosecond-induced emission are awaiting better experimental realizations. Comparative FE with and without hydrogen adsorption may provide an easier connection between theory and experiment.

In the conclusion, these more controlled experiments and particularly the energy spectra can now guide the scientific community to better understand electron and ion emission from graphene edges.

Acknowledgments

The authors acknowledge the contributions of M. Choueib, R. Dielhl, and N. Blanchard. This work was carried out within the framework of the Plateforme Nanofils et Nanotubes Lyonnaise de l'Université Lyon 1.

References

1. Geim, A. K., & Novoselov, K. S. (2007). The rise of graphene. *Nat. Mater.*, **6**, 183–196.
2. Castro Neto, A. H., Guinea, F., Peres, N. M. R., Novoselov, K. S., & Geim, A. K. (2009). The electronic properties of graphene. *Rev. Mod. Phys.*, **81**, 109–162.

3. Abergela, D. S. L., Apalkovb, V., Berashevicha, J., Zieglerc, K., & Chakrabortya, T. (2010). Properties of graphene: A theoretical perspective. *Adv. Phys.*, **59**, 261–482.
4. Tada, K., & Watanabe, K. (2002). Ab initio study of field emission from graphitic ribbons. *Phys. Rev. Lett.*, **88**, 127601.
5. Wu, Z.-S., Pei, S. Ren, W., Tang, D., Gao, L., Liu, B., Li, F., et al. (2009). Field emission of single-layer graphene films prepared by electrophoretic deposition. *Adv. Mater.*, **21**, 1756–1760.
6. Shaw, J. L., Boos, J. B., Kong, B. D. Robinson, J. T., & Jernigan, G. G. (2019). Field emission energy distribution and three-terminal current-voltage characteristics from planar graphene edges. *J. Appl. Phys.*, 125, 054502-9.
7. Chen, L., Yua, H., Zhong, J., Song, L., Wuc, J., & Su, W. (2017). Graphene field emitters: A review of fabrication, characterization and properties. *Mater. Sci. Eng. B*, **220**, 44–58.
8. Obraztsov, A. N., & V. Kleshch, V. I. (2009). Cold and laser stimulated electron emission from nanocarbons. *J. Nanoelectron. Optoelectron.*, **4**, 1–13.
9. Miller, R., Lau, Y. Y., & Booske, J. H. (2007). Electric field distribution on knife-edge field emitters. *Appl. Phys. Lett.*, **91**, 074105.
10. Qin, X., Wang, W., Xu, N., Li, Z., & Forbes, R. (2011). Analytical treatment of cold field electron emission from a nanowall emitter, including quantum confinement effects. *R. Soc. A*, **467**, 1029–1051.
11. Minoux, E. (2010). Sources électroniques à base de nanotubes de carbone - Application aux tubes amplificateurs hyperfréquence. *Thesis Online*, https://tel.archives-ouvertes.fr/pastel-00001904/, p. 62.
12. Watcharotonea, S., Ruoff, R. S., & Read, F. H. (2008). Possibilities for graphene for field emission: Modeling studies using the BEM. *Proc. Seventh Int. Conf. Charged Particle Optics, Phys. Procedia*, **1**, 71–75.
13. Lado, J. L., Garcia-Martinez, N., & Fernandez-Rossier, J. (2015). Edge states in graphene-like systems. *Synth. Metals*, **210**, 56–67.
14. Shimomura, Y., Takane, Y., & Wakabayashi, K. (2011). Electronic states and local density of states in graphene with a corner edge structure, *J. Phys. Soc. Jpn.*, **80**, 054710-9, DOI: 10.1143/JPSJ.80.054710.
15. Stein, S. E., & Brown, R. L. (1987). Electron properties of large condensed polyaromatic hydrocarbons. *J. Am. Chem. Soc.*, **109**, 3721–3729.

16. Nakada, K., Mitsutaka, F., Dresselhaus, G., & Dresselhaus, M. S. (1996). Edge state in graphene ribbons: Nanometer size effect and edge shape dependence. *Phys. Rev. B*, **54**, 17954–17961.

17. Araidai, M., Nakamura, Y., & Watanabe, K. (2004). Field emission mechanisms of graphitic nanostructures. *Phys. Rev. B*, **70**, 245410.

18. Young, R. D., & Muller, E. W. (1959). Experimental measurement of the total-energy distribution of field-emitted electrons. *Phys. Rev.*, **113**, 115–120.

19. Huang, S. F., Leung, T. C., Li, B., & Chan, C. T. (2006). First-principles study of field-emission properties of nanoscale graphite ribbon arrays. *Phys. Rev. B.*, **72**, 035449.

20. Wang, W., Qin, X., Xu, N., & Li, Z. (2011). Field electron emission characteristic of graphene. *J. Appl. Phys.*, **109**, 044304.

21. Saroka, V. A., Batrakov, K. G., Demin, V. A., & Chernozatonskii, L. A. (2015). Band gaps in jagged and straight graphene nanoribbons tunable by an external electric field. *J. Phys.: Condens. Matter*, **27**, 145305.

22. Muller,E.W.(1950).SichtbarmachungderPhthalocyaninmolekelmitdem Feldelektronenmikroskop. *Naturwissenschaften*, **37**, 333–334.

23. Saito, Y., Hata, K., & Murata, T. (2000). Field emission patterns originating from pentagons at the tip of a carbon nanotube. *Jpn. J. Appl. Phys.*, **39**, L271–L272.

24. Luo, M., & Li, Z. (2016). Multi-field electron emission pattern of 2D emitter: Illustrated with graphene. *J. Appl. Phys.*, **120**, 204304-5.

25. Li, Z., Xu, N., & Kreuzer, H. J. (2012). Coherent field emission image of graphene predicted with a microscopic theory. *Phys. Rev. B*, **85**, 115427–115435.

26. Wang. W., & Li, Z. (2012). Graphene with the secondary amine-terminated zigzag edge as a line electron emitter. *Appl. Phys. A*, **109**, 353–356.

27. Chen, X., & Li. B. (2015). First-principles study of field emission from zigzag graphene nanoribbons terminated with ether groups. *Chinese J. Chem. Phys.*, **28**, 573–578.

28. Gao, Y., & Okada, S. (2019). Field emission properties of edge-functionalized graphene. *Carbon*, **142**, 190–195.

29. Sun, S., Ang, L. K., Shiffler, D., & Luginsland, J. W. (2011). Klein tunneling model of low energy electron field emission from single-layer graphene sheet. *Appl. Phys. Lett.*, **99**, 013112.

30. Utsugi, D., Hu, C., & Watanabe, K. (2012). Femtosecond laser-driven field emission from graphene nanoribbons: Time-dependent density-functional theory simulations. *Appl. Phys. Express*, **5**, 105101.

31. Ang, Y. S., Liang, S.-J., & Ang, L. K. (2018). Theoretical modeling of electron emission from graphene. *MRS Bull.*, **42**, 505–510.

32. Xiao, Z., She, J., Deng, S., Tang, Z., Li, Z., Lu, J., & Xu, N. (2010). Field electron emission characteristics and physical mechanism of individual single-layer graphene. *ACS Nano*, **4**, 6332–6336.

33. Santandrea, S., Giubileo, F., Grossi, V., Santucci, S., Passacantando, M., Schroeder, T., Lupina, G., & Di Bartolomeo, A. (2011). Field emission from single and few-layer graphene flakes. *Appl. Phys. Lett.*, **98**, 163109.

34. Nakakubo, K., Asaka, K., Nakahara, H., & Saito, Y. (2012). Evolution of field electron emission pattern from multilayered graphene induced by structural change of edge. *Appl. Phys. Express*, **5**, 055101.

35. Yokoyama, N., Nakakubo, K., Iwata, K., Asaka, K., Nakahara, H., & Saito, Y. (2016). Field emission patterns showing symmetry of electronic states in graphene edges. *Surf. Interface Anal.*, **48**, 1217–1220.

36. Kleshch, V. I., Bandurin, D. A., Orekhov, A. S., Purcell, S. T., & Obraztsov, A. N. (2015). Edge field emission of large-area single layer graphene. *Appl. Surf. Sci.*, **357**, 1967–1974

37. Zhao, L., Luo, G., Cheng, Y., Li, X., Zhou, S., Luo, C., Wang, J., Liao, H.-G., Golberg, D., & Wang, M.-S. (2020). Shaping and edge engineering of few-layered freestanding graphene sheets in a transmission electron microscope. *Nano Lett.*, **20**, 2279–2287.

38. Diehl, R., Choueib, M., Choubak, S., Martel, R., Perisanu, S., Ayari, A., Vincent, P., Purcell, S. T., & Poncharal, P. (2020). Narrow energy distributions of electrons emitted from clean graphene edges. *Phys. Rev. B*, **102**, 035416 (7p).

39. Swanson, L., & Schwind, G. (2009). A review of the cold field electron cathode. *Adv. Imag. Elect. Phys.*, **159**, 63–98.

40. Swanson, L. W., & Crouser, L. C. (1970). The effect of polyatomic adsorbates on the total energy distributions of field emitted electrons. *Surf. Sci.*, **23**, 1–29.

41. Choubak, S., Biron, M., Levesque, P. L., Martel, R., & Desjardins, P. (2013). No graphene etching in purified hydrogen. *J. Phys. Chem. Lett.*, **4**, 1100–1103.

42. Choubak, S., Levesque, P. L., Gaufres, E., Biron, M., Desjardins, P., & Martel, R. (2014). Graphene CVD: Interplay between growth and

etching on morphology and stacking by hydrogen and oxidizing impurities. *J. Phys. Chem. C*, **118**, 21532–21540.35.

43. Poncharal, P., Ayari, A., Michel, T., & Sauvajol, J.-L. (2008). Raman spectra of misoriented bilayer graphene. *Phys. Rev. B*, **78**, 113407-4.
44. Descombin, A., Poncharal, P., Pascale-Hamri, A., Choueib, M., Diehl, R., Vincent, P., Purcell, S. T., Ayari, A., & Perisanu, S. (2019). Giant, voltage tuned, quality factors of single wall carbon nanotubes and graphene at room temperature. *Nano Lett.*, **19**, 1534–1538.
45. Binh, V. T., Purcell, S. T., & Garcia, N. (1993). Reply to comment by N. Ernst et al. *Phys. Rev. Lett.*, **70**, 2504.
46. Purcell, S. T., Vincent, P., Rodriguez, M., Journet, C., Vignoli, S., Guillot, D., & Ayari, A. (2006). Evolution of the field-emission properties of individual multiwalled carbon nanotubes submitted to temperature and field treatments. *Chem. Vap. Deposition*, **12**, 331–344.
47. Purcell, S. T., Binh, V. T., & Garcia, N. (1995). 64 meV measured energy dispersion from cold field emission nanotips. *Appl. Phys. Lett.*, **67**, 436–438.
48. Zhang, H., Tang, J., Yuan, J., Yamauchi, Y., Suzuki, T. T., Shinya, N., Nakajima, K., & Qin, L.-C. (2015). An ultrabright and monochromatic electron point source made of a LaB6 nanowire. *Nat. Nanotech.*, **11**, 273–279.
49. Chong, M. C., Afshar-Imani, N., Scheurer, F., Cardoso, C., Ferretti, A., Prezzi, D., & Schull, G. (2018). Bright electroluminescence from single graphene nanoribbon junctions. *Nano Lett.*, **18**, 175–181.

Chapter 7

FEM and FIM of Graphene

Yahachi Saito
Toyota Physical and Chemical Research Institute,
Yokomichi 41-1, Nagakute, Aichi 480-1192, Japan
ysaito@toyotariken.jp

Graphene is a two-dimensional material with atomic thickness and displays intriguing field-emission microscopy (FEM) and field ion microscopy (FIM) images that reflect electronic states at the edges. This chapter presents the FEM and FIM patterns peculiar to graphene edges and their interpretations.

7.1 Introduction

Graphene is now attracting considerable attention as a new material with two-dimensional structure and a peculiar electronic structure. In Table 7.1, electrical, mechanical, and thermal properties of graphene are summarized. Graphene with a finite size has edges inevitably, and electronic states at the edges exhibit peculiar properties, e.g., localized states depending on the types of edge (zigzag or armchair) [1] and spin polarization (ferromagnetic order) at the zigzag edge [2, 3]. A study on spin-polarized electrons using field-emission

Nanostructured Carbon Electron Emitters and Their Applications
Edited by Yahachi Saito

ISBN 978-981-4877-62-6 (Hardcover), 978-1-003-14199-0 (eBook)
www.jennystanford.com

technique is presented in Chapter 8. Experimental and theoretical studies on field-emission (FE) from graphene are reviewed in the other chapters of this book. Here, field-emission microscopy (FEM) and field ion microscopy (FIM) measurements of graphene, focusing on emission microscope images of graphene edges, are presented.

Table 7.1 Electrical, mechanical, and thermal properties of graphene

Property	Value	Reference
Work function cf. graphite	4.32 eV (layer number, N =1), 4.4 eV (N =2), 4.55 eV (N =3), 4.58 eV (N =4) on SiC 4.6 eV	H. Hibino et al., *Phys. Rev. B* **79** (2009), 125437 T. Takahashi et al., *Phys. Rev. B* 32 (1985) 8317
Electrical resistivity	~10^{-4} Ω-cm (few-layer graphene)	R. Murali et al., *Appl. Phys. Lett.* **94** (2009), 243114
Current-carrying capability	10^9 A/cm^2 (few-layer graphene)	ibid.
Carrier mobility	3000–10,000 cm^2/Vs ~40,000 cm^2/Vs at RT (on SiO_2) 200 cm^2/Vs	K. S. Novoselov et al., *Science* **306** (2004), 666 J.-H. Chen et al., *Nat. Nanotech.* **3** (2008), 206 X. Wang et al., *Phys. Rev. Lett.* **100** (2008) 206803
Young's modulus	0.5 TPa 1.0 TPa	I. W. Frank et al., *J. Vac. Sci. Tech. B* **25** (2007), 2558 C. Lee et al., *Science* **321** (2008), 385
Breaking strength	130 GPa	ibid.
Thermal conductivity at RT Single-layer graphene Few-layer graphene	 ~4840–5300 W/Km ~3080–5150 W/Km ~2800–1300 W/Km (decrease with N from 2 to 4)	 A. A. Balandin et al., *Nano Lett.* **8** (2008), 902 S. Ghosh et al., *Appl. Phys. Lett.* **92** (2008), 151911 S. Ghosh et al., *Nat. Matter.* **9** (2010), 555

7.2 FEM of Graphene

7.2.1 Preparation of Graphene Emitter

Three methods for preparing graphene emitters are described here, though various methods can be proposed. First one is scratching method: A piece of graphene is mounted on a tip of an electropolished tungsten (W) needle by rubbing the tip on the surface of highly oriented pyrolytic graphite (HOPG) under a scanning electron microscope by using piezo-driven micromanipulators [4]. The W needle was prepared by electrochemical polishing of a W wire (e.g., 0.15 mm in diameter). The obtained graphene emitter was multilayered graphene consisting of several to a few tens of layers, as shown in Fig. 7.1. The W needle attaching a piece of graphene is spot-welded to a W filament (0.15 mm in diameter), which can be used for flash-heating, being the same as that used for carbon nanotube (CNT) emitters.

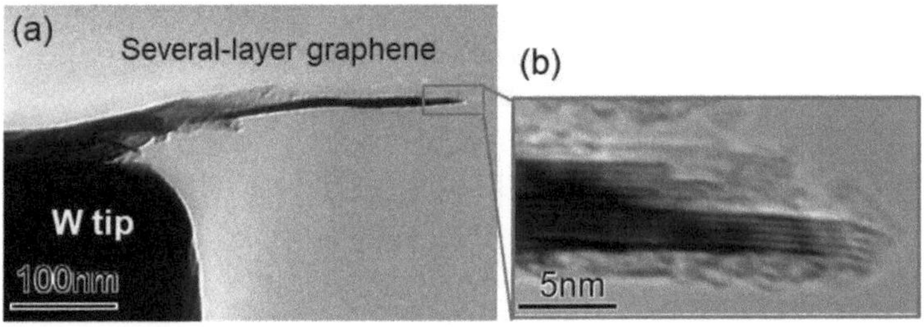

Figure 7.1 TEM pictures of a graphene emitter prepared by the scratching method. (a) Low- and (b) high-magnification images of five-layered graphene before FE measurement.

Second method is electrophoretic deposition of graphene oxide (GO): A flake of GO is attached to the tip of a W needle by applying electric voltage (DC 4–6 V) between the W needle and a counter electrode dipped in GO solution (5.0×10^{-3} wt% in water). Third one is dropping and drying of GO solution: A drop of GO solution is trickled down on a tip of the W needle heated on an aluminum foil to dry. The experimental apparatus for FEM/FIM of graphene is the same as that for CNT emitters described in Chapter 1.

7.2.2 "Lip" Pattern Typical of Graphene Field Emitter

Figure 7.2a shows a typical FEM image of graphene, manifesting unique features: Striated (or oval) spots are aligned in parallel with each other, and a dark line (node) running through the pattern perpendicular to the striation direction of spots [4]. As a whole, the pattern has a mirror symmetry with respect to the dark line. The direction of the dark line (node) is parallel to the graphene edge. The experimental pattern looks like a lip, thus this pattern is called a "lip" pattern [5].

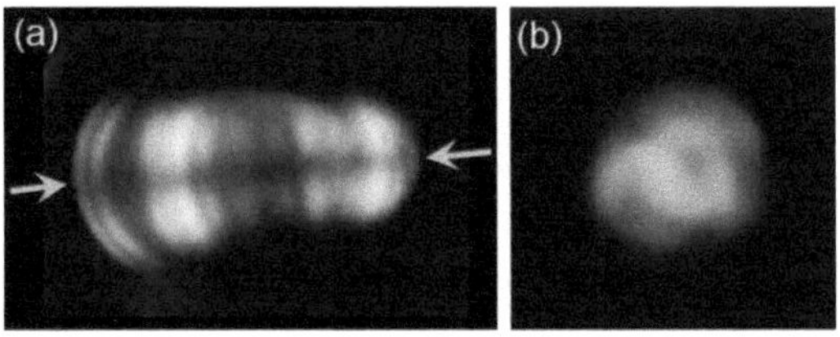

Figure 7.2 (a) FEM image of a graphene with open edges, showing a "lip" pattern. Two arrows indicate the location of a node. (b) FEM image of graphene from looped edges after intensive FE, showing a "dim" pattern.

Theoretically, taking into account only π-orbitals, Li et al. predicted dragonfly-shaped FEM patterns from a graphene emitter [6]. In their predicted patterns, the central node (dark line) parallel to the graphene edge is reproduced, which is due to the destructive interference of electron waves emitted from π-orbitals which have two lobes with a phase of difference π on either side of the graphene plane.

The origin of the arrayed oval spots, on the other hand, is not clear. Yamaguchi et al. [7] ascribed the array of streaked spots in their experimental FEM pattern from reduced GO (rGO) to an interference of electrons emitted from a few emission sites, but it is difficult to explain the characters of the lip pattern by the simple interference effect. Instead, each oval spot in our FEM may corresponds to respective emission site along the graphene edge. More specifically, we conjecture that oval spots originate from π-electron orbitals which have high density of states at each carbon atom on the graphene edge. Experimental and theoretical supports for this hypnosis are described in the next section.

In the present interpretation of FEM patterns, we postulate that these emission patterns came from a single sheet of graphene even though the emitter used was multilayered graphene. The reason of the postulation is as follows. Since graphite has the ABAB stacking sequence, atom positions in adjacent layers being shifted with each other. Thus, the edges of the multilayered graphene are indented layer by layer. The offsets between the edges of two adjacent layers have been actually observed by TEM [8], the distances between edges of adjacent layers being measured to be 0.34 and 0.37 nm for zigzag and armchair edges. So, it is expected that the lip patterns originate from an effectively single layer of graphene, which is most protruded at the side of a sheet of multilayer graphene.

7.2.3 Change of FEM Images from "Lip" to Dim Pattern

When the FE measurement was prolonged and the emission currents exceed a certain value (several tens μA, depending on emitter), the lip pattern changed to a dim pattern as shown in Fig. 7.2b. The dim pattern shows the same feature as that observed for single-wall CNTs (SWCNTs) and thin CNTs with small diameter (see Chapter 1). The change of FEM patterns results from the structure change of graphene edges, as has been revealed by in situ TEM of FE from graphene emitters [4]. Figure 7.3 shows a series of TEM images of a multilayer graphene edge before and after intense FE measurements. Before FE, the tip of the multilayer graphene was broken (each layer of graphene is open) and covered with thin amorphous carbon (Fig. 7.3a). After the intense FE with emission currents of 35 μA (Fig. 7.3b), free edges of graphene layers were connected to those of adjacent layers (sometimes skipping one or two layer) to form loops (capped structure), and also amorphous carbon at the tip was removed. A subsequent FE experience with emission currents of 75 μA (Fig. 7.3c) brought about the reduction of the number of constituent graphene layers, suggesting evaporation of graphene layers, while the tip shows the looped structure (closed ends). The structure change from the open to the closed ends is caused by the Joule heating of the graphene emitter itself during the intense FE experience. Similar structure change of graphene edges from open to closed ends was reported to occur by heat treatment above 1500 °C [9].

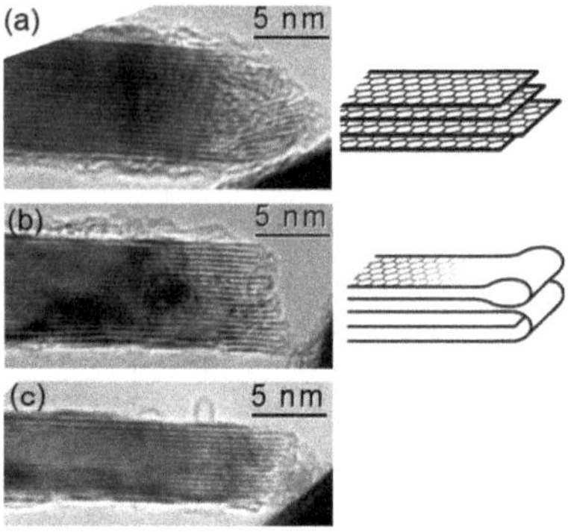

Figure 7.3 TEM images showing the structure change of a graphene edge. (a) Before FE, (b) and (c) after FE with emission currents 35 and 75 μA, respectively. Schematic drawings illustrating structures of the edges (a) and (b) are shown on the right of each picture.

7.2.4 Origin of "Lip" Pattern

The "lip" pattern exhibits following features: An array of streaks (or oval spots) is elongated in the direction perpendicular to the graphene plane. Each stripe is divided into two wings by a central dark band running through the pattern. The direction of the dark band (node) is parallel to the graphene edge. The node is considered to be due to the destructive interference of electrons emitted from two lobes of π-orbitals with the phase difference π, thus pairs of streaks (or oval spots) on both sides of the node correspond to pairs of π-orbitals on both sides of a graphene layer. The reason for streaking of FEM images perpendicular to the graphene layer is the strong anisotropy of magnification for two-dimensional emitters like graphene. The magnification of FEM in the direction perpendicular to the graphene layer is much larger than that in the direction parallel to the graphene layer because the electric field lines near the graphene edge are strongly diverging toward the direction perpendicular to a graphene plane, compared to the direction parallel to the edge, as illustrated in Fig. 7.4.

Information on the spacing between adjacent streaks in the lip pattern is vital for the interpretation of the pattern. Some useful information is obtained by observing FEM images of adsorbates on the graphene edges. When a residual gas molecule adsorbed on the graphene emitter, it made an appearance as a bright spot [5] like the phenomena observed in CNT emitters. The molecular species, though not identified, was possibly hydrogen or carbon monoxide molecule because they are the most abundant residual gas species in the ultrahigh vacuum. The size of the bright spot originating from an adsorbate is similar to the width of each striated spot, indicating that the origin of each striated spot has a size comparable with that of gas molecules.

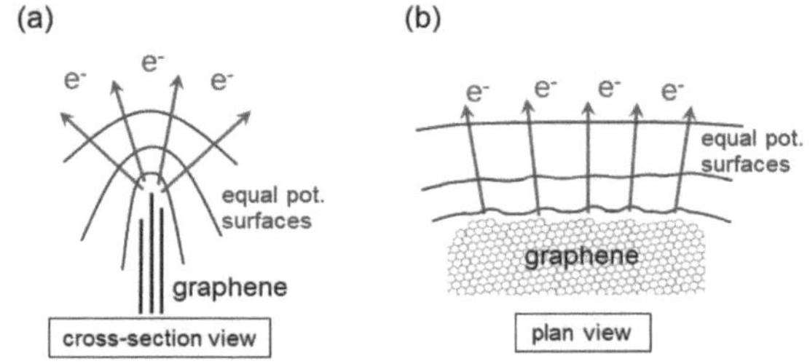

Figure 7.4 Schematics illustrating the electric field distribution near a graphene edge: (a) cross-section view and (b) plane view.

FEM images of aluminum (Al) clusters deposited on the graphene emitter also gave similar information on the spacing between streaks in the lip pattern. Using an atomic resolution image of the Al cluster, the spacing between adjacent streaked spots for clean graphene emitters was estimated to be in a range between 0.14 and 0.23 nm [5]. The derived distances of 0.14 and 0.23 nm in the lip pattern are very close to the C–C bond length (0.144 nm) and the distance between carbon atoms at the zigzag edge (0.249 nm). This indicates that the streaked spots correspond to emission sites with atomic size, supporting the above conjecture that the lip patterns reflect the symmetries of π orbitals localized at the graphene edge.

7.2.5 Frequent Encounter of "Lip" Patterns in Graphene-Related Materials

The lip patterns characteristic of graphene (including several-layered graphene) are sometimes observed in nanocarbon materials related to graphene. For example, FEM patterns observed from bamboo-shaped CNTs [10] bear a close resemblance to the lip pattern, or rather, are the very same as the lip pattern. It is inferred that the tips of bamboo-shaped CNTs were open, and thus free edges of graphene layers were exposed. Pt/C composite emitters produced by electron beam-induced deposition (EBID) of Pt-containing organic molecules showed FEM [11] patterns which also exhibit the selfsame features as those of the lip pattern. Since the Pt/C composite emitters were subjected to heat treatment prior to or during FE, amorphous carbon in the emitters was presumably transformed to graphite layers by a catalytic action of Pt.

7.3 FIM of Graphene

7.3.1 FIM Images of Graphene

Figure 7.5a,b shows an FIM image of graphene scratched from HOPG and the corresponding FEM image, respectively. In the FIM image, arrays of streaked spots are observed. Careful inspection of the image finds that some streaks are split into two spots (dipole spots) as indicated by arrows and some ones are arced as indicated by two-headed arrows. The dipole and arced patterns are encountered in FIM images of open-ended CNT and DWCNTs, respectively, as described in Chapter 1. The arced streaks sometimes look like a spindle shape with a hollow center. White broken lines in the FIM image (Fig. 7.5a) stand for the locations of graphene edges. At least four edges contribute to this FIM image. The direction of elongation and splitting of spots in FIM is perpendicular to the edge direction, being consistent with the streaking direction in the "daisy-like" FIM pattern of open-ended CNT (see Chapter 1). The FEM image in Fig. 7.5b from the same graphene emitter shows the "lip" pattern characteristic of graphene. The number of graphene edges contributing to the FEM image is only one or two, being smaller than that in the FIM image. The reason of this difference in the number of observed edges is not clear, though differences in field strength and image formation mechanism between FEM and FIM may be responsible.

Figure 7.6a,b shows an FIM image of a flake of GO and the corresponding FEM image, respectively. The FIM image, similarly that of the scratched graphene (Fig. 7.5a), exhibits an array of streaked spots including dipole and arced patterns (indicated by arrows and two-headed arrows). The corresponding FEM image shows a typical "lip" pattern.

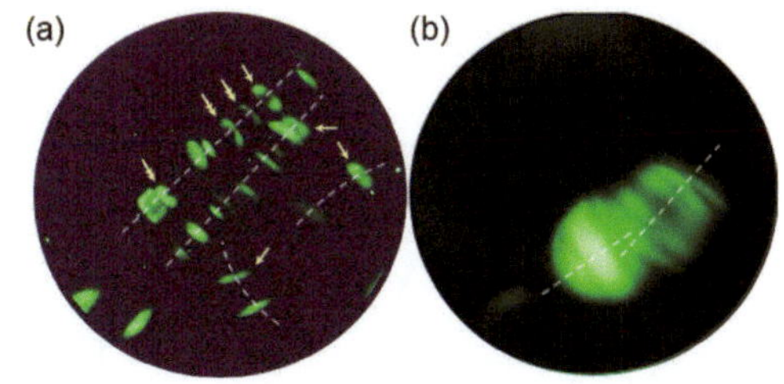

Figure 7.5 (a) FIM image of graphene scratched from HOPG (emitter voltage V_E = 2.5 kV, Ne = 1 × 10^{-3} Pa, at 80 K), and (b) the corresponding FEM image (V_E = −1.5 kV, at 80 K). White broken lines indicate the positions of edges of graphene. Split spots (dipole spots) and arced streaks are indicated by arrows and two-headed arrows, respectively.

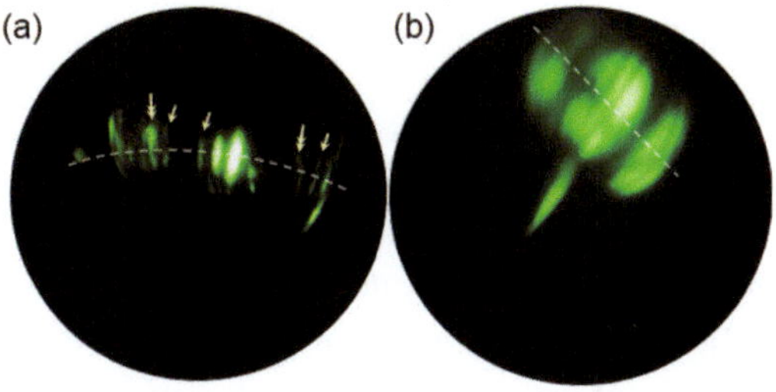

Figure 7.6 (a) FIM images of GO (V_E = 2.5 kV, Ne = 3 × 10^{-3} Pa, at 100 K) and (b) the corresponding FEM image (V_E = −0.8 kV, at 90 K). White broken lines indicate the positions of edges of graphene. Arrows and two-headed arrows in (a) indicate split spots (dipole spots) and arced streaks, respectively.

FIM images (Figs. 7.5a and 7.6a) show more finely resolved structures than the corresponding FEM images. The striation of spots perpendicular to the graphene layer is due to the strong anisotropy of magnification, as explained in the previous section. It should be noted that pairs of bright spots (dipole spots) extending to the opposite side of graphene sheet are reminiscent of two lobes of π-orbitals in graphene. Taking into account the image formation mechanism of FIM, the paired spots represent the locations where imaging gas atoms (Ne) are ionized, as illustrated in Fig. 7.7. An electron in a Ne atom tunnels to an empty orbital (π^*-orbital) at graphene edge, and the thus-formed Ne^+ ion is repelled from the emitter tip and fly to the observation screen. So, the bright spots in the FIM are considered to represent a spatial distribution of the unoccupied π^*-orbital at the graphene edge.

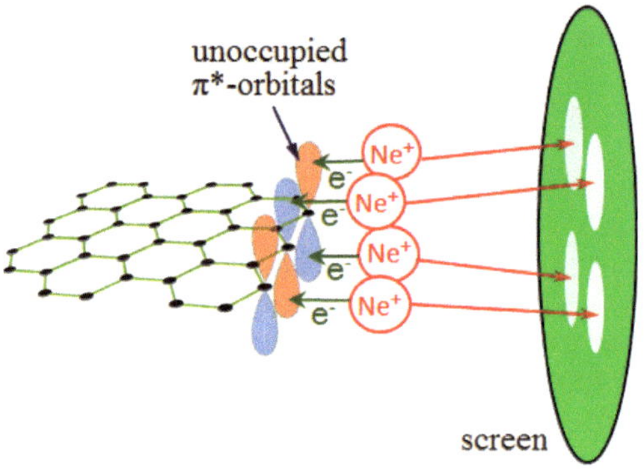

Figure 7.7 Schematic illustrating the origin of double spots in FIM of graphene. Ne is the imaging gas.

7.3.2 Historic Survey of Graphite FIM

An FIM study on a graphite crystal, which was reported in 1968 by Williams [12], clearly showed arrays of streaks running transverse to the layer plane of graphite. The streak pattern is quite similar to those shown in the previous section. Dipole spots can also be found in the reported FIM image by a careful survey of the image, though he did not give any comment on these. Even a pencil lead emitter was shown to provide the similar streaked FIM pattern by Khaimar et al. [13]. Murr and Inal [14] reported the double spots (dipole spots) pattern in FIM images from graphite fiber emitters. The appearance of the dipole spots is quite similar to those of graphene described above. They interpreted the double spots as representing carbon atoms in the corner of a single graphite platelet, i.e., each spot corresponds to an atom located at either end of a graphite platelet. Their interpretation on the dipole spots pattern is different from ours. The dipole pattern has the following features: The intensity, the shape, and the size of constituent spots are almost equal with each other, and each spot is always paired exactly on both sides of the edge with the mirror symmetry. It is difficult to explain these characters by the corner atoms at the edge of a graphite platelet. Rather, these characters are persuasively explained by the hypothesis that the paired spots are originated from one entity, e.g., an electron orbital extending to the opposite side of a graphene layer.

7.4 Conclusion

FEM and FIM images characteristic of graphene emitters, e.g., "lip" pattern consisting of streaked lobes, double (dipole) spots, etc., are summarized and their origins are discussed. It is shown that these unique patterns can be found even in graphite and multilayered graphene, because the most extruding graphene layer at the edge of a graphite platelet contributes to FE. A new interpretation of double (dipole) spots in FIM of graphene and related emitters is presented, i.e., the paired spots split on either side of the graphene edge originate from empty π^*-orbitals of carbon atoms at the edge.

Acknowledgments

This work was financially supported, in part, by JSPS KAKENHI grant number JP17K19053.

References

1. Fujita, M., Wakabayashi, K., Nakada, K. and Kasukabe, K. (1996). Peculiar localized state at zigzag graphite edge, *J. Phys. Soc. Jpn.*, **65**, pp. 1920–1923.
2. Magda, G. Z., Jin, X., Hagymási, I., Vancsó, P., Osváth, Z., Nemes-Incze, P., Hwang, C., Biró, L. P. and Tapasztó, L. (2014). Room-temperature magnetic order on zigzag edges of narrow graphene nanoribbons, *Nature*, **514**, pp. 608–611.
3. Yazyev, O. V., Capaz, R. B. and Louie, S. G. (2011). Theory of magnetic edge states in chiral graphene nanoribbons, *Phys. Rev. B*, **84**, pp. 115406-1–115406-5.
4. Nakakubo, K., Asaka, K., Nakahara, H. and Saito, Y. (2012). Evolution of field electron emission pattern from multilayered graphene induced by structural change of edge, *Appl. Phys. Exp.*, **5**, pp. 055101-1–055101-3.

5. Yokoyama, N., Nakakubo, K., Iwata, K., Asaka, K., Nakahara, H. and Saito, Y. (2016). Field emission patterns showing symmetry of electronic states in graphene edges, *Surf. Interface Anal.*, **48**, pp. 1217–1220.
6. Li, Z., Xu, N. and Kreuzer, H. J. (2012). Coherent field emission image of graphene predicted with a microscopic theory, *Phys. Rev. B*, **85**, pp. 115427-1–115427-9.
7. Yamaguchi, H., Murakami, K., Eda, G., Fujita, T., Guan, P., Wang, W., Gong, C., Boisse, J., Miller, S., Acik, M., Cho, K., Chabal, Y. J., Chen, M., Wakaya, F., Takai, M. and Chhowalla, M. (2011). Field emission from atomically thin edges of reduced graphene oxide, *ACS Nano*, **5**, pp. 4945–4952.
8. Jia, X., Hofmann, M., Meunier, V., Sumpter, B. G., Dergado, J. C., Herrera, J. M. R., Hsieh, Y. P., Reina, A., Kong, J., Terrones, M. and Dresselhous, M. S. (2009). Controlled formation of sharp zigzag and armchair edges in graphitic nanoribbons, *Science*, **323**, pp. 1701–1705.
9. Campos-Delgado, J., Kim, Y. A., Hayashi, T., Gómez, A. M., Hofmann, M., Muramatsu, H., Endo, M., Terrones, H., Shull, R. D., Dresselhaus, M. S. and Terrones, M. (2009). Thermal stability studies of CVD-grown graphene nanoribbons: Defect annealing and loop formation, *Chem. Phys. Lett.*, **469**, pp. 177–182.
10. Wu, P., Deng, S. Z., Chen, J. and Xu, N. S. (2007). Fine-structured field emission images originating from coherently scattering of electrons within a multi-walled carbon nanotube, *Appl. Surf. Sci.*, **254**, pp. 1389–1393.
11. Murakami, K., Wakaya, F. and Takai, M. (2007). Observation of fringelike electron-emission pattern in field emission from Pt field emitter fabricated by electron-beam-induced deposition, *J. Vac. Sci. Technol. B*, **25**, pp. 1310–1314.
12. Williams, W. S. (1968). Field-ion microscopy of graphite, *J. Appl. Phys.*, **39**, pp. 2131–2132.
13. Khaimar, R. S., Dharmadhikari, C. V. and Joag, D. S. (1989). Pencil lead tips: A field ion and field electron emission microscopic study, *J. Appl. Phys.*, **65**, pp. 4735–4738.
14. Murr, L. E. and Inal, O. T. (1971). Field ion microscopy of graphite fibers, *J. Appl. Phys.*, **42**, pp. 3487–3493.

Chapter 8

Spin-Polarized Field-Emitted Electrons from Graphene Oxide Edges

Shigekazu Nagai
Graduate School of Engineering, Mie University,
1577 Kurima-machiya Tsu, Mie 514-8507, Japan
nagai@elec.mie-u.ac.jp

This chapter presents experimental evidence for the theoretical prediction of spin-polarized π-electrons in graphene and related materials by measuring the spin polarization of the field-emitted electrons from graphene oxide edges. From the findings, the spin polarization of the field-emitted electrons from graphene oxide, which show lip patterns reflecting the interference of π-electrons, is up to 67%. However, stepwise changes in the spin polarization of the field-emitted electrons were observed in a time trace, which is caused by the adsorption of residual gases such as hydrogen. These results verify the theoretical prediction that the localized electrons at the graphene oxide edge are spin-polarized.

Nanostructured Carbon Electron Emitters and Their Applications
Edited by Yahachi Saito

ISBN 978-981-4877-62-6 (Hardcover), 978-1-003-14199-0 (eBook)
www.jennystanford.com

8.1 Introduction

Graphene, which has hexagonally arranged carbon atoms, is a two-dimensional material with a specific band structure owing to its in-plane motion of π-electrons [1]. The spin-polarized electronic structure has been studied in both theoretical and experimental aspects [2–5] after Fujita et al. predicted the localization of spin-polarized electrons at the zigzag edges of graphite [6]. Recently, the localization of spin-polarized electrons at the zigzag and armchair edges of graphene as well as graphene oxide has also been predicted based on first-principle calculations. Particularly, the spin-polarized π-electrons localized at the zigzag edges of nonmagnetic graphene are assumed to be a type of spintronics material. However, to analyze the spin polarization of graphene edges, they either need to be grown or transferred on the substrate; therefore, interaction with the substrate material is unavoidable. Consequently, it is difficult to experimentally observe the intrinsic spin-polarized electronic state of the graphene edges.

Because of the chemical stability of graphene, the use of graphene edges as a field-emission type electron source has been actively studied [7–9]. Many research groups have reported that field-emission microscopy (FEM) images of graphene and graphene oxide show a characteristic striated pattern with a dark line in the center (i.e., a lip pattern). The origin of the lip pattern is theoretically assumed to be caused by the interference of π-electrons at the graphene edge [10, 11].

These results suggest that the field-emitted electrons from graphene edges, which form the lip pattern, may be spin-polarized. Because the field-emitted electrons are directly extracted into the vacuum by the tunneling effect, their spin polarization may reflect that of the graphene edges.

Our recent work [12, 13] on spin polarization of field-emitted electrons from graphene oxide edges is presented in this chapter. These edges were attached to electropolished tungsten tips to extract field-emitting electrons from graphene oxide. The spin polarization of the field-emitted electrons from these edges was measured using a Mott spin polarization analyzer [14]. The experimental findings demonstrated the existence of spin-polarized electrons at the graphene edges and revealed their potential application in spin-polarized field emitters.

8.2 Experimental Method

Graphene oxide flakes of approximately 10 μm in size were attached as field emitters to a tungsten needle tip by electrophoresis. Graphene oxide was dispersed in water at a concentration of 5.0×10^{-3} wt%, and a tungsten needle was dipped in the mixture to approximately 7 mm. The graphene oxide was bonded to the apex of the tungsten needle by applying 10 V between the counter electrode and the tungsten needle as a cathode and anode, respectively. The scanning electron microscope (SEM) image of the prepared graphene oxide sample is shown in Fig. 8.1.

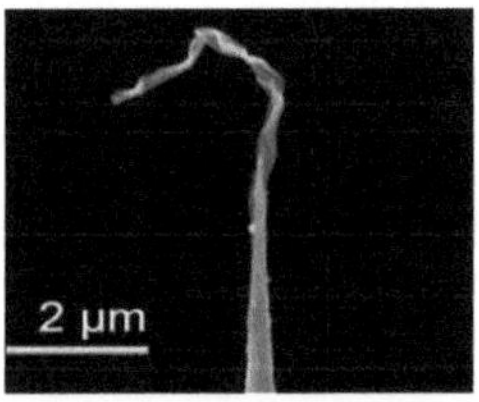

Figure 8.1 Example of SEM image of graphene oxide field emitter [13].

The spin polarization of the field-emitted electrons was measured using a field-emission microscope (FEM) equipped with a 90° sector type spin rotator and a retarding potential Mott spin polarization analyzer [14, 15]. The emitter was attached to the cold finger of a cryostat that could be cooled down to 47 K, and a negative high voltage was applied. An extraction electrode was then placed 5 mm in front of the emitter; the emission current could be controlled by applying a positive high voltage to the extraction electrode. The spin polarization of the electrons passing through the central probe hole on a fluorescent screen was measured. Because a component of the spin polarization of electrons parallel to the electron beam trajectory could not be measured using the Mott scattering spin detector, the spin direction was controlled by the spin rotator. The emitter axial component was defined as the *z*-direction, and the horizontal and vertical directions on the fluorescent screen were defined as the *x* and y components, respectively. The operating voltage of the Mott spin polarization analyzer was 25 kV. An effective Sherman function of 0.14 calibrated by the Fe/W field emitter was used for converting scattering asymmetries to the spin polarizations.

8.3 Spin Polarization at Edges of Graphene Oxide

To confirm the FEM image of the lip pattern observed from the prepared graphene edge, it is necessary to remove the residual impurities on the graphene oxide edge. Therefore, the impurities at the edges were removed by continuous field electron emission at an applied voltage of −440 V and emission current of 1 μA, as reported in the previous study [9]. The FEM image observed afterward is shown in Fig. 8.2. A striated electron emission pattern with a dark line crossing the center, i.e., lip pattern, was observed successfully. This pattern is caused by the interference of the electron orbitals of π-electrons at the zigzag edges of graphene oxide, which is consistent with the theoretical predictions.

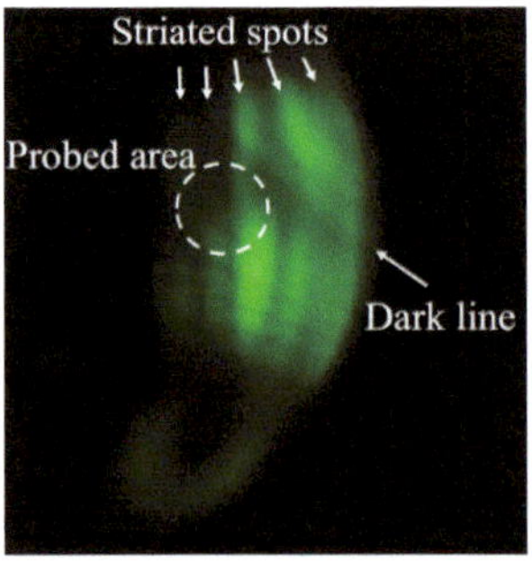

Figure 8.2 FEM image of the lip pattern observed from graphene oxide emitter.

The spin polarization components measured from an area within the lip pattern, indicated by a dotted circle in Fig. 8.2, are listed in Table 8.1. In this measurement, the emitter was at room temperature and the probe current was 110 nA. The spin vectors of the emitted electrons are aligned in the x- and z-directions whereas the spin polarization for the y-direction is relatively weak. The

x- and y-directions lie the horizontal and vertical axes on the FEM image, respectively, and z-direction is along the emitter axis. Each component of the spin polarization was calculated by the following equation:

$$P = \frac{1}{S_{\text{eff}}} \frac{N_{\text{R}} - N_{\text{L}}}{N_{\text{R}} + N_{\text{L}}} \tag{8.1}$$

where S_{eff} is effective Sherman function and N_{R} and N_{L} are the number of electrons detected by electron multipliers located at right and left side, respectively. The total spin polarization P_{total} calculated by the vector summation is approximately 57% as per the following equation:

$$P_{\text{total}} = \frac{N_{\uparrow} - N_{\downarrow}}{N_{\uparrow} + N_{\downarrow}} \tag{8.2}$$

where $N_{\uparrow}$ and $N_{\downarrow}$ are the number of electrons with up and down spin, respectively. The spin polarization of the field-emitted electrons varied between 20 and 60% per sample; however, it was found that the emitted electrons from graphene oxide were significantly spin-polarized. Additionally, in many samples, the alignment of the spin vector with the emitter axis in the z-direction was observed. Because the emitted electrons reflecting the lip pattern may be caused by the π-electrons localized at the graphene oxide edge, the experimental results are evidence of the localization of spin-polarized electrons at the graphene oxide edge. The spin polarization of the field-emitted electrons increased to 67% from 57% when graphene oxide was cooled to 45 K. The increase in polarization is considered to be caused by the suppression of the fluctuation of the spin vector owing to the cooling of the sample. The emission pattern with the high polarization (67%) is shown in Fig. 8.3a. The spin polarization, meanwhile, decreased to 23% when the FEM image changed, as shown in Fig. 8.3b. The change in the electron emission pattern may be attributed to the structural change of the graphene edges caused by the evaporation of the edge atoms, or the adsorption of residual gas molecules. These are known to induce not only a change in the electron emission pattern but also a decrease in the spin polarization of the localized electrons, i.e., the magnetic moment. To elucidate this factor, the spin polarization of the field-emitted electrons was measured as a function of the emission current and elapsed time.

Table 8.1 Spin polarization of field-emitted electrons from graphene oxide

Component	*x*	*y*	*z*	Total
Spin polarization (%)	32	13	48	57

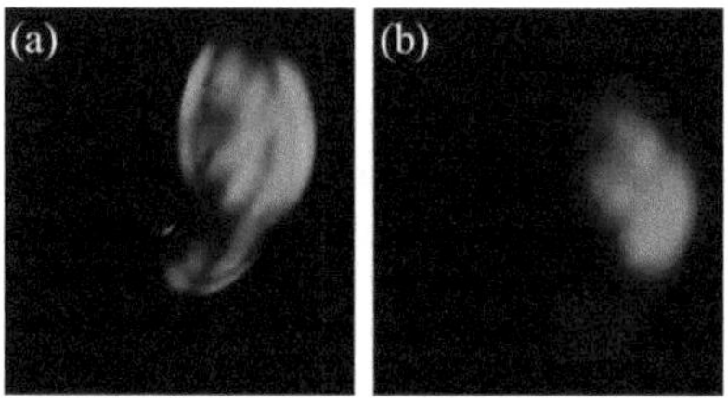

Figure 8.3 Change of FEM images of the graphene oxide emitter with spin polarization of 67% (a) and 23% (b).

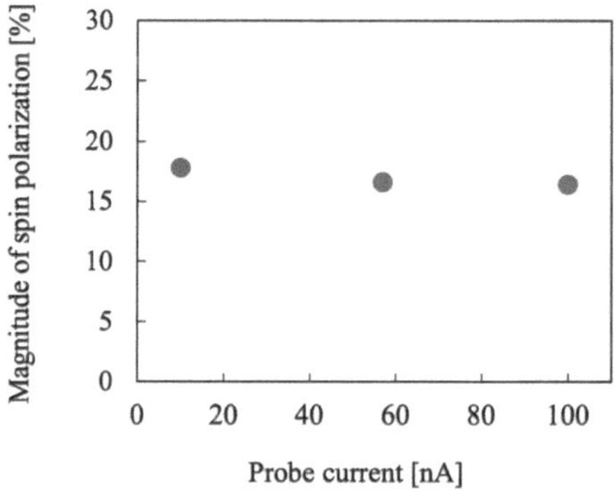

Figure 8.4 Magnitude of the spin polarization of field-emitted electrons from graphene oxide as a function of probe current.

The spin polarization was measured as a function of probe current at room temperature. Figure 8.4 shows the spin polarizations measured with a probe current ranging from 10 to 100 nA. The magnitude of the spin polarization is plotted as values by the vector sum of the components in all directions. Consequently, a significant change was not observed in the spin polarization, in comparison to the statistical error of the measurement by the Mott spin polarization analyzer, which was approximately 5%, with respect to the change in the probe current. Therefore, the effect of the emission current and electric field strength is negligible in the spin polarization measurements of the field-emitted electrons from the graphene oxide emitters. Alternatively, the spin polarization at the edges may be promoted by the effect of the electric field applied to the graphene oxide [16].

8.4 Change in Spin Polarization Due to Adsorption

To verify the decrease in the spin polarization of the field-emitted electrons, the spin polarization was measured as a function of elapsed time at room temperature, as shown in Fig. 8.5. The polarization measurements shown in Fig. 8.5 were calculated from the vector sum of all the components in the spin direction. During the measurement, the applied voltage was adjusted to keep the emission current constant. The polarization values ranged from 20 to 25% up to 60 min after the start of the measurement. And then, an abrupt decrease in polarization was observed at 75 and 110 min. Figure 8.6 shows the I–V characteristics before the measurement and after 110 min. After the decrease in polarization at 110 min, the threshold voltage for field-emission was observed to be lower than that before the measurement. This suggests the structure change of the graphene edge. Interestingly, the spin polarization suddenly decreased at 70 min and then restored to its original value at 90 min. The recovery of the spin polarization indicates that the polarization change is not attributed to the electric field strength associated with the structural change of the graphene edge. However, it is possible that residual gases were adsorbed on the graphene edges and affected the spin-states at the edges.

Theoretical calculations show that the 5.43 eV work function of clean graphene edges changes to 4.29 and 7.74 eV due to adsorption of H and O atoms, respectively [17]. In the field-emission phenomenon, a lower work function indicates a lower threshold of the applied voltage. Thus, it is more likely that the adsorption of hydrogen atoms on the graphene oxide edge affects the spin polarization of field-emitted electrons. The stepwise change in polarization is also probably caused by adsorption and desorption of hydrogen, because hydrogen is the main residual gas in an ultrahigh vacuum of the order of 10^{-8} Pa.

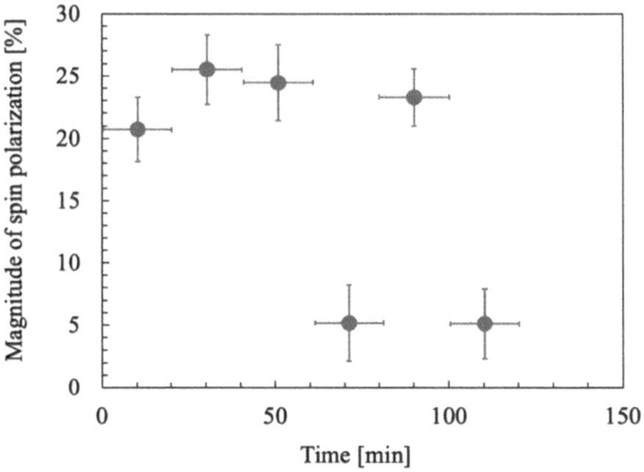

Figure 8.5 Magnitude of spin polarization of field-emitted electrons from graphene oxide as a function of elapsed time.

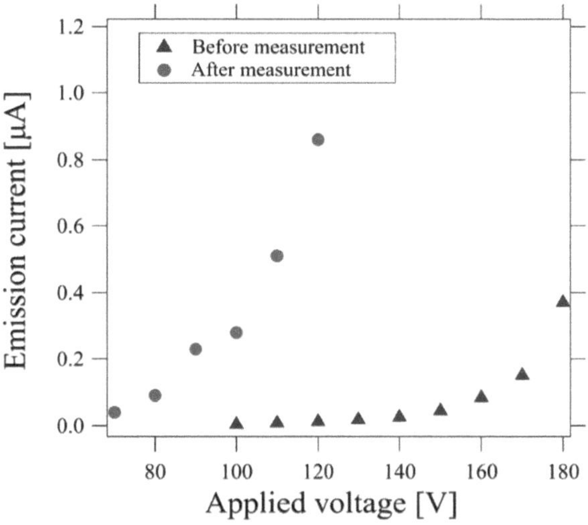

Figure 8.6 Current–voltage characteristics before and after measurements of spin polarization shown in Fig. 8.5.

Antiferromagnetism is predicted in the graphene nanoribbon (GNR) region flanked by zigzag edges [16]. In other words, at each zigzag graphene edge, the electrons in the π-orbitals of carbon atoms take a ferromagnetic spin arrangement. Theoretical studies have reported the absence of magnetization in the center of graphene nanoribbons without hydrogen terminations and a magnetization of $1.3\mu_s$ at each edge atom. Meanwhile, if all the dangling bonds are terminated by a single hydrogen atom, the magnetization is theoretically reduced to $0.3\mu_B$ [18]. This is because the electrons localized at the spin-polarized edges are derived from the presence of flat bands, which depend on the number of edge atoms terminated by hydrogen. The spin polarization of an electron localized at an edge is associated with the number of hydrogen atoms adsorbed on a dangling bond at a single edge atom. Therefore, from this theory, it can be concluded that hydrogen adsorption on the edge of graphene reduces the localization of the edge electrons and the spin polarization. To clarify this, it is necessary to evaluate the electron emission sites using techniques with high spatial resolution such as field ion microscopy.

8.5 Conclusion and Remarks

In this chapter, the spin polarization of field-emitted electrons from graphene oxide attached to the tip of a tungsten needle by electrophoresis was presented. It was observed that the spin polarization of field-emitted electrons showing the lip pattern due to π-electrons achieved a maximum of 67%, proving that the electrons localized at the graphene oxide edge are significantly spin-polarized. Although spin polarization is dependent on the probe current, stepwise change in the spin polarization is observed in the time trace. This change may be caused by the adsorption and desorption of residual gas such as hydrogen. The change in the electron emission pattern was also observed, which was presumed to be caused by the rearrangement of the edge atoms. In order to clarify the cause of the polarization change, field ion microscopy of the edge structure with atomic resolution is desired in the future, in addition to the preparation of graphene emitters with well-defined edge termination.

Acknowledgments

The author thanks Prof. Y. Saito for fruitful discussions as well as Prof. Nishina for supplying graphene oxide samples. The author also acknowledges the financial support from the Ministry of Education, Culture, Sports, Science and Technology of Japan (Challenging Research (Exploratory) No. 17K19053).

References

1. Novoselov K. S., Geim A. K., Morozov S. V., Jiang D., Katsnelson M. I., Grigorieva I. V., Dubonos S. V. and Firsov A. A. (2005). Two-dimensional gas of massless Dirac fermions in graphene, *Nature*, **438**, pp. 197–200.
2. Acik M. and Chabal Y. J., (2011). Nature of graphene edges: A review, *Jpn. J. Appl. Phys.*, **50**, p. 070101 17.
3. Huang B., Liu F., Wu J., Gu B. L. and Duan W. (2008). Suppression of spin polarization in graphene nanoribbons by edge defects and impurities, *Phys. Rev. B*, **77**, p. 153411.
4. Qin S., Guo X., Cao Y., Ni Z. and Xu Q. (2014). Strong ferromagnetism of reduced graphene oxide, *Carbon*, **78**, pp. 559–565.
5. Zhang Y.-T., Jiang H., Sun Q.-F. and Xie X. C. (2010). Spin polarization and giant magneto resistance effect induced by magnetization in zigzag graphene nanoribbons, *Phys. Rev. B*, **81**, p. 165404.
6. Fujita M., Wakabayashi K., Nakada K. and Kusakabe K. (1996). Peculiar localized state at zigzag graphite edge, *J. Phys. Soc. Jpn.*, **65**, pp. 1920–1923.
7. Nakakubo K., Asaka K., Nakahara H. and Saito Y. (2012) Evolution of field electron emission pattern from multilayered graphene induced by structural change of edge, *Appl. Phys. Express*, **5**, p. 055101.
8. Yamaguchi H., Murakami K., Eda G., Fujita T., Guan P., Wang W., Gong C., Boisse J., Miller S., Acik M., Cho K., Chabal Y. J., Chen M., Wakaya F., Takai M. and Chhowalla M. (2011). Field emission from atomically thin edges of reduced graphene oxide, *ACS Nano*, **5**, pp. 4945–52.
9. Yokoyama N., Nakakubo K., Iwata K., Asaka K., Nakahara H. and Saito Y. (2016). Field emission patterns showing symmetry of electronic states in graphene edges, *Surf. Interface Anal.*, **48**, pp. 1217–1220.
10. Chen J., Li Z. and Wang W. (2014). Manifesting pseudo-spin polarization of graphene with field emission image, *J. Appl. Phys.*, **115**, p. 053701.

11. Li Z., Xu N. and Kreuzer H. J. (2012). Coherent field emission image of graphene predicted with a microscopic theory, *Phys. Rev. B*, **85**, p. 115427.

12. Kunoh K., Nagai S., Kishi T., Uchikoshi H., Iwata T., Hata K. and Saito Y. (2019). Evidence of spin-polarized electronic state at edges of graphene oxide studied by field emission spin polarimetry, *Proc. 12th International Symposium on Atomic Level Characterization '19*, pp. 395–396.

13. Nagai S., Ikemizu H., Kunoh K., Hata K., Watanabe Y., Hoshino T., Irita M., Nakahara H. and Saito Y. (2017). Spin polarization of field emitted electrons from graphene edges, *Proc. 11th International Symposium on Atomic Level Characterization '17*, pp. 288–290.

14. Burnett G. C., Monroe T. J. and Dunning F. B. (1994). High-efficiency retarding-potential Mott polarization analyzer, *Rev. Sci. Instrum.*, **65**, pp. 1893–1896.

15. Nagai S., Sakakibara H., Hata K., Okada M. and Mimura H. (2011). Measurement of z-direction component of electron spins field-emitted from a single-crystal magnetite whisker, *Ultramicroscopy*, **111**, pp. 405–408.

16. Son Y.-W., Cohen M. L. and Louie S. G. (2006). Half-metallic graphene nanoribbons, *Nature*, **444**, pp. 347–349.

17. Wang W. and Li Z. (2011). Potential barrier of graphene edges, *J. Appl. Phys.*, **109**, p. 114308.

18. Tada K., Haruyama J., Yang H. X., Chshiev M., Matsui T. and Fukuyama H. (2011). Ferromagnetism in hydrogenated graphene nanopore arrays, *Phys. Rev. Lett.*, **107**, p. 217203.

Chapter 9

Theoretical Coherent Field Emission of Graphene

Zhibing Li
State Key Laboratory of Optoelectronic Materials and Technologies, Guangzhou, 510275, P. R. China
Guangdong Province Key Laboratory of Display Material and Technology, Guangzhou, 510275, P. R. China
School of Physics, Sun Yat-Sen University, Guangzhou, 510275, P. R. China
stslzb@mail.sysu.edu.cn

The theme of this chapter is to discuss the possibility of extracting quantum mechanical phase information of an emitter from cold field-emission patterns. The theory is illustrated with coherent cold field emission (CCFE) of graphene.

9.1 Coherent Cold Field Emission

Let us first explain what do we mean by coherent field emission. Adopting the single-electron picture, a quasi-particle with the same charge as an electron in an emitter is described by a state density operator

Nanostructured Carbon Electron Emitters and Their Applications
Edited by Yahachi Saito

ISBN 978-981-4877-62-6 (Hardcover), 978-1-003-14199-0 (eBook)
www.jennystanford.com

$$\hat{\rho} \sim \sum_{n,i} f_n \,|\, \psi_{n,i} \rangle \langle \psi_{n,i} \,| \tag{9.1}$$

where $|\psi_{n,i}\rangle$ and $\langle\psi_{n,i}|$ are the Dirac ket-vector and bra-vector of the i-th degenerated quasi-eigenstate with energy E_n. Each quasi-eigenstate contributes to the state with the weight given by the occupation distribution f_n. In thermal equilibrium at temperature T with Fermi level E_F, the occupation distribution is the Fermi–Dirac distribution $f_n^{(0)} = [e^{(E_n - E_F)/(k_B T)} + 1]^{-1} \cdot |\psi_{n,i}\rangle$ is not an exact eigenstate of energy due to tunneling, electron–phonon scattering, electron–electron scattering, and so on, and thus E_n has a small imaginary part $-i\Gamma$, which accounts for the instability of the state. The exact value of Γ is hard to know. Empirically, $2\Gamma \sim 0.005 - 0.05$ eV for typical emitters.

At a long distance away from the emitter, the state of emitted electron can be written as

$$\hat{\rho}_{\text{out}} \sim \sum_{n,i} D_{n,i} f_n \,|\, \phi_{n,i} \rangle \langle \phi_{n,i} \,| \tag{9.2}$$

where $|\phi_{n,i}\rangle$ is the outgoing wave of electron emitted from $|\psi_{n,i}\rangle$. The weight of the outgoing wave is proportional to the transmission coefficient $D_{n,i}$ and the occupation distribution f_n. Again, $|\phi_{n,i}\rangle$ cannot be an exact eigenstate of energy. Following Heisenberg's uncertainty principle, the energy uncertainty of the state approximated by Γ must be the same order as $\frac{\hbar}{2\tau_0}$, where τ_0 is the emission time of the state. Therefore, $|\phi_{n,i}\rangle$ is actually a wave packet in the coordinate representation. A schematic diagram of cold field emission (CFE) is sketched in Fig. 9.1(left). The emitter is the dark region on the left. In equilibrium, electrons basically occupy the quasi-particle states below the Fermi level. The solid curve is the potential energy of electron. The emitted electrons are free electrons in the vacuum region, described by wave packets.

Like a classical wave, the wave packet representing the probability density amplitude in quantum mechanics can be separated into several sub-waves by an object and would exhibit interference phenomenon when the sub-waves rejoin again. Roughly speaking, the electron beam is coherent when the relative phase of the wave packet is observable. The coherence conditions somehow depend on the interests: the characteristic size of object, the time scale, the measurement apparatus, and so on. Coherence length is the maximum distance in which the relative phase can be measured. For instance, in the setup of double-slit interference (as sketched in Fig. 9.1(right)), one can define a longitudinal coherence length l_l as the width of the wave packet in the direction of propagation, and define a transversal coherence length l_t as the maximum slit spacing with which the interference is still observable. In Fig. 9.1(right), l_{AB} is the slit spacing, l_{AC} is the distance between A and C, and l_{BC} the distance between B and C. The interference fringe at C can be observed if condition $l_l > |l_{AC} - l_{BC}|$ and $l_t > l_{AB}$ are fulfilled. Thus, both coherence lengths can be determined by adjusting the slit spacing and watching the change of interference region.

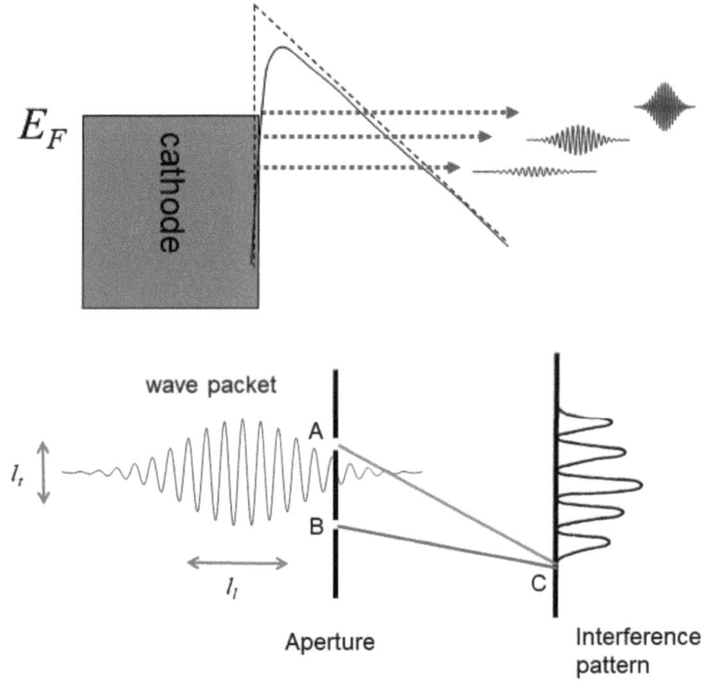

Figure 9.1 (Above) Schematic picture of cold field emission. (Below) Schematic picture of double slit interference.

For the interest of measuring relative phase of different times, a coherence time τ_c is used to describe the coherence ability of two (or more) sub-waves. Since τ_c has the same order as τ_0, τ_c - 2L - 10^{-13} - 10^{-12} s for typical emitters. On the other hand, $l_l = \upsilon\tau_c$ with

υ the group velocity of the wave packet. In a conventional CFE experiment, the distance between the cathode (emitter) and the anode is much longer than the size of emitter. The group velocity may be approximated as $\upsilon = \sqrt{\frac{2eV_a}{m_e}}$ with V_a the anode voltage. For the typical voltage of $V_a = 10^4$ V, $\upsilon = 0.2c$, the longitudinal coherent length l_l is about 60–600 nm.

Yet according to the Pauli exclusive principle, two electrons cannot occupy the same state. Hence the interference is in fact due to each individual electron, although the interference fringes appear as a collection of point-like images of many electrons. This is very different from a classical wave. Therefore, the wave packets of two successively emitted electrons should not overlap, otherwise the interference pattern will become blurred. Accordingly, the interference requires an additional condition: the spacing of two wave packets $l_s > l_l$, i.e., the average time interval between two wave packets $\tau_s > \tau_c$, noting that $l_s = v\tau_s$. The time interval can be estimated by the CFE current as $\tau_s = e/I$. For the turn-on current of $I = nA$, $\tau_s = 1.6 \times 10^{-10}$ s, that leads to $l_s = 1$ cm. They should be compared with τ_c and l_l.

Finally, one should note that the total emission beam described by the state operator Eq. 9.2 is non-coherent superpositions of wave packets (non-pure state). In other words, states of different (n,i) do not interfere each other. The interference happens within each outgoing wave that is described by the wave function $\phi_{n,i}$. The total interference pattern is the summation of the interference patterns of individual wave packets with weights given by the non-pure state. Therefore, only when the beam is dominated by wave packets of energies in a narrow range one could observe significant interference, even if each individual wave packet fulfills all coherence conditions. The emission energy range can be estimated by the Fowler–Nordheim theory as [Fowler and Nordheim, 1928; Modinos, 1984]

$$-\frac{e\hbar\gamma' V_a}{2r_t\sqrt{2m_e W_0}} < E - E_F < k_B T \tag{9.3}$$

where W_0 is the work function with respect to E_F, r_t the curvature of the emitting surface, and γ' a constant depending on the geometric structure of the setup. In a typical CFE experiment, the local field

$\frac{\gamma' V_a}{r_t}$ is about 1 V/nm, then the deviation of emission energy with respect to the Fermi level is within 0. 05 eV at room temperature. That is about half of the upper bound of Γ. The narrow energy distribution is a merit of CFE for coherent emission.

In principle, coherent interference of emitted electron wave contains information of the geometric structure, the electron energy band (long wavelength property), the local atomic orbitals (short wavelength property), and dynamic interactions in the emitter. Excellent coherence is inherent in CFE for two reasons: the emission energy is restricted within a narrow range about the Fermi level by the tunneling barrier, and thermal fluctuation is almost irrelevant for the tunneling process. The quantum mechanical size effect may further enhance the coherence of CFE from a nano-emitter. As an extreme example, the spatial coherence of electrons emitted from a nano-tip terminated by a single atom has been proven succinctly in inline electron holography [Horch and Morin, 1993]. Combining holography and coherent diffraction, angstrom resolution has been achieved with a CCFE e-beam of a single atom tip [Latchevskaia et al., 2015]. The laser-assisted FE (LAFE), which is controllable temporally, is an attractive technique toward 4D electron holography [Zewail and Thomas, 2010]. Furthermore, the LAFE will have good coherent properties, as it is dominated by the emission of optical resonant energy levels. Emission patterns inheriting features of the quantum states of nano-emitters could lead to a novel spectroscopy for nano-structures, as a complementarity of the angle-resolved photoemission spectroscopy [Kordruk, 2014].

As the first synthesized material of monolayer of atoms, graphene is interesting on its own for its rich novel properties, and has been extensively studied [Brey and Fertig, 2006; Geim, 2009; Neto et al., 2009; Das Sarma, et al., 2011]. Among those, the linear dispersion, pseudo-spin symmetry, and relativity Hall effect [Novoselov, 2005; Zhang et al., 2005] are symbolic. The CCFE would provide a novel method to probe the quantum states (especially the edge states) of graphene once we have a theory to connect CCFE patterns to the quantum states. On the experimental aspect, as have been demonstrated in other chapters of this book and many papers [Xiao et al., 2010; Jiang et al., 2013; Tang et al., 2020], low-dimensional carbon emitters have promising CFE properties such as low threshold field and large emission current density. Observations of CCFE patterns of CNTs and graphene are accumulating [Oshima et al., 2002; Wu et al., 2007; Zajec et al., 2011]. Due to the perfect monolayer atomic structure, electrons in graphene have long coherent lengths (a few micrometers at room temperature) [Geim, 2009]. Interference fringes produced by Saito's group have shown the excellent coherence of the e-beam emitted from graphene [Yamaguchi et al., 2011; Nakakubo et al., 2012; Yokoyama et al., 2016].

This chapter is devoted to a theoretical study of CCFE from graphene. Going beyond the Fowler–Nordheim theory of CFE, we will show that emission electrons carry the quantum mechanical features of graphene into vacuum. In order to have a compact analytical solution, only emission from a uniform one-dimensional edge of graphene will be considered. This setup has two remarkable features which are commonly shared by two-dimensional emitters. First, it preserves the translational symmetry along the edge when the applied electric field is normal to the edge and parallel to the plane of graphene. Due to this symmetry, the lattice momentum in the direction parallel to the edge is conserved and encoded in the emission orientation. Second, the tunneling barrier has a minimum ridge in front of the edge. It leads to the self-focusing effect of a two-dimensional emitter: the emission dispersion in the vertical direction is much smaller than a spherical emitter. Therefore, the emission may have good momentum resolution along the edge direction and a good position resolution in the direction vertical to the edge.

The electronic structure of graphene is associated to the 2p carbon orbitals. The low-energy excitations can be appropriately described by two Weyl spinors originated from the honeycombed atomic structure. Graphene has a number of properties which are important for CCFE: (i) the linear dispersion of low-energy quasi-particles (two Dirac cones); (ii) zero density of states at the Fermi level of neutral graphene (at the intrinsic Fermi level); (iii) the edge-dependent effective barrier height of forward emission. The linear dispersion is significantly different from the parabolic dispersion, which leads to different electron supply for emission. Because of the low density of states in the vicinity of the Fermi level, field penetration in graphene is important. The vacuum barrier of electron tunneling depends on the edge structure severely. For instance, the hydrogen terminated armchair edge has the effective work function of 4.3 eV, while it is 7.1 eV for the zigzag edge. Therefore CFE from the former is much easier than the latter.

In the next section, we will first discuss a classical model of two-dimensional emitters. The vacuum potential energy barrier and the non-coherent emission current will be calculated. In Section 9.3, we will introduce the path-decomposition approach (PDA) to CCFE. The PDA is applied to CCFE of graphene in Section 9.4, where the CCFE patterns are calculated. In a magnetic field, Landau levels emerge. We will see that the emission patterns can reflect the feature of Landau levels.

9.2 Graphene Emitter with a Uniform Edge

As a simple model, graphene is represented by a metallic two-dimensional thin film of small thickness w. In the CFE setup, a metallic thin film of height h stands vertically on a flat cathode. Let the uniform edge coincide with the y-axis. An anode with y-translational invariance locates at $x = x_a(\gg h)$, having electrostatic potential V_a, as sketched in Fig. 9.2(a, b). The atomic structure of graphene emitter is sketched in Fig. 9.2(c), where the black balls are carbons with interatomic distance a = 0.142 nm. The emission edge is the armchair edge terminated by hydrogens (white balls). We assume that the emitter has L_x columns and L_y rows of carbons, and both L_x and L_y are large, and $h = \sqrt{3a}\, L_x$. Let F_a denote an applied electric field, B a magnetic field, and Δ a gate voltage on the edge region. The rectangular lattice plotted in red dashed lines defines the unit cells that preserve the y-translational invariance, which is convenient for the study of CFE. The honeycomb lattice can be divided into two triangle sublattices. Each carbon of the A-sublattice is neighboring with three carbons of the B-sublattice, and vice versa.

For a two-dimensional CFE setup having translational invariance in one direction, the electrostatic potential satisfies a two-dimensional Poisson equation, which may be solved with a powerful mathematical method named as conformal transformation: $z + ix \to \tilde{z} + i\tilde{x}$, defined by $\tilde{z} + i\tilde{x} = \sqrt{(z + ix + ih)^2 + h^2}$ [Li, 2010].

The cathode and emitter are transformed into a flat plane at $\tilde{x} = 0$, and the anode is transformed into a flat plane at $\tilde{x}_a = \sqrt{2(x_a^2 - h^2)}$. Hence the electrostatic potential energy in the transformed space is $\tilde{u}_0(\tilde{x}) = -\frac{\tilde{x}}{\tilde{x}_a} eV_a$.

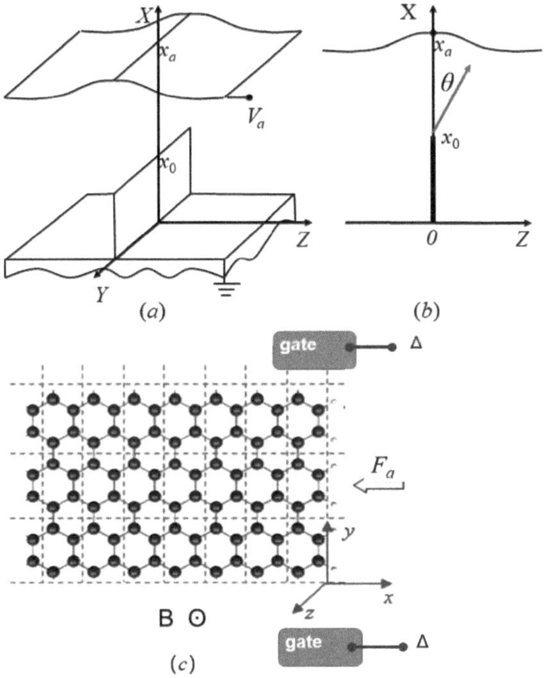

Figure 9.2 (a) Metallic thin film emitter stands vertically on the cathode plane. The upper surface is the anode at voltage V_a. (b) The side view of the setup. (c) Graphene emitter. Black (white) balls are carbon (hydrogen) atoms. F_A, B, and Δ are electric field, magnetic field, and gate voltage, respectively. The red rectangles are unit cells.

The stationary Schrodinger equation for the emission wave of energy E in the transformed space reads

$$\left[-\frac{\hbar^2}{2m_e}\left(\partial_{\tilde{x}}^2 + \partial_{\tilde{z}}^2\right) + \frac{(W_0 - W + \tilde{u}_0(\tilde{x}))(\tilde{z}^2 + \tilde{x}^2)}{\sqrt{(\tilde{z}^2 + \tilde{x}^2 - h^2)^2 + 4h^2\tilde{x}^2}}\right]\psi = 0 \qquad (9.4)$$

where $W = E - \frac{\hbar^2 k_y}{2m_e}$ is the energy of motion in the x–z plane. As obvious in Fig. 9.3(left), the potential valley along the x-axis is the most probable emission path. The emission is confined by the valley to a small disperse angle $\left(\theta \equiv \frac{z}{x}\right)$. Since z and so $\tilde{z}$ are small in the barrier region, one may neglect the $\tilde{z}$-dependence in the effective potential of (9.4). In this approximation, the equation is solvable. With the solution to emission wave function, one can easily obtain the emission current $\mathbf{J} = (j_x, j_y, j_z)$, where $j_y \sim ek_y\,|\psi|^2$. In a long distance from the edge,

$$\frac{|j_z|}{|j_x|} = \frac{h^2}{x_a^2}\sin\theta\cos^3\theta \tag{9.5}$$

As expected, the forward emission current is dominant. The vertical current maximized at $\theta = \frac{\pi}{3}$ is tiny because $\frac{h^2}{x_a^2} \ll 1$. This self-orientation phenomenon is universal for two-dimensional emitters.

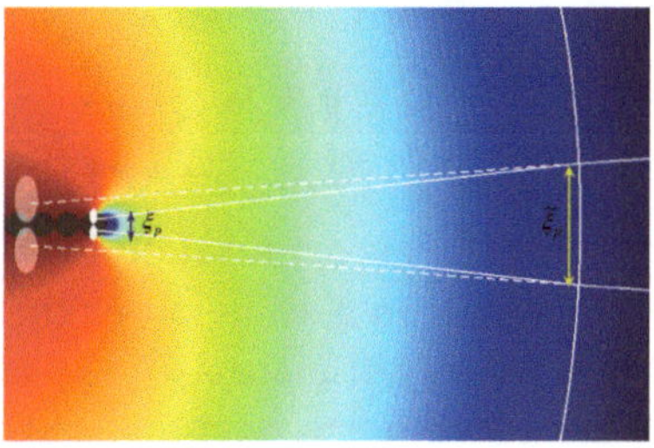

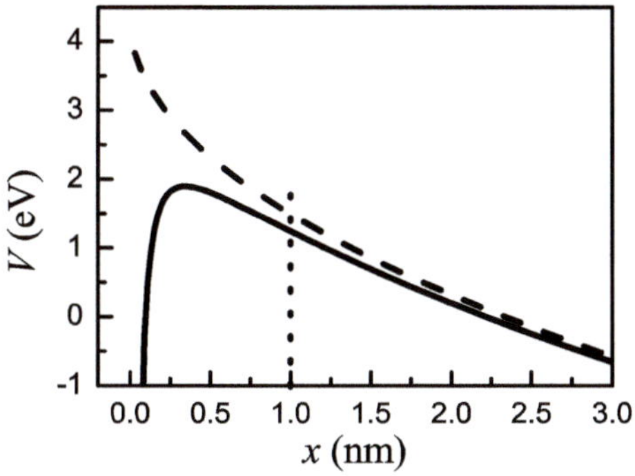

Figure 9.3 (Above) The vacuum energy barrier in the vicinity of graphene emitter [Luo and Li, 2016]. (Below) The vacuum barrier along x-axis. The dotted line indicates the potential energy where the inside and outside wave functions are matched (see Section 9.3) [Li, Xu, and Kreuzer, 2012].

The finite thickness effect has been included in an improved model [Qin et al., 2011]. If the thickness w of the nano-wall is sufficiently small, the total emission current depends on the applied field $F_{\mathrm{a}} = \frac{V_{\mathrm{a}}}{x_{\mathrm{a}}}$ as

$$J = a^{\mathrm{avf}} \left(\frac{h^{3/2}}{w} \right) W_0^{-9/4} F_{\mathrm{a}}^3 \exp\left(-\frac{2bW_0^{5/2}}{5ehF_{\mathrm{a}}^2} \right) \tag{9.6}$$

where $b = \left(\frac{2\sqrt{2m_{\mathrm{e}}}}{3e\hbar} \right)$ is the second FN constant, and a^{avf} the other universal constant in place of the first FN constant. Obviously, Eq. 9.6 is different from the Fowler–Nordheim theory.

In the metallic model, the electric field is screened completely inside the nano-wall. The charge density can be found from the uniform charge density $\tilde{\lambda}_0(\tilde{x},\tilde{z}) = \frac{\varepsilon_0 V_{\mathrm{a}}}{\sqrt{2} x_{\mathrm{a}}}$ by the reverse conformal transformation. For $x < 0$, it is

$$\lambda_0(x) = -\frac{\varepsilon_0 \sqrt{2} V_{\mathrm{a}}}{x_{\mathrm{a}}} \frac{h + x}{\sqrt{-x(x + 2h)}} \tag{9.7}$$

It is divergence at $x = 0$. This artificial difficulty arises from the fact that $\lambda_0(x)$ is only defined in a scale larger than the lattice spacing. To get around it, one can estimate $\lambda_0(0)$ by an extrapolation from the densities of the next and the next-next to the last column of unit cells, resulting $\lambda_0(0) = -\frac{0.22\varepsilon_0 V_{\mathrm{a}}}{x_{\mathrm{a}}}\sqrt{\frac{h}{a}}$. A physical error of Eq. 9.7 is the in-equipotential which contradicts to the complete screening assumption.

In fact, most nano-emitters have non-negligible field penetration because of the finite size and finite density of states. The field penetration is crucial for CFE because it changes the effective height of the tunneling barrier and the emission energy distribution. However, one can use Eq. 9.7 as the zero-order approximation to obtain a correction to the electrostatic potential energy if the density of states (DOS) of an emitter is known from a microscopic theory. Denote the potential energy due to the field penetration by $u(x)$. The excessive charge in the emitter are electrons occupying energy levels between the intrinsic Fermi level [$E_{\mathrm{F0}} = E_{\mathrm{F}} + u(x)$] and the Fermi level ($E_F$). Adopting the graphene DOS of tight-binding model $n(E)=\frac{2|E-E_{F0}|}{\pi(\hbar v_F)^2}$, one can obtain the first order correction of $u(x)$ as $u_1(x)=-\hbar v_F\sqrt{\frac{\pi}{e}|\lambda_0(x)|}$, where $v_F \sim \frac{c}{330}$ is the Fermi velocity. Furthermore, a nano-edge would have significant zero-field polarization that cannot be accounted by the work function as it could be in a flat emitter. The zero-field polarization has been calculated with density functional theory [Wang and Li, 2011a], resulting in an edge-effective dipole-potential energy $-\frac{f_d x}{\rho^2}$ with $\rho=\sqrt{x^2+z^2}$ and f_d = 0.244 eV nm.

Including both the dipole-potential energy and the electrostatic potential energy $u_0(x,z)$ (that can be obtained from $\tilde{u}_0(\tilde{x})$ by the reverse transformation) in the vacuum, the vacuum potential energy of graphene emitter is given by $U(x,z)=W_0-\frac{f_d x}{\rho^2}+u_0(x,z)+\Delta$, where the gate voltage Δ includes the potential drop $u_1(0)$ originated from the field penetration. For typical parameters W_0 = 4.32 eV [Wang et al., 2011b], h =10 μm, x_a = 1 mm, and V_a = 20 kV, $U(x, z)$ is shown in Fig. 9.3(left), with the gate voltage Δ = 0. The dark circles on the left represent the carbons of graphene, the smaller pair of white spots represent the pi-orbital of the edge carbon while the larger one is its image seen from the outer boundary of the barrier for electron at the Fermi level with zero y-directional kinetic energy (white arc on the right). The solid lines are the profile of emission beam. The vacuum barrier along x-axis is plotted in Fig. 9.3(right), where the solid(dash) line is the potential energy including (excluding) the effective dipole-potential energy.

9.3 Path-Decomposition Approach

The basic microscopic process of the CCFE may be understood as a quasi-particle in an eigenstate of the emitter being annihilated and then created as a free electron in an outgoing state in the vacuum. The major difficulty is to find the phase connection between the eigenstate of the emitter and the emitted wave. The path-decomposition approach (PDA) provides a method for this task, thus it is endowed with the ability to predict CCFE patterns [Li, Xu, and Kreuzer, 2012]. Remarkably, the approach preserves the phase relation while the exact atomic potential, which is usually unknown, is not required, therefore, it is very efficient. We will illustrate the PDA with graphene emitter.

Let us consider the simplest model that the emitted electrons reach the detector directly, passing no sample object during the propagation. Thus, the image of CCFE is the interference pattern of $\phi_{\mathrm{n,i}}$ with proper weights as given by Eq. (9.2). To find the connection between quantum states of a nano-emitter to the emitted state, one ought to match $\psi_{\mathrm{n,i}}$ with $\phi_{\mathrm{n,i}}$ at a emission surface of the emitter smoothly. There is no real problem for a metallic emitter as the free electron model also works well for the emitter. In a nano-emitter, however, electrons would be strongly correlated, $\psi_{\mathrm{n,i}}$ should be understood as collective mode which is only defined in a scale larger than the lattice spacing, therefore one could not match it to $\phi_{\mathrm{n,i}}$ directly. One has to go down to the scale of atomic orbital if one wants to read out the information of such scale from CCFE patterns.

In the PDA, $\psi_{\mathrm{n,i}}$ is expanded in compositions of atomic orbitals. Denote the wave function of the a-th atomic orbital of the *j*-th atom (or unit cell) by $\varphi_j^{\alpha}(\mathbf{r}-\mathbf{r}_j)$, with $\mathbf{r}_{\mathrm{j}}$ the position of the atom. The atomic orbital in the emission region will be referred to as edge atomic orbitals (EAOs). Expanding $\psi_{\mathrm{n,i}}$ in atomic orbitals,

$$\psi_n(\mathbf{r})=\sum_{\alpha,j} C_{n,j}^{\alpha}\varphi_j^{\alpha}(\mathbf{r}-\mathbf{r}_j) \tag{9.8}$$

Here and after, the quantum number *i* for the degeneracy is skipped. The subscript *j* implies that the form of function $\varphi_j^{\alpha}(\mathbf{r})$ would depend on the position of atom. By diagonalizing the Hamiltonian of the emitter in the bases of atomic orbitals, one can obtain the amplitudes $C_{n,j}^{\alpha}$.

Each φ_j^{α} that appears in the expansions of the eigenstate ψ_{n} emits an outgoing sub-wave $\varphi_{j,n}^{\alpha}$. In the vacuum barrier region where the atomic potential is smooth, one can match the EAO with the sub-wave from it without much difficulty. The outgoing wave ϕ_{n} from ψ_{n} is coherent superpositions of the outgoing sub-waves.

Using the language of the path-integral, the wave of frequency *w* emitted from the eigenstate ψ_{n} for large times is strictly given by the path-decomposition formula [Auerbach and Kivelson, 1984]:

$$\tilde{\Phi}_n(\mathbf{r},\omega)=\int_{\Omega}G^r(\mathbf{r},\mathbf{r}')\frac{i\hbar\overleftrightarrow{\partial_n}}{2m_e}\int_{\mathbf{r}''}G(\mathbf{r}',\mathbf{r}'')\psi_n(\mathbf{r}'') \tag{9.9}$$

where $\int_{\Omega}$ is the surface integral over an emission surface Ω that separates the emitter and the vacuum, and $\overleftrightarrow{\partial_n}$ is the right normal derivative minus the left one on Ω. The G is the retarded fixed energy Green's function and the restricted Green's function G^{r} is defined by a path integral over energy-fixed paths that stay strictly outside Ω and do not touch it except at the initial point. The physical meaning of Eq. (9.9) is clear. The emission electron must go through the emission surface Ω on which the amplitude $G\psi_{\mathrm{n}}$ at each point acts as a point source. The operator $-\frac{i\hbar}{2m}\overleftrightarrow{\partial_n}$ represents the mean flux through the surface. The restricted Green's function G^{r} propagates the wave from Ω to the screen. Thus the surface integral gives the emission wave as coherent superpositions of secondary waves from all points on Ω. We will choose Ω as a surface of constant potential in the tunneling barrier region such that only the least action path has significant contribution to G^{r} and the semi-classical approximation (JWKB) is applicable. We also require that Ω locates in the barrier region dominated by the atomic potential so that the wave functions of EAOs are still feasible on it. In Fig. 9.3(right), intersectional point of Ω and the x-axis is indicated by the dotted line.

Using the spectrum representation for $G(\mathbf{r}', \mathbf{r}'')$ in the reverse Laplace transformation of Eq. (9.9), one obtains,

$$\phi_n(\mathbf{r})=-\int_{\Omega}G^r(\mathbf{r},\mathbf{r}')\frac{\hbar\overleftrightarrow{\partial_n}}{2m_e}\sum_{\alpha,j}C^{\alpha}_{n,j}\varphi^{\alpha}_j(\mathbf{r}'-\mathbf{r}_j) \tag{9.10}$$

Because φ^{α}_j $(\mathbf{r}' - \mathbf{r}_j)$ decreases with $|\mathbf{r}' - \mathbf{r}_j|$ exponentially, only those orbitals close to the surface Ω, i.e., EAOs, have significant contribution to Eq. (9.10). It could be further simplified if the emission surface Ω has certain symmetry.

9.4 CFE Patterns of Graphene

In this section, we will illustrate the PDA by applying it to graphene CCFE. We have seen in Section 9.2 that the two-dimensional feature of graphene enables us to find the tunneling barrier and penetration field reliably. The simple emission surface enables us to have a complete analytical solution of graphene emitter, with and without a vertical magnetic field. In the magnetic field, the quasi-eigenstates are Landau states. Characteristic features of these fantastic quantum states can manifest in the CCFE patterns.

9.4.1 Quantum States

Relevant electronic properties of graphene can be attributed to the π-orbital of carbon. At the intrinsic Fermi level E_{F0}, there are two Bloch wave vectors, known as Dirac points, $\mathbf{K}_{-} = (-K, 0)$ and $\mathbf{K}_{+} = (K, 0)$, with $K = \dfrac{4\pi}{3\sqrt{3}a}$. A Bloch wave with energy in the vicinity of E_{F0} can be denoted by a relative wave vector $\mathbf{k} = (k_x, k_y)$ and a valley spin $\sigma = \pm$, such that $\mathbf{K}_{\mathbf{k}} = \mathbf{K}_{\sigma} + \mathbf{k}$. In the absence of magnetic field, a bulk eigenstate labeled by p has energy $E_{\mathrm{p}} = s\hbar v_{\mathrm{F}}|\mathbf{k}|$, where E_{F0} has been set to zero, thus the bands with $s = -1$ are filled while $s = 1$ bands are empty at the zero temperature.

In the magnetic field B along the z-direction and vanishing beyond graphene, an eigenstate of a given k_y is localized in the x-direction when the Landau gauge $\mathbf{A} = (0, Bx, 0)$ is chosen. Without losing generality, we assume $B > 0$. The magnetic field forms a harmonic potential well for each k_y. The bottom of the potential well locates at $x_{\mathrm{B}} = k_y l_B^2$ of the x-axis, with the magnetic length $l_{\mathrm{B}} \sqrt{\dfrac{\hbar c}{eB}}$. Bulk Landau states have energies $E_{\mathrm{p}}\, s\sqrt{\lambda_n}\,\varepsilon_0$, where $\varepsilon_0 = \dfrac{\sqrt{2}\hbar v_{\mathrm{F}}}{l_{\mathrm{B}}}$. The parameter λ_{n} depends on x_{B} with the subscript n being a non-negative integer, and $\lambda_n \to n$ for large negative k_y.

The Bloch waves are amplitudes $C_{n,j}^{\alpha}$ of the atomic orbitals in the expansions (9.8). Recalling the y-translational invariance, $C_{n,j}^{\alpha}$ can be written as $\tilde{C}_{p,x_{j_1}} \dfrac{1}{\sqrt{L_y}} e^{ik_y y_{j_2}}$, where (x_{j1}, y_{j2}) are the coordinates of carbon of the j_1-th column and the j_2-th row (see Fig. 9.2c). Amplitudes $\tilde{C}_{p,x_{j_1}}$ can be found by diagonalizing the Hamiltonian of graphene with the armchair edge.

Without the B field, the amplitude at the edge is [Li, Xu, and Kreuzer, 2012]

$$\tilde{C}_{p,0} \sim \sin(\gamma_p) \tag{9.11}$$

where $\gamma_p = \frac{k_y}{k_x}$ is the phase difference between the A-sublattice and the B-sublattice, called pseudo-spin phase difference. Remarkable, γ_p has a geometric origin, deeply rooted in the definition of spinor. Under the B field, it is [Chen, Li, Wang, 2014]

$$\tilde{C}_{p,0} \sim D_{\lambda_p}(\sqrt{2}k_y l_B)[1-\sin(\eta s\gamma_p)] \tag{9.12}$$

where $D_{\lambda p}$ is the parabolic cylinder function, and $\lambda_p = k_y a + \lambda_0$. The parameter γ_0 is called the pseudo-spin polarization angle.

The boundary conditions for the armchair edge breaks the x-translational invariance and mixes the states of different valley spins, hence k_x and σ are not conserved in CFE. But there are still two remarkable symmetries: (i) invariance in y-translational transformation; (ii) invariance in the time-y reflection defined as the combination transformation of time reverse and the y-directional reflection. Due to (i), k_y is conserved, hence the wave functions of eigenstates may be written as $\psi_p(\mathbf{r}) = \psi_p(x,z)\frac{1}{\sqrt{L_y}}e^{ik_y y}$. Due to (ii), the parity of the time-y reflection, denoted by η, is conserved. Hence an eigenstate under the CFE condition can be denoted by $p = \{s, |k_x|, k_y, \eta\}$ in the absence of B field, or by $p = \{s, n, k_y, \eta\}$ in the present of B field.

9.4.2 Emission Waves and Patterns

Expanding the eigenstate $\psi_p(x, z)$ in atomic orbitals,

$$\psi_p(x,z) = \sum_{m,j_1} \tilde{C}_{p,x_{j_1}} \varphi_{p,j_1}(x - x_{j_1}, z) \tag{9.13}$$

where x_{j1} is the x-coordinate of carbon of the j_1-th column, and $\varphi_{p,j1}$ is the y-directional Fourier component of the atomic orbitals of the same column. Because the tunneling probability of an atomic orbital is exponentially decreasing versus the distance between the atom and the edge, only the EAOs $\varphi_{p,0}(x,z)$ have significant contribution to CFE. Therefore the summation in the integrand of (9.10) is reduced to $\tilde{C}_{p,0}\varphi_{p,0}$, with $\tilde{C}_{p,0}$ given by (9.11) (without B) and (9.12) (with B).

In the PDA as discussed in Section 9.3, the EAO should smoothly match ϕ_p (the wave function of emitting electron) on Ω. In fact, all required information about the atomic wave function is its behavior in the vicinity of Ω. Because the potential energy is smooth in the vicinity of Ω, both $\varphi_{\mathrm{p},0}(x, z)$ and ϕ_{p} have simple forms in the matching region. The former can be modeled by a simple exponential function multiplied by an odd function of z to account the symmetry of the π-orbital, having the form $\varphi_{p,0}(\rho,z) \sim z\sqrt{\rho}e^{-\rho/\xi_{\mathrm{p}}}$ where $\xi_{\mathrm{p}} = \dfrac{\hbar}{\sqrt{2m_e W_p}}$ is the effective size at the x–z plane when the edge carbon is in the p-state [see Fig. 9.3(left)].

With the restricted Green's function G^{r} calculated by the JWKB approximation [Li, Xu, and Kreuzer, 2012], the asymptotic outgoing wave function (9.10) can be obtained as

$$\phi_p(\mathbf{r}) = N_0 \tilde{C}_{p,0}\left(\frac{W_{\mathrm{p}} S_{\mathrm{p}}^5}{eF_{\mathrm{a}} x}\right)^{\frac{1}{4}} e^{-\frac{b_0}{2}S_{\mathrm{p}}} \chi_{\mathrm{p}}(x,z) e^{i(k_x x + k_y y)} \tag{9.14}$$

where N_0 is a normalization constant, $b_0 = 1 + \dfrac{4}{5\sqrt{3}}$ and

$$S_{\mathrm{p}} = \frac{4}{9\xi_{\mathrm{p}}\hbar}\left(\frac{W_{\mathrm{p}}}{eF_{\mathrm{a}}}\right)^2 \tag{9.15}$$

It is a large number in CFE, thus we only keep the leading terms in S_{p}. The off-axis fading function is

$$\chi_{\mathrm{p}}(x,z) = \mu_{\mathrm{p}}^{3/4}\sqrt{\frac{\pi}{x}}\frac{z}{x} e^{-\frac{\mu_{\mathrm{p}}}{2}\left(\frac{z}{x}\right)^2} \tag{9.16}$$

where $\mu_{\mathrm{p}} = \alpha\xi_{\mathrm{p}}\hbar k_{\mathrm{a}}^2$ with $\alpha = 0.45$ and $k_{\mathrm{a}} = \dfrac{1}{\hbar}\sqrt{2m_{\mathrm{e}}V_{\mathrm{a}}}$.

The emission wave arriving on the anode plane is the analogue of the two-slit Young's interference with an effective slit constant $z_{\mathrm{s}} = \pi\sqrt{\alpha\hbar\xi_{\mathrm{p}}}$ and the effective wave number k_{a}. However, only two fringes are observable while others (ignored in (9.14)) are heavily suppressed by the fading factor in the z-direction.

With the emission wave function obtained, it is straightforward to find the emission current density of each individual eigenstate of graphene. The total emission current density is a non-coherent summation of the individual current densities weighted by the occupation distribution f_{p}. If the screen plane and the anode plane coincide at x_{a} and both perpendicular to the x-axis, the electric current density on the screen is

$$j_a(y,z) \sim \sum_p{}' \frac{e|\tilde{C}_{p,0}\chi_p(x_a,z)|^2}{2\grave{A}a^2\xi_p t_a} e^{-b_0 S_p} f_p \tag{9.17}$$

where $t_a = \frac{2m_e x_a}{\hbar k_a}$ is the flying time. The summation is over eigenstates of graphene with $k_y = \frac{m_e y}{\hbar t_a}$ fixed. In the absence of magnetic field, the summation over k_x can be turned into integral. Under the (strong) magnetic field, k_x is not a proper quantum number. Instead, one should sum over n that labels the Landau levels.

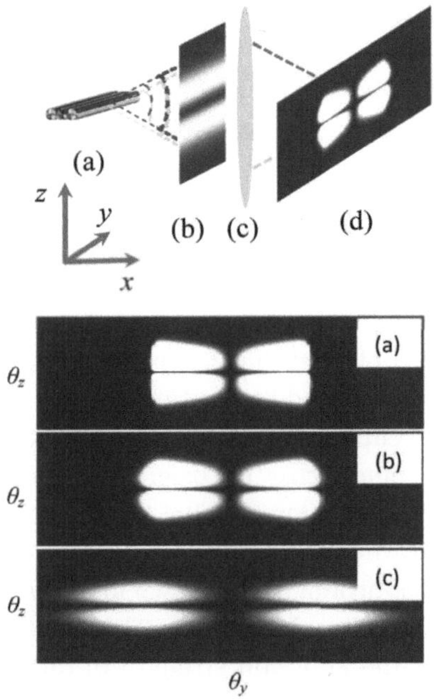

Figure 9.4 (Above) (a) The graphene edge. (b) The emission image near the anode plane. (c) The electric objective lens. (d) The image on the back focal plane of the objective lens. (Below) The image on the back focal plane for a graphene height h = 10 μm and an anode distance x_a = 1 μm in the field F_A = 20 V/μm. The region shown are given by $|\theta_y| < 0.005$ and $|\theta_z| < 0.00015$. The panels (a), (b), and (c) are for temperatures 100 K, 300 K, and 1000 K, respectively [Li, Xu, and Kreuzer, 2012].

In the proposed setup, Fig. 9.4(left), the electric objective lens is placed right behind the anode plane and the screen is on the back focal plane of the lens. Denote the focal length with x_f. For large x_f

and small emission angles, the wave function on the back focal plane is basically the Fourier transformation of the wave function on the anode plane. The emission angles with respect to the y and z axes are $\theta_y = \frac{y_f}{x_f}$ and $\theta_z = \frac{z_f}{x_f}$, respectively. The angular distribution of the emission current on the back focal plane reads

$$g(\theta_y, \theta_z) \sim \sum_p'' \frac{\sqrt{x_{\mathrm{f}}}\,\mu_{\mathrm{p}}^{3/2} \,|\tilde{C}_{\mathrm{p},0}|^2}{2\grave{\mathrm{A}} a^2 \xi_{\mathrm{p}} t_{\mathrm{a}}} \theta_z^2 e^{-b_0 S_{\mathrm{p}} - \mu_{\mathrm{p}} \theta_z^2} f_{\mathrm{p}} \tag{9.18}$$

where the summation is over eigenstates of graphene with $k_y = \theta_y k_{\mathrm{a}}$ fixed.

Eqs. (9.17) and (9.18) are rather general. They can be applied to the cases without or with magnetic field. All quantities with subscript p contain information of the quantum states. It is obvious that S_{p} governs the tunneling probability and μ_{p} governs the disperse angle in the z-direction. Both of them as well as the effective size of edge carbons ξ_{p} are W_{p}-dependent, and $W_{\mathrm{p}} = W_0 - E_{\mathrm{p}} + \frac{\hbar^2 k_y^2}{2m_{\mathrm{e}}} - \Delta$. The shift of Fermi level at the edge given by Δ is partly due to the field penetration as discussed in Section 9.2. It can also be controlled by a gate voltage applied at the edge region. The states contributing to the emission are filtered by the tunneling probability (the lower bound) and the occupation function f_{p} (the upper bound).

Because the angle factor θ_z^2 originated from the symmetry of π-orbital, there is a dark slit parallel to the graphene edge with $\theta_z = 0$. It is caused by the destructive interference of two partial waves emitted from the π-orbital that has two lobes with phase difference π on either side of the graphene plane. The y-directional distribution is determined by the emission strength of different k_y that enters the formula through W_{p} and, remarkably, through $|C_{\mathrm{p},0}|^2$ also. The latter depends on the pseudo-spin phase difference!

In the case of no magnetic field, the pseudo-spin phase difference γ_{p} appears in $|\tilde{C}_{\mathrm{p},0}|^2$ in the form $\sin^2(\gamma_{\mathrm{p}})$. States of different γ_{p} are emitted to different θ_y because of $|\gamma_{\mathrm{p}}| = |\theta_y| \frac{\hbar v_{\mathrm{F}} k_{\mathrm{a}}}{|E_{\mathrm{p}}|}$. It predicts a dark slit vertical to graphene edge with $\theta_y = 0$. The pattern dims in both sides of large y-directional angles due to suppression of the occupation distribution at large k_y. Therefore the image on the back focal plane is like a dragonfly [Fig. 9.4(right)] that is dark in the center cross with $\theta_y\theta_z = 0$. Theoretically, the dragonfly pattern can be observed at temperatures as high as 1000 K. Before the lens where the waves of different k_y have not been dispersed, one will see two bright bands parallel to the y-axis [Fig.9.4b(left)].

Under the magnetic field, $|\tilde{C}_{p,x_j}|^2$ is given by (9.12). The relative phase of pseudo-spins is $\gamma_P = k_y a + \gamma_0 = \theta_y k_a a + \gamma_0$ with γ_0 an unknown constant phase. The contribution of a Landau band to the total emission may be controlled by a gate voltage at the edge, which shifts up or down the Fermi level with respect to the Landau level. However, there is no simple method to break the n-degeneracy. Now we see that the CCFE pattern can in principle separate the emission of $\eta = 1$ from $\eta = -1$, and vise versa: their maxima and minima appearing alternatively and periodically in θ_y. Because of the magnetic field, the current density is not symmetrical for $y \rightarrow -y$. Fig. 9.5 gives an example of current density on the anode plane, using (9.18) with following parameters: $F_a = 20$ V/μm and $B = 15$ Tesla, $\Delta = 0$, $\gamma_0 = 0$, $h = 10$ μm, $x_a = 1$ mm, and the room temperature. Two main peaks separating in z direction are caused by interference of two lobes of the π-orbital.

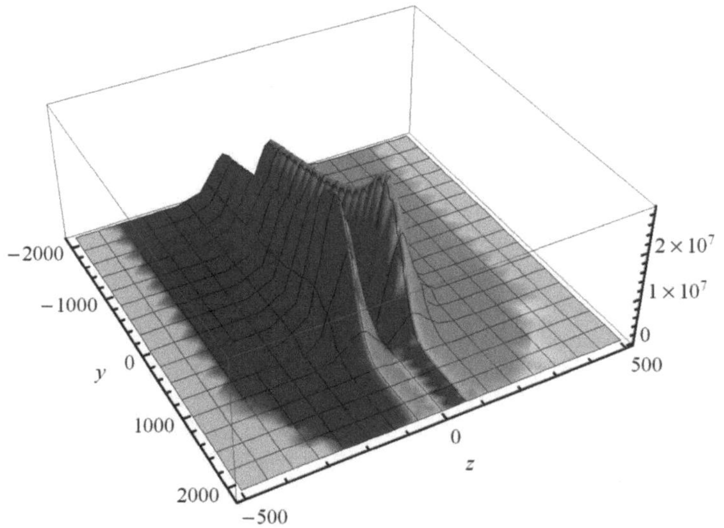

Figure 9.5 The emission pattern for $F_a = 20$ V/μm and $B = 15$ Tesla, at room temperature, with $\Delta = 0$ and $\gamma_0 = 0$. The vertical axis is the emission current density on the screen (mA/cm^2). The y and z coordinates are in unit of nm [Chen, Li, and Wang, 2014].

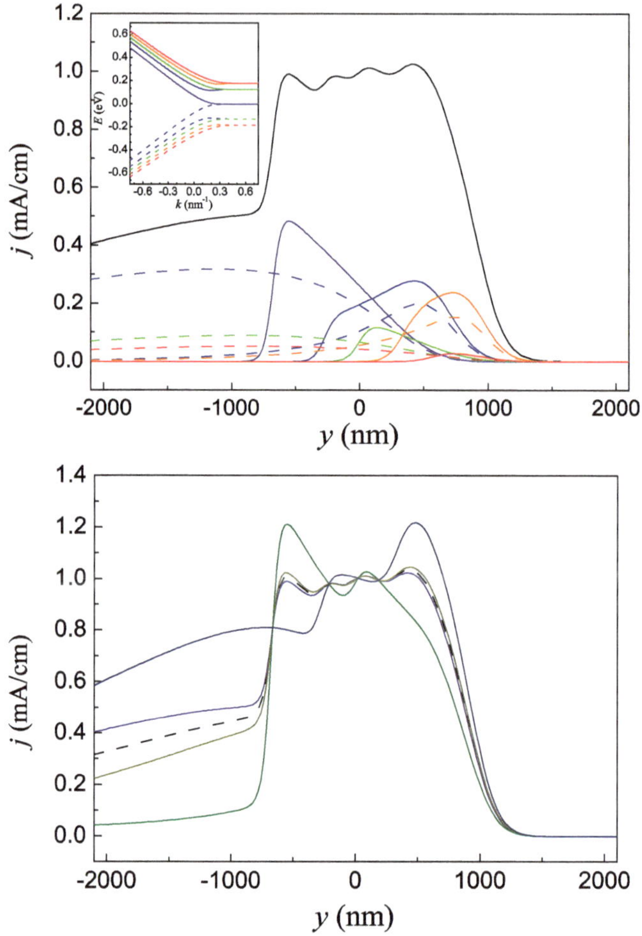

Figure 9.6 (Above) The y-directional current line-density on the screen for F_a = 20 V/mm and B = 15 Tesla, T = 77 K, Δ = 0.2, and γ_0 = 0. Black solid curve is the total line-density. Each colored curve is the partial line-density corresponding to the Landau level as shown in the inset in the same color. (Below) Dependence of the current line-density on the pseudo-spin relative phase, with Δ = 0. The black dashed curve is the un-polarized line-density with γ_0 = 0. The solid curves with their left wings from higher to lower are corresponding to $\gamma_0 = -\pi/2, 0, \pi, \pi/2$ [Chen, Li, and Wang, 2014].

Integration of Eq. (9.17) over z gives the y-directional current line-density, as shown by the solid black curve in Fig. 9.6 (left) for the temperature of liquid nitrogen (77 K). Colored curves are partial current line-densities from individual Landau levels close to the Fermi level (see inset of the same figure). Each partial current line-density is colored with the same color of its corresponding Landau level. Solid (dashed) curves are for positive (negative) levels. The Landau levels exhibit in the emission image as the multi-peak structure along the y-direction.

The effect of the pseudo-spin polarization angle is shown in Fig. 9.6 (right) for F_a = 20 V/μm, B = 15 Tesla, T = 77 K, and $\Delta = -0.2$. The pattern of $\gamma_0 = 0$ and π almost coincide with the un-polarized pattern (black dashed) both in the middle region and on the right wing. However, the left wings of them are clearly distinguishable. The patterns of $\gamma_0 = \pm\frac{\pi}{2}$ are dramatically different from those of $\gamma_0 = 0$ and π. The pattern of $\gamma_0 = \frac{\pi}{2}$ has the main peak on the left and the second peak on the right. Its left wing is the lowest. On the other hand, the pattern of $\gamma_0 = -\frac{\pi}{2}$ has the main peak on the right, the second peak on the left, and a fat tail on the left wing. In conclusion, the CCFE current distribution has distinct structure determined by the pseudo-spin polarization.

9.5 Discussions and Summary

Coherent cold field emission (CCFE) is a developing field, promising in applications of electron holography, parallel e-beam lithography, high-resolution holographic microscopy, and so on. Nano-emitters of various shapes and materials have been used for CCFE aiming to construct a practical coherent electron source and to reveal the physics behind the phenomenon. For all these purposes, it is important to know how the phase of an emission wave is related to quantum states of an emitter. Graphene as a unique two-dimensional emitter with well-defined structure is an ideal platform for developing CCFE theory.

We try to make the concept of CCFE clearer by a general discussion on electron coherence. We also explain the longitudinal and transversal spatial coherence lengths as well as the temporal coherence length.

To predict the field emission quantitatively, it is crucial to calculate the vacuum potential energy barrier reliably. This is done by a close investigation into the metallic model of two-dimensional emitters. The excessive charge distribution, the vacuum potential energy barrier, and the field penetration are calculated with the mathematical trick of conformal transformation, which works only for two-dimensional systems. As a bi-product, the model predicts self-focusing as a universal phenomenon of 2D emitters.

Going further to the atomic scale, we introduce the path-decompose approach (PDA) and show how it works with graphene emitter. The PDA provides a bridge connecting the quantum states of a solid emitter to the emission wave function. In this method, CCFE process of a nano-emitter is viewed as coherent decay of surface atomic orbitals whose relative phase is specified by the eigenstate of the emitter. Therefore, the approach can extract the quantum mechanical phase information of the emitter from CCFE patterns in principle.

Applying the PDA to a graphene emitter, one can obtain analytical solution to emission wave function in three-dimensions. The symbolic dragonfly-like emission pattern of graphene is explained by the symmetry of the π-orbital of carbon as well as the peculiar quantum geometric phase of graphene. According to the theory, the interference patterns may remain to a temperature as high as 1000 K.

Analytical solution is also available when a vertical magnetic field is added. The PDA predicts that the emission patterns can reflect the features of Landau levels. Remarkably, the patterns are sensitive to the phase difference of two sublattices, providing a novel method to probe the pseudo-spin of graphene.

A research interest would be the laser-assisted CCFE. The PDA has been applied to the CCFE of graphene in a laser field [Luo and Li, 2016]. It is shown that the occupations in Landau levels may be manipulated by laser controlling resonant transitions. One may read out the dynamic information of an emitter from the time-depending CCFE pattern.

Alhough considerable progress has been made, there is still room for improving the scattering theory of CCFE [Lepetit, 2019; Marquez-Mijares and Lepetit, 2019]. Recently, time-dependent density functional theory has been successfully applied to the CCFE of a long (5,5) carbon nanotube [Wang, Yang, Li, 2018]. The first principle approach requires huge computational resources, which is its major obstacle to general applications [Li, 2015]. Therefore, a method combining the PDA and the first principle approach would be attractive.

Finally, reliability of a theory and efficiency of a method can only be confirmed by experiments. Naturally, more experiments on CCFE of nano-emitters are appealing.

Acknowledgments

I am grateful to Profs. N. S. Xu, J. H. Kreuzer, S. Z. Deng, J. C. She, J. Chen, R. Forbes, and Dr. W. L. Wang, J. K. Chen, and M. Luo for many helpful discussions and fruitful collaborations. Thanks to D. Y. Li for amending the text. The project is supported by the Key Area Research and Development Program of Guangdong Province (Grant No. 2019B030330001) and the National Key Research and Development Project of China (Grant No. 2016YFA0202001).

References

Fowler, R. H. and Nordheim, L. W. (1928). Electron emission in intense electric fields. *Proc. R. Soc. A*, 119, 173–181 (doi:10.1098/rspa.1928.0091).

Modinos, A. (1984). *Field, Thermionic, and Secondary Electron, Emission Spectroscopy.* Plenum Press.

Horch, S. and Morin, R. (1993). *Field emission from atomic size sources. J. Appl. Phys.*, 74, 3652.

Latychevskaia, T., Longchamp, J-N., Es-cher, C., Fink, H.-W. (2015). *Holography and coherent diffraction with low-energy electrons: a route towards structural biology at the single molecule level. Ultramicroscopy*, 159, 395–402.

Zewail, A. H. and Thomas, J. M. (2010). *4D Electron Microscopy: Imaging in Space and Time.* Imperial College Press.

Kordruk, A. A. (2014). *ARPES experiment in fermiology of quasi-2D metals. Low. Temp. Phys.*, 40, 286–296.

Geim, A. K. (2009). *Graphene: status and prospects. Science*, 324, 1530–1534.

Brey, L. and Fertig, H. A. (2006). *Electronic states of graphene nanoribbons studied with the Dirac equation. Phys. Rev. B*, 73, 235411(5).

Neto, A. H. C., Guinea, F., Peres, N. M. R., Novoselov, K. S., and Geim, A. K. (2009). The electronic properties of graphene. *Rev. Mod. Phys.*, 81, 109–162.

Das Sarma, S., Adam, S., Hwang, E. H., and Rossi, E. (2011). Electronic transport in two-dimensional graphene. *Rev. Mod. Phys.*, 83, 407–470.

Novoselov, K. S., Geim, A. K., Morozov, S. V., Jiang, D., Katsnelson, M. I., Grigorieva, I. V., Dubonos, S. V., and Firsov, A. A. (2005). Two-dimensional gas of massless Dirac fermions in graphene. *Nature* (London), 438, 197–200.

Zhang, Y., Tan, Y.-W., Stormer, H. L., and Kim, P. (2005). Experimental observation of the quantum Hall effect and Berry's phase in graphene. *Nature*, 438, 201–204.

Xiao, Z. M., She, J. C., Deng, S. Z, Kang, T. Z., Li, Z. B., Lu, J. M., Xu, N. S. (2010). Field electron emission characteristics and physical mechanism of individual single-layer graphene. *ACS Nano*, 4, 6332–6336.

Jiang, L., Yang, T., Liu, F., Dong, J., Yao, Z., Shen, C., et al. (2013). Controlled synthesis of large-scale, uniform, vertically standing graphene for high performance field emitters. *Adv. Mater.*, 25(2), 250–255.

Tang, S., Deng, S. Z., Zhao, P., Zhan, R. Z., Chen, J., and Zhang, Y. (2020). Self-optimizing effect of a few-Layer graphenes top-edge structure during field electron emission observed by in situ TEM. *ACS Appl. Mater. Interfaces.*, 12, 16815–16821.

Oshima, C., Mastuda, K., Kona, T., Mogami, Y., Komaki, M., Murata, Y., Yamashita, T., Kuzumaki, T., Horiike, Y. (2002). Young's interference of electrons in field emission patterns. *Phys. Rev. Lett.*, 88, 038301(4).

Wu, P., Deng, S. Z., Chen, J., Xu, N. S. (2007). Fine-structured field emission images originating from coherently scattering of electrons within a multi-walled carbon nanotube. *Appl. Surf. Sci.*, 254, 1389–1393.

Zajec, B., Nemanic, V., Zumer, M., Bryan, E. N., Nemanich, R. J. (2011). Ring-shaped field emission patterns from carbon nanotube films. *Carbon*, 49, 3332–3339.

Yamaguchi, H. et al. (2011). Field emission from atomically thin edges of reduced graphene oxide. *ACS Nano.*, 5, 49450–4952.

Nakakubo, K., Asaka, K., Nakahara, H., and Saito, Y. (2012). *Evolution of* field electron emission pattern from multilayered graphene induced by structural change of edge. *Appl. Phys. Express*, 5, 055101(3).

Yokoyama, N., Nakakubo, K., Iwata, K., Asaka, K., Nakahara, H., and Saito, Y. (2016). Field emission patterns showing symmetry of electronic states in graphene edges. *Surf. Interface Anal.*, 48, 1217–1220.

Li, Z. B. (2010). Field electron emission from two-to three-dimensional space: an instance of electrodynamic and quantum-mechanical interdisciplinary study. *College Physics*, 29(7), 1–4 (Chinese).

Qin, X. Z., Wang, Xu, W. L., N. S., Li, Z. B., and Forbes, R. G. (2011). Analytical treatment of cold field electron emission from a nanowall emitter, including quantum confinement effects. *Proc. R. Soc. A*, 467, 1029–1051.

Wang, W. L. and Li, Z. B. (2011). Potential barrier of graphene edges. *J. Appl. Phys.*, 109, 114308(6).

Wang, W. L., Qin, X. Z., Xu, N. S., and Li, Z. B. (2011). Field electron emission characteristic ofgraphene. *J. Appl. Phys.*, 109, 044304(7).

Li, Z. B., Xu, N. S., and Kreuzer, H. J. (2012). Coherent field emission image of graphene predicted with a microscopic theory. *Phys. Rev. B*, 85, 115427(9).

Auerbach, A. and Kivelson, S. (1984). Path decomposition for multidimensional tunneling. *Phys. Rev. Lett.*, 53, 411–414.

Chen, J. K., Li, Z. B., and Wang, W. L. (2014). Manifesting pseudo-spin polarization of graphene with field emission image. *J. Appl. Phys.*, 115, 053701(4).

Luo, M. and Li, Z. B. (2016). Multi-field electron emission pattern of 2D emitter: illustrated with graphene. *J. Appl. Phys.*, 120, 204304(5).

Lepetit, B. (2019). A three dimensional numerical quantum mechanical model of electronic field emission from metallic surfaces with nanoscale corrugation. *J. Appl. Phys.*, 125, 025107.

Marquez-Mijares, M., and Lepetit, B. (2019). A three dimensional numerical quantum mechanical model of field electron emission from metallic surfaces covered with carbon adsorbates. *J. Appl. Phys.*, 126, 065107.

Li, Z. B. (2015). Density functional theory for field emission from carbon nano-structures. *Ultramicroscopy*, 159, 162–172.

Wang, W. L., Yang, W. T., and Li, Z. B. (2018). Field electron emission images far away from a semi-infinitely long emitter: a multiscale simulation. *J. Phys. Chem. C*, 122, 27754–27760.

Chapter 10

Influence of Edge Structures of Graphene on Field-Emission Properties

Yanlin Gao and Susumu Okada
Department of Physics, University of Tsukuba, Japan
ylgao@comas.frsc.tsukuba.ac.jp

Theoretical studies on electrostatic potential properties of graphene edges in terms of their shapes and functionalization under an external electric field were addressed by using the density functional theory. The shapes and attached functional groups of the edges cause substantial variation of their electrostatic potential outside the edges. The emission current from graphene edges is also investigated based on the Fowler–Nordheim equation, corroborating that the current is sensitive to their shape and functional groups.

10.1 Introduction

Electrons in materials are strongly bound by the deep ionic potential. By applying external physical stimuli, electrons overcome or go through the potential, emitting from their surfaces. Depending on the physical stimuli, the electron emission can be roughly classified

Nanostructured Carbon Electron Emitters and Their Applications
Edited by Yahachi Saito

ISBN 978-981-4877-62-6 (Hardcover), 978-1-003-14199-0 (eBook)
www.jennystanford.com

into thermal electron emission, photoelectron emission, and field-induced electron emission. For the thermal electron emission and photoelectron emission, electrons get adequate energy from external stimuli to overcome the potential barrier. For the field emission, an external electric field narrows and lowers the potential barrier outside the materials, enabling the electrons to tunnel through the barrier. Field emission is a quantum mechanical process of electrons tunneling through the potential barrier of surfaces of materials (metal or semiconductor) to vacuum under a strong electric field of the order of 10^6–10^7 V/cm. The simplified Fowler–Nordheim theory gives the relation among the field-emission current, I, electric field, F, and the work function, ϕ, of the material surface with the following formula:

$$I = \frac{A a \beta^2 F^2}{\phi} \exp\left(-\frac{b\phi^{3/2}}{\beta F} \right) \tag{10.1}$$

$$\beta = \frac{-b\phi^{3/2}}{m}, \tag{10.2}$$

where $a = 1.54 \times 10^{-6}$ A · eV/V^2, $b = 6.83 \times 10^9$ eV$^{-3/2}$ · V/m, β is the field enhancement factor at the surfaces, and m is the slope of the Fowler–Nordheim plot, obtained by $\ln(I/F^2)$ as a function of $1/F$. Several materials, such as molybdenum, silicon, and diamond, have been reported to work as the electron-emission sources of field-emission devices. Among such materials, low-dimensional materials cause substantial emission current owing to their structural morphology. Two-dimensional (2D) materials have been attracting much attention as constituent materials for designing field-emission devices because of their geometric and electronic structures [Bando *et al.* (2006); Terrones *et al.* (2008); Chen and Zou (2011); Late *et al.* (2014); Erande *et al.* (2015); Suryawanshi *et al.* (2015); Ahmadi *et al.* (2018)]. Erande *et al.* fabricated a nanosheet emitter consisting of few-layered black phosphorus, which shows a turn-on electric field of 4.2 V/μm at the current of 10 μA/cm^2. Suryawanshi *et al.* fabricated WS_2 planar emitters and provided the mechanism of the current fluctuation [Suryawanshi *et al.* (2015)]. It was reported that the BN sheets protruding from the ZnS nanowire shows superior field-emission properties with a turn-on electric field of 1.9 V/μm, field enhancement factor of 1600, and small-field-emission current fluctuation [Chen and Zou (2011)]. The turn-on electric field of the BN sheets with zigzag edges was reported to be approximately 1 V/μm [Terrones *et al.* (2008)].

Among 2D materials, graphene has been attracting a great deal of attention because of its excellent electron emission properties such as low turn-on electric field, high emission-current density, and long-term emission stability. Graphene-based emitters showed different field-emission efficiency depending on the substrate and graphene morphology. It was experimentally reported that edge atomic sites of suspended graphene preferentially emit electrons compared with the other atomic sites deposited on supporting substrates [Xiao *et al.* (2010); Tsai *et al.* (2012)]. A graphene foam emitter derived from highly porous reduced graphene oxide (rGO) showed outstanding field-emission properties with a turn-on electric field of 1.06 V/μm and stable current of 9.2 mA/cm^2 for 22 hours. Jeong *et al.* demonstrated that the length of rGO arrays affects its field-emission property and ZnO coating layers may further improve the device stability [Jeong *et al.* (2012)]. Zhang *et al.* investigated the effect of the morphology of few-layer graphene on the field-emission property leading to the conclusion that the optimal morphology of few-layer graphene for the field emission is a sharp corner, large height, and less amorphous carbon [Zhang *et al.* (2012)].

Although these experimental works showed that graphene is applicable to field-emission devices, microscopic insight into the physical mechanisms of field emission from the graphene edge is still unclear. Several theoretical works give knowledge about the emission from limited cases of graphene edges modeled by graphene nanoribbons (GNRs) [Tada and Watanabe (2002); Huang *et al.* (2005)]. Thus, in this chapter, we demonstrate the electrostatic properties of GNRs in terms of the edge shape [Gao and Okada (2018)] and functionalization [Gao and Okada (2019)], for providing microscopic insight into the field-emission mechanism of graphene-based electron emitters.

10.2 Theoretical Procedures

All calculations in this chapter were based on density functional theory (DFT) [Hohenberg and Kohn (1964); Kohn and Sham (1965)]. The exchange-correlation potential among interacting electrons was treated using the generalized gradient approximation with the Perdew–Burke–Ernzerhof functional [Perdew *et al.* (1997)]. The interaction between electrons and ions was described by ultrasoft pseudopotentials generated by the Vanderbilt scheme [Vanderbilt (1990)]. The valence wave function and deficit charge densities were expanded in terms of plane-wave basis sets with cut-off energies of 25 and 225 Ry, respectively. All atomic structures were optimized until the force on each atom was below 5 mRy/Å under the zero–electric field condition. Integration over the Brillouin zone was carried out using equidistant 4-*k* points in one-dimensional Brillouin zone along ribbon direction, which enabled sufficient convergence in the total energy and electronic structures of graphene and other carbon materials [Yamanaka and Okada (2016)]. All atomic structures were fixed as those under zero–electric field during the calculations of electronic properties under an external electric field.

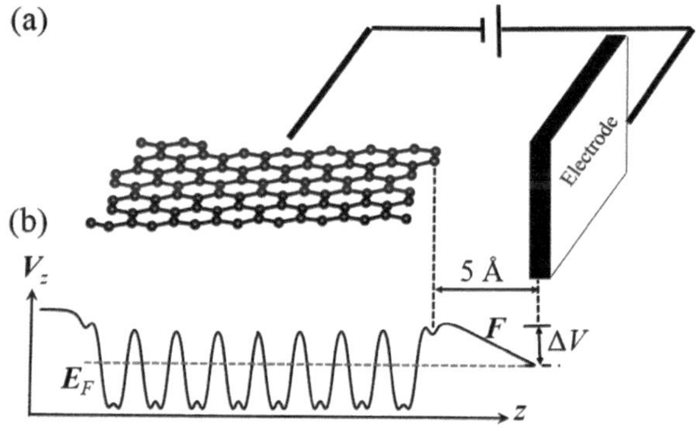

Figure 10.1 (a) Structural model and (b) schematic diagram of the plane-averaged electrostatic potential of a GNR under a lateral electric field. The dashed horizontal line is the Fermi energy, and ΔV denotes the electrostatic potential barrier for electron emission from the edges.

The effective screening medium (ESM) method was used to investigate the electrostatic potential properties of graphene edges modeled by functionalized GNRs with various edge shapes under a lateral electric field [Otani and Sugino (2006)]. To apply an electric field to the graphene edge, we considered a planar counter metal electrode described by an ESM with an infinite relative permittivity, which is separated by 5 Å vacuum spacing from the functionalized GNR edge [Fig. 10.1(a)]. In contrast, an open boundary condition is imposed at the opposite cell boundary described by a relative permittivity of 1 with the vacuum spacing of 5 Å from the other GNR edge, where the C atom is terminated by the H atom. Figure 10.1(b) shows the schematic diagram of the electrostatic potential for the edge-functionalized GNR under the electric field.

10.3 Structural Models

To study the effect of the edge shape of graphene on its field-emission property, we considered the clean and edge-hydrogenated GNRs' edges with edge angles of 0° for the armchair edge, 8°, 16°, and 23° for the chiral edges, and 30° for the zigzag edge (Figs. 10.2 and 10.3).

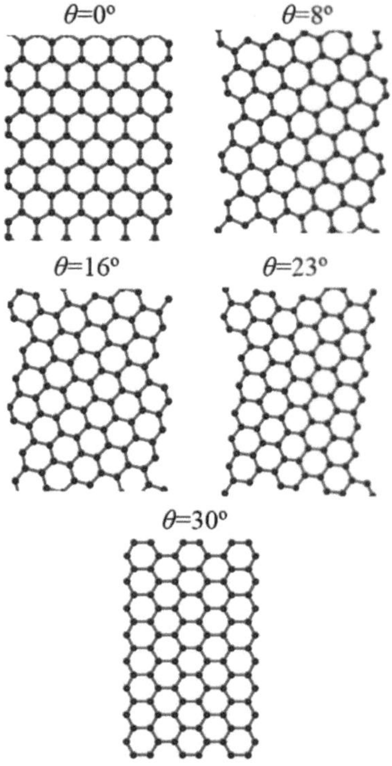

Figure 10.2 Optimized geometries of clean GNRs with edge angles of 0°, 8°, 16°, 23°, and 30°.

In addition to the edge shape, as for the effect of edge functionalization, we considered the armchair GNRs and zigzag GNRs, one of whose edges facing the electrode is terminated by O, CHO, COOH, H, OH, and NH functional groups, and the other is terminated by H atoms. Figure 10.4 shows the optimized geometries of edge-functionalized armchair GNRs and zigzag GNRs. To avoid steric hindrance between functional groups on adjacent edge atomic sites for the CHO and COOH cases, the edge atomic sites are alternately terminated by the H atom and the functional groups.

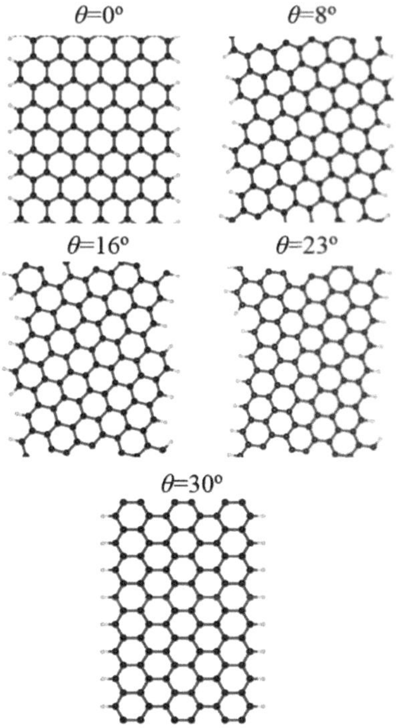

Figure 10.3 Optimized geometries of hydrogenated GNRs with edge angles of 0°, 8°, 16°, 23°, and 30°.

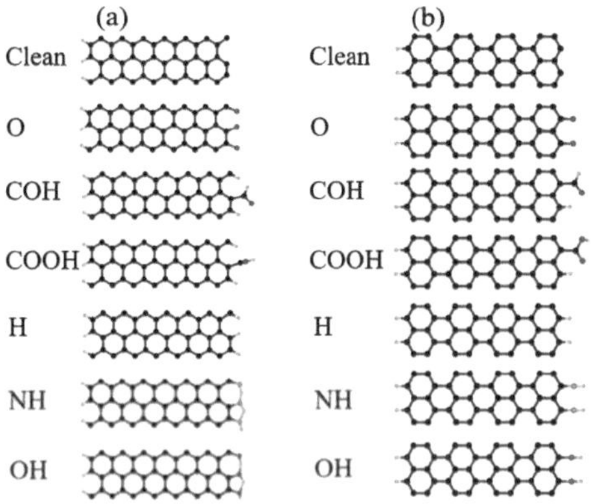

Figure 10.4 Optimized geometries of functionalized (a) armchair GNRs and (b) zigzag GNRs. The C, O, N, and H atoms are colored in black, red, grey, and pink, respectively.

10.4 Edge-Shape Effect on the Field-Emission Property of Graphene

Figures 10.5(a) and 10.5(b) show the electrostatic potential barrier for the electron emission from the clean and hydrogenated edges, respectively, as a function of the electric field. The potential barrier monotonically decreases with increasing electric field, irrespective of edge shape and termination. For clean edges, the potential barrier in the armchair edge is the lowest among the edges studied here for all electric fields because of its smallest work function. The potential barrier in the chiral edges (edge angles of 8°, 16°, and 23°) rapidly decreases with increasing electric field compared with those in the armchair and zigzag edges. Hence, the zigzag edge possesses the highest potential barrier for all electric fields. For the hydrogenated edges, the potential barrier monotonically decreases with the increasing electric field. The barrier strongly depends on the edge angle, although the work function is insensitive to the edge angle: the armchair edge has the lowest barrier, whereas the zigzag edge has the highest barrier for all electric fields.

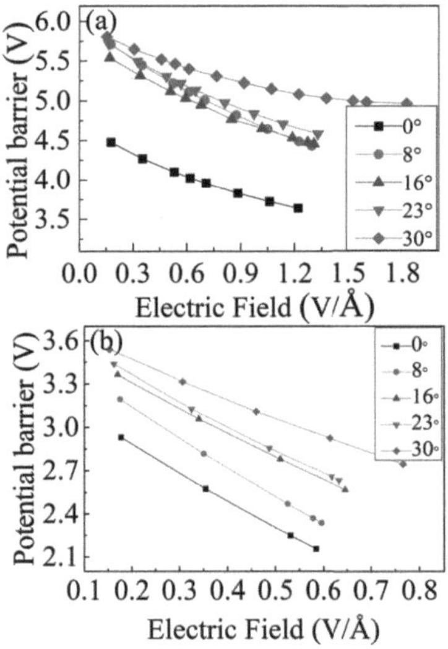

Figure 10.5 Electric-field dependence of the potential barrier for the field-emission in (a) clean and (b) hydrogenated GNRs with the edge shape ranging from armchair to zigzag.

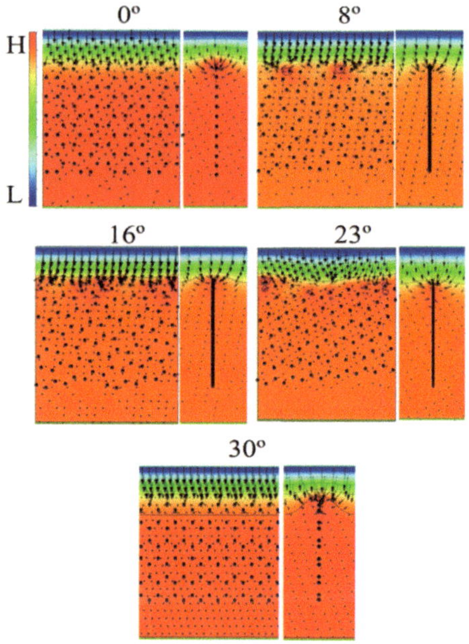

Figure 10.6 Contour and vector plots of the electrostatic potential and the electric field of clean GNRs with edge angles of 0°, 8°, 16°, 23°, and 30° subject to the critical electric field for the electron emission. The left and right panels for each angle correspond to parallel and vertical planes, respectively, of the GNRs. Black dots denote the atomic position of GNRs. Red, green, and blue indicate the high, moderate, and low potential regions, respectively.

To clarify the influence of the edge shape on electrostatic properties of the GNRs, we analyzed the electrostatic potential and electric field outside the GNRs with different edge angles under an external electric field. Figures 10.6 and 10.7 present contour plots of the electrostatic potential and vector plots of electric field for the GNRs with clean and hydrogenated edges, respectively, under the critical electric field at which the electrostatic potential on the electrode crosses the E_F [Fig. 10.1(b)]. The electrostatic potential under the external electric field is calculated by subtracting the electrostatic potential in the absence of an external field from that with the critical electric field to remove the deep potential valleys arising from the ions. In all cases, the electrostatic potential monotonically decreases on approaching the electrode, indicating a constant electric field in the vacuum between the graphene edge and the electrode. In contrast, the field concentrates near the edge atomic sites because of the atom-thick electron distribution in the GNRs. In addition, the electrostatic potential and the electric field are spatially modulated along the edge direction in some instances, by reflecting the detailed atomic structures of the edges: for the armchair and zigzag edges, the electrostatic potential and electric field are uniform along the edge direction, irrespective of the termination, suggesting that electrons are homogeneously distributed along the edge. In contrast, for clean chiral edges, we found the potential peaks at the edge atomic sites belonging to zigzag portions, leading to the concentration of the electric field. The fact indicates that zigzag sites are preferred in electron emissions than armchair sites in graphene with the chiral edges because of the electron concentration induced by the dangling bond. For chiral GNRs with hydrogenated edges, the potential peaks and field concentrations occur at the zigzag sites, even though the edges do not possess a dangling bond. In this case, the potential peak is ascribed to the higher charge density around the zigzag sites induced by the edge states.

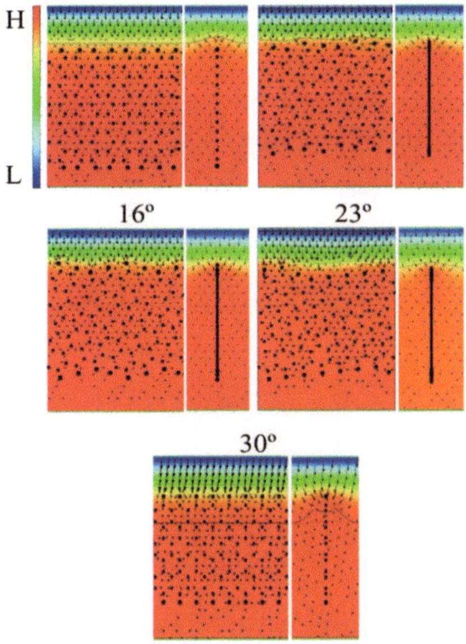

Figure 10.7 Contour and vector plots of the electrostatic potential and the electric field of edge-hydrogenated GNRs with edge angles of 0°, 8°, 16°, 23°, and 30° subject to the critical electric field for the electron emission. The left and right panels for each angle correspond to parallel and vertical planes, respectively, of the hydrogenated GNRs. Black dots denote the atomic position of hydrogenated GNRs. Red, green, and blue indicate the high, moderate, and low potential regions, respectively.

We analyzed the emission current density as a function of the plane-averaged external electric field. The current density, I, is calculated from the relation

$$I = \lambda v T, \tag{10.3}$$

where λ is the averaged electron charge density accumulating near the edge by the external electric field, and v is the collision frequency of electrons estimated by $v = E_k/h$ with the kinetic energy E_k. Since the electron emission is associated with the electron near the E_F, the emission current may be overestimated by the definition of v under the low-temperature condition. The transmission coefficient T is evaluated using the equation

$$T = \exp\left[\frac{-4\pi}{h}\int\sqrt{2m(V(z)-E_F)dz}\right], \tag{10.4}$$

where $V(z)$ is the plane-averaged electrostatic potential.

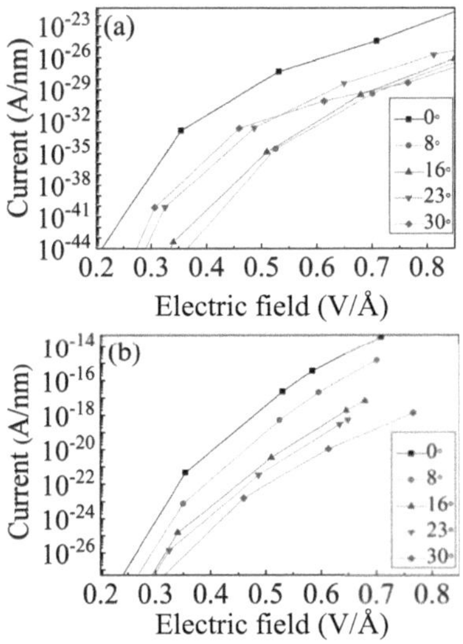

Figure 10.8 Field-emission current density as a function of the plane-averaged electric field for (a) clean and (b) hydrogenated GNRs.

The emission current density from clean and hydrogenated edges as a function of the electric field is shown in Figs. 10.8(a) and 10.8(b), respectively. The emission current density increases with increasing the electric field. It is also sensitive to the edge shape and termination. The armchair edge generates a larger current density than the other edges for both clean and hydrogenated cases. The edge hydrogenation substantially enhances the emission current density for all edge shapes. These behaviors are attributed to a decrease in the electrostatic potential barrier outside the ribbon edge upon edge hydrogenation.

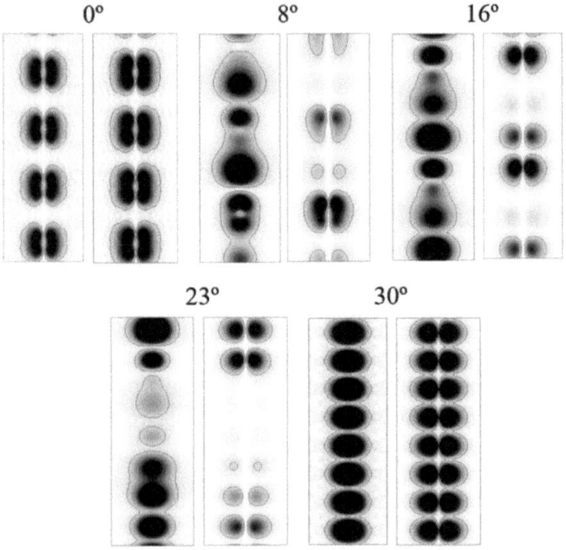

Figure 10.9 Contour plots of LDOS of GNRs, with various edge angles under the critical electric field for the electron emission. The contours are depicted on the plane parallel to the electrode situated at the 1 Å from the outermost edge atomic site. In each figure, the left and right panels correspond with LDOS of clean and hydrogenated edges.

Figures 10.9 shows the local density of states (LDOS) outside the GNRs, with clean and hydrogenated edges near the E_F, to simulate projected images by the emitted electrons from the edges since the electron tunneling from the edge to the electrode is attributed to the electron states near the E_F. The LDOS is defined by

$$\rho(r) = -\frac{1}{\pi}\mathrm{Im}\sum_{n,k}\int_{E_F-\Delta}^{E_F} dE \frac{|\phi_{\mathrm{n,k}}|^2}{E-\varepsilon_{\mathrm{n,k}}+i\delta}, \tag{10.5}$$

where Δ is 1 eV. The LDOS depends on the edge shape and termination. Thus, the edge shape and termination affect the field-emission patterns from the graphene edges. For clean and hydrogenated armchair edges, the LDOS is uniformly extended along the direction of the edge with bonding π-state nature. For clean chiral and zigzag edges, the LDOS does not have a nodal line with respect to the graphene layer, indicating their dangling σ-bond state nature. The LDOS also contains edge states because it is slightly elongated along the direction normal to the graphene layer. In contrast, for hydrogenated chiral and zigzag edges, the LDOS image exhibits the nonbonding π-state nature, which also implies the fact that the edge state at or near the zigzag portions of the hydrogenated chiral or zigzag edges significantly contributes to the electron emission.

10.5 Edge-Functionalization Effect on the Field-Emission Property of Graphene

Since the electrostatic potential barrier outside a material is a key factor in determining its field emission, we calculate this barrier for electron emission from edges with OH, H, O, CHO, COOH, and NH functional groups. Figure 10.10 shows the potential barriers of GNRs with clean and functionalized edges as functions of the electric field. Each potential barrier depends on the edge shape and termination. For the armchair edge, the potential barrier decreases monotonically with increasing electric field for all functional groups. Moreover, the potential barrier depends on the edge functionalization: the armchair GNR with O, NH, or CHO edge has a larger potential barrier than that with a clean edge, while the armchair GNR with COOH, H, or OH has a smaller potential barrier than that with a clean edge [Fig. 10.10(a)]. Therefore, we expect that COOH, H, and OH improve the field emission of graphene with the armchair edge. In particular, the edge with the OH group will have the largest field-emission current among all armchair edges because it has the lowest potential barrier among all edge-functionalized armchair GNRs.

For the zigzag edge, the potential barrier also monotonically decreases with increasing electric field for all functionalizations except for OH, whose potential barrier increases on increasing the field [Fig. 10.10(b)]. Even when the external electric field increases, the potential barrier of the OH edge remains the lowest for all zigzag edges studied here. As in the case of armchair edges, zigzag edges with O and CHO groups have higher potential barriers compared with other functionalized zigzag edges. By summarizing these facts, graphene with an OH-functionalized zigzag edge is the most efficient for electron emission among the structures studied here.

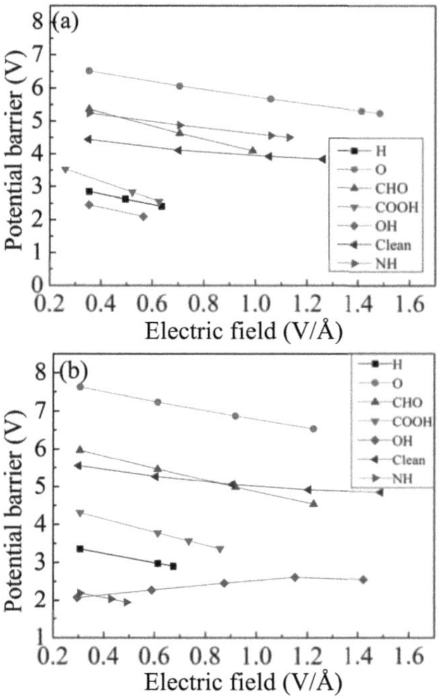

Figure 10.10 Potential barriers for electron emission from functionalized (a) armchair and (b) zigzag edges as functions of the external electric field.

To clarify the physical origin of this anomalous potential barrier associated with the field for the zigzag GNR terminated by the OH group, we investigate the electrostatic potential and electric field of functionalized zigzag GNRs under the critical external electric field (Fig. 10.11). The electrostatic potential shown in Fig. 10.11 is calculated by taking the difference between the electrostatic potential for the zigzag GNRs under the electric field and those under the zero electric field to remove the deep potential arising from the ions. The electrostatic potential peak emerges in the vacuum region outside the OH-terminated edge and extends along the edge. This potential peak outside the OH-terminated edge leads to the opposite electric field to the external electric field by the electrode at the vicinity of the edge atomic site, which suppresses the electron emission from the OH-terminated edge. The potential peak is ascribed to the spilled electron in the vacuum region owing to the nearly free electron (NFE) state, which causes the repulsive potential for the electrons. Therefore, the spilled electron due to the NFE state causes the anomalous potential barrier for electron emission from the OH-terminated edges. For the edges with H, O, CHO, COOH, and NH groups, the electrostatic potential monotonically decreases on approaching the electrode, leading to a constant electric field except at the vicinity of the edge atomic site. At the vicinity of the edges, the field concentration occurs owing to the ultimate thickness of the electron distribution on zigzag GNRs. In addition, along the edge direction, we also found that the spatial modulation of the electrostatic potential, owing to the functional groups, leads to further field concentration around them. Thus, zigzag GNRs with H, O, CHO, COOH, and NH edges may efficiently emit the electron from their edges, and the electric field required for the electron emission depends on their work function.

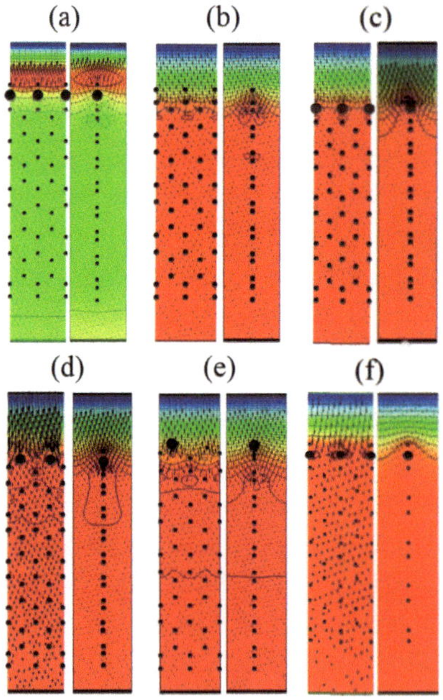

Figure 10.11 Contour and vector plots of electrostatic potential and the electric field, respectively, of zigzag GNR functionalized by (a) OH, (b) H, (c) O, (d) CHO, (e) COOH, and (f) NH functional groups under the critical electric field for the electron emission. The left and right panels in each figure correspond to parallel and vertical planes, respectively, of zigzag GNRs. Black dots denote the atomic position of edge-functionalized zigzag GNRs. Red, green, and blue indicate the high, moderate, and low potential regions, respectively.

We investigate the emission current from functionalized edges as a function of the external electric field. Figures 10.12(a) and 10.12(b) show calculated emission current densities for the clean and functionalized armchair and zigzag edges, respectively, as functions of the external electric field. Each emission current density *I* depends on the external electric field, edge shapes, and edge functional groups. Overall, the emission current monotonically increases with increasing electric field. The increase of the current strongly depends on the edge shape and functional group. For the armchair edge [Fig. 10.12(a)], OH, H, or COOH group enhances the emission current, while O, CHO, or NH suppresses the current. The emission current strongly correlates with the potential barrier outside the edge: a small potential barrier leads to a large emission current from the functionalized armchair edges.

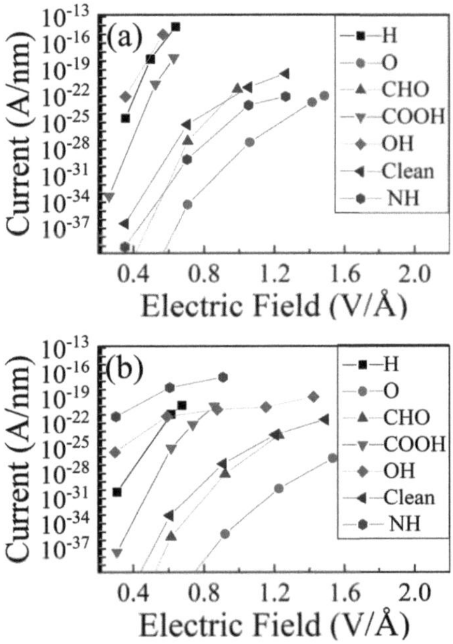

Figure 10.12 Field-emission current densities from functionalized (a) armchair and (b) zigzag edges as functions of the external electric field.

For the zigzag edge, the O functionalization substantially suppresses the emission current, while other groups except CHO enhance the current. The NH functionalization provides the largest current because this zigzag edge has a low potential barrier. The zigzag edge with an OH group also has a large emission current because of the low potential barrier outside the edge. Note that the emission current from the OH edge depends less on the external electric field than that from the other functionalized edges. This implies that the edge with the OH group is a stable emission source with respect to the electric field. For H, COOH, CHO, OH, and O groups, the armchair edge provides a larger emission current than that with the zigzag edge, as in the case of clean edges.

Figure 10.13 shows LDOS for functionalized zigzag GNRs near the E_F and at the vacuum space between the functional edge and electrode. From the LDOS images at the plane in vacuum parallel to the electrode, the LDOS depends on the functional group attached to the edges. The fact indicates that the functional group affects the field-emission pattern from the functionalized edges of graphene. For H, O, CHO, COOH, and NH edges, the LDOS possess symmetric distribution with respect to the ZGNR plane, indicating the p_z nature of C and O/N atoms. In contrast to these functionalized edges, LDOS of the OH-terminated edge shows different characteristics from the other edges: the LDOS does not exhibit p_z orbital nature but exhibits extended nature along the edge direction, leading to approximately uniform distribution.

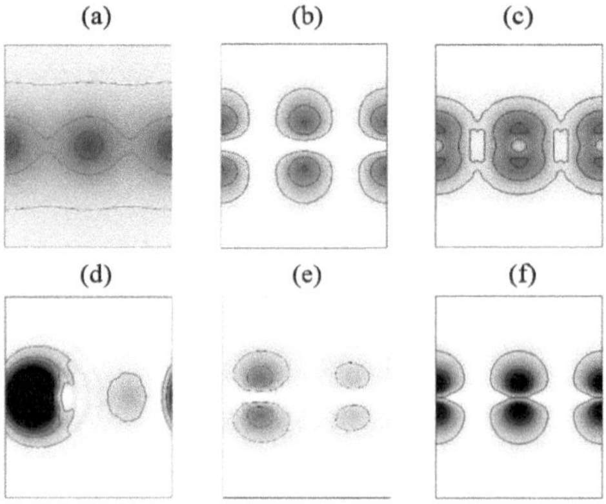

Figure 10.13 Contour plots of LDOS of zigzag GNRs functionalized by (a) OH, (b) H, (c) O, (d) CHO, (e) COOH, and (f) NH functional groups under the critical electric field for the electron emission. The contours are depicted at the plane parallel to the electrode situated 1 Å from the outermost edge atomic site.

10.6 Conclusion

We investigated the electronic properties of edges of clean and hydrogenated GNRs with respect to their edge shapes under the lateral electric fields, using the DFT with the ESM method. The electrostatic potential properties of the GNR were found to depend on the edge shape and termination. The armchair edge possesses lower potential barrier and larger field-emission current than the other edges, irrespective of edge termination. Furthermore, hydrogenation decreases the potential barrier by introducing a dipole, leading to an increase in the field-emission current. In addition, for chiral GNRs, the electron emission is primarily attributed to the zigzag portion of edges because of the concentration of electrons arising from either the dangling bond state or the edge state.

We also investigated the electrostatic potential properties of GNRs with zigzag and armchair edges functionalized by O, CHO, COOH, H, NH, and OH groups under the external electric field. Our calculations revealed that the work functions and potential barriers for the electron emission from these edges depend on the attached functional groups. Accordingly, the emission current depends not only on the applied electric field but also on the functional group. GNRs with H, OH, and COOH functional groups exhibit remarkable field-emission currents from their edges, owing to their low potential barriers. In contrast, edges with O termination have lower emission currents compared with the other functionalized edges, owing to their higher electrostatic potential barriers. The edges functionalized by the NH group show unusual behavior whereby the NH group enhances the field-emission current from the zigzag edge but suppresses that from the armchair edge because the dipole moment of the NH in the zigzag edge is larger than that in the armchair edge, arising from the conformation difference of the NH group.

References

Ahmadi, S., Nezhad, P. D. K., Hosseinian, A. and Vessally, E. (2018). A computational study on tuning the field emission and electronic properties of BN nanocones by impurity atom doping, *Physica E*, 100, pp. 63–68, doi:10.1016/j.physe.2018.03.010.

Bando, Y., Yin, L. and Golberg, D. (2006). Field nanoemitters: ultrathin BN nanosheets protruding from Si_3N_4 nanowires, *Nano Lett.*, 6, pp. 2982–2986, doi:10.1039/c0jm02955f.

Chen, Z.-G. and Zou, J. (2011). Field emitters: ultrathin BN nanosheets protruded from BN fibers, *J. Mater. Chem.*, 21, pp. 1191–1195, doi:10.1039/C0JM02955F.

Erande, M. B., Suryawanshi, M. A., Sachin R. More and Late, D. J. (2015). Electrochemically exfoliated black phosphorus nanosheets-prospective field emitters, *Eur. J. Inorg. Chem.*, 2015, pp. 3102–3107, doi:10.1002/ejic.201500145.

Gao, Y. and Okada, S. (2018). Electrostatic properties of graphene edges for electron emission under an external electric field, *Appl. Phys. Lett.*, 112, p. 163105, doi:10.1063/1.5023845.

Gao, Y. and Okada, S. (2019). Field emission properties of edge-functionalized graphene, *Carbon*, 142, pp. 190–195, doi:10.1016/j.carbon.2018.10.019.

Hohenberg, P. and Kohn, W. (1964). Inhomogeneous electron gas, *Phys. Rev.*, 136, p. B864, doi:10.1103/PhysRev.136.B864.

Huang, S. F., Leung, T. C., Li, B. and Chan, C. T. (2005). First-principles study of field-emission properties of nanoscale graphite ribbon arrays, *Phys. Rev. B*, 72, p. 035449, doi:10.1103/PhysRevB. 72.035449.

Jeong, H. J., Kim, H. Y., Jeong, H. D., Jeong, S. Y., Han, J. T. and Lee, G.-W. (2012). Arrays of vertically aligned tubular-structured graphene for flexible field emitters, *J. Mater. Chem.*, 22, p. 11277, doi:10.1039/c2jm31263h.

Kohn, W. and Sham, L. J. (1965). Self-consistent equations including exchange and correlation effects, *Phys. Rev. A*, 140, p. 1133, doi: 10.1103/PhysRev.140.A1133.

Late, D. J., Shaikh, P. A., Khare, R., Kashid, R. V., Chaudhary, M., More, M. A. and Ogale, S. B. (2014). Pulsed laser-deposited MoS_2 thin films on W and Si: field emission and photoresponse studies, *Appl. Mater. Interfaces*, 6, pp. 15881–15888, doi:10.1021/ am503464h.

Otani, M. and Sugino, O. (2006). First-principles calculations of charged surfaces and interfaces: a plane-wave nonrepeated slab approach, *Phys. Rev. B*, 73, p. 115407, doi:10.1103/PhysRevB.73.115407.

Perdew, J. P., Burke, K. and Ernzerhof, M. (1997). Generalized gradient approximation made simple [Phys. Rev. Lett. 77, 3865 (1996)], *Phys. Rev. Lett.* 78, 7, p.139, doi:10.1103/PhysRevLett.78.1396.

Suryawanshi, S. R., Kolhe, P. S., Rout, C. S., Late, D. J. and More, M. A. (2015). Spectral analysis of the emission current noise exhibited by few layer WS_2 nanosheets emitter, *Ultramicroscopy*, 149, pp. 51–57, doi:10.1016/j.ultramic.2014.10.004.

Tada, K. and Watanabe, K. (2002). Ab initio study of field emission from graphitic ribbons, *Phys. Rev. Lett.*, 88, p. 127601.

Terrones, M., Charlier, J.-C., Gloter, A., Cruz-Silva, E., Terres, E., Li, Y. B., Vinu, A., Zanolli, Z., Dominguez, J. M., Terrones, H., Bando, Y. and Golberg, D. (2008). Stacking-dependent optical absorption in multilayer graphene, *Nano Lett.* 8, pp. 1026–1032, doi:10.1021/nl072713m.

Tsai, J. T. H., Chu, T. Y. E., Shiu, J. Y. and Yang, C. S. (2012). Field emission from an individual freestanding graphene edge, Small 8, pp. 3739–3745.

Vanderbilt, D. (1990). Soft self-consistent pseudopotentials in a generalized eigenvalue formalism, *Phys. Rev. B*, 41, p. 7892, doi: 10.1103/PhysRevB.41.7892.

Xiao, Z., She, J., Deng, S., Tang, Z., Li, Z., Lu, J. and Xu, N. (2010). Field electron emission characteristics and physical mechanism of individual single-layer graphene, *ACS Nano*, 4, pp. 6332–6336.

Yamanaka, A. and Okada, S. (2016). Energetics and electronic structure of graphene nanoribbons under a lateral electric field, *Carbon*, 96, pp. 351–361.

Zhang, Y., Du, J., Tang, S., Liu, P., Deng, S., Chen, J. and Xu, N. (2012). Optimize the field emission character of a vertical few-layer graphene sheet by manipulating the morphology, *Nanotechnology*, 23, p. 015202, doi:org/Nano/23/015202.

Chapter 11

Theory of Thermionic Electron Emission for 2D Materials

Y. S. Ang and L. K. Ang
Science, Mathematics, and Technology,
Singapore University of Technology and Design,
8 Somapah Road, Singapore 487372, Singapore
yeesin_ang@sutd.edu.sg, ricky_ang@sutd.edu.sg

This chapter is intended to provide a brief overview on the recent development of the thermionic electron emission models for two-dimensional (2D) materials, such as graphene. The overview is not meant to be comprehensive. Its main objective is for the readers to appreciate that the traditional models developed decades ago, like Richardson–Dushman (RD) law for thermionic emission and Fowler–Nordheim (FN) law for field emission, are no longer valid due to the unique physical properties of 2D materials. The focus is to highlight the different temperature scaling laws for thermionic emission from 2D materials at different conditions such as lateral or vertical emission and electron momentum conservation. A recently developed universal thermionic emission model for 2D electronic systems is introduced, which offers a unifying theoretical framework to understand the thermionic emission physics in 2D materials.

Nanostructured Carbon Electron Emitters and Their Applications
Edited by Yahachi Saito

ISBN 978-981-4877-62-6 (Hardcover), 978-1-003-14199-0 (eBook)
www.jennystanford.com

11.1 Introduction

Electron emission from a material or charge injection from one medium to another medium is a fundamental process in cathode, diode, gas ionization, electrical contact, and many other areas. Depending on the energy used to release the electrons, it can be broadly characterized into three different processes known as thermionic emission (TE, by thermal energy), field emission (FE, by quantum tunneling), and photoemission (PE, by absorption of photons or optical tunneling of the light). The fundamental models for these emissions (TE, FE, PE) have been formulated many decades ago, known respectively, as the Richardson–Dushman (RD) law, Fowler–Nordheim (FN) law, and Fowler Dubrige (FB) law or Keldysh model.

In this chapter, we will focus on the new developments of electron thermionic emission theory for two-dimensional (2D) atomically thin nanomaterials, such as graphene, which has attracted tremendous research attentions in recent years. For a comprehensive introduction on the traditional electron emission models from three-dimensional (3D) or bulk materials, the readers are recommended to refer to a recent book by K. L. Jensen [1]. Carbon nanotube (CNT), a nanomaterial, which was widely studied before graphene, was intensively studied as field electron emitters for various applications [2] and new formulation of field-emission theory for CNT had been developed [3].

We first start with an introduction of the well-known traditional emission models (of 3D materials) broadly used by the research community, supplemented by discussions on selected topics focused on some new developments and key applications. The three key electron emission processes (without light illumination), namely the (1) thermionic emission, (2) thermal field emission, and (3) cold field emission, together with (4) space-charge-limited transport, are illustrated in Fig. 11.1. We then present the motivations and justifications why these traditional models should be revised to account for the unique physical properties of 2D materials in which the key assumptions used in the derivation of the traditional models are no longer valid. Finally, we illustrate some recent revised thermionic emission models for 2D materials using graphene as an example. These models are important not only for thermionic emitters but also for 2D-material-based electrical contact. The chapter concludes with an outlook and discussion of future works required for further expanding the theory of electron emission to the realms of field- and photoemission.

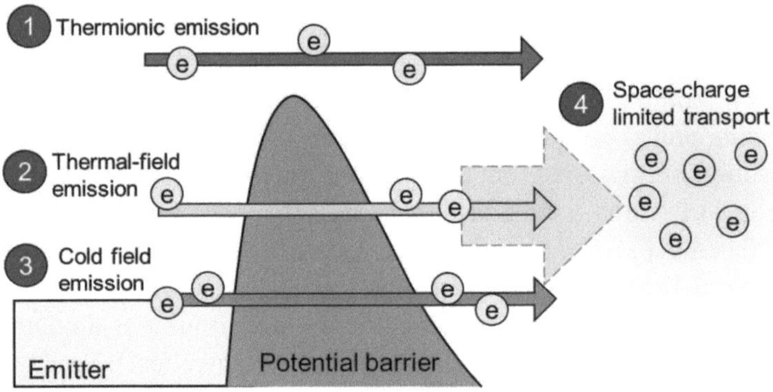

Figure 11.1 Schematic drawing illustrating the three electron emission processes (i.e., thermionic emission, thermal field emission, and cold field emission) and the space-charge-limited current.

11.2 Thermionic Emission

In 1901, Owen Richardson reported the experimental results of thermionic emitted electrons from a heated metallic surface [4]. He received a Nobel Prize in Physics in 1928 for his work on the thermionic emission with the emission law named after him. Subsequently, the thermionic emission from metals was intensively studied by many scientists including Owen Richardson himself, Saul Dushman, and others. Now the thermionic emission from metallic materials is well described by the Richardson–Dushman (RD) law [4, 5], which provides the amount of the thermionic emitted current density J_{RD} as a function of the emitting surface's temperature (T) and the work function of the material (Φ_{B}):

$$J_{\mathrm{RD}} = A_{\mathrm{RD}} T^2 \exp(-\Phi_{\mathrm{B}} / kT)\,. \tag{11.1}$$

Here, $A_{\mathrm{RD}} = 4\pi e m_{\mathrm{e}} k^2/h^3$ is a constant ($\approx 1.2 \times 10^6$ A m^{-2} K^{-2}), where k is the Boltzmann constant, e and m_{e} are the electron charge and free electron mass, respectively, and Φ_{B} is the work function in eV. Note Eq. (11.1) is in its simplest form and further refinements include the modification of the constant A_{RD} by adding various nonideal effects, such as replacing $A_{RD} \to \lambda(1 - R)A_{RD}$, where R is the reflection probability of electron wave function on the emitter surface and λ is a material-specific correction factor.

In spite of the uncertainties in the pre-factor A_{RD}, the temperature scaling of T^2 in Eq. (11.1) remains valid for many 3D bulk materials. By using experimental results to plot a graph of $\ln(J_{RD}/T^2)$ as a function of $1/T$ (commonly known as the *Arrhenius plot*), the linearity of the data points with a negative slope provides a direct evidence of thermionic emission. From the plot, the slope of the linear fit, M, can be used to estimate the work function as $\Phi_B = Mk_B$ based on the RD law (see Eq. (11.1)). Equation (11.1) thus provides a useful tool for extracting the work function of a cathode.

If a local surface electrostatic field of magnitude F is applied to reduce the surface barrier (due to image charge effect), Eq. (11.1) becomes

$$J_{RD} = \gamma A_{RD} T^2 \exp[-\left(\Phi_B - c\sqrt{F}\right)/kT], \tag{11.2}$$

where $c = \sqrt{e^3/4\pi\varepsilon_0}$ is the Schottky constant, ε_0 is the permittivity of free space, and γ is a fitting factor that compactly contains the effect of quantum mechanical reflections of electron sat the surface, the quality of the interface, and other factors.

One of the key applications of the RD law is in the Shockley diode equation named after the transistor coinventor William Shockley in 1949 [6]. The Shockley diode equation basically gives the I–V (current–voltage) characteristic of an idealized diode (or p-n junction) with an applied voltage in either forward or reverse bias, given by

$$I = I_s\left[\exp\left(\frac{qV}{\eta kT}\right) - 1\right]. \tag{11.3}$$

Here, I_s is the reverse bias saturated current, which equals to the RD law times the emitting cross-section area (A), namely $I_s = J_{RD} \times A$, when the carrier transport undergoes dominantly thermionic emission across the diode. Note that η is an *ideality factor* fitted from the experimental data and typically ranges from 1 to 2 in most Schottky diode devices. For an ideal diode, we have $\eta = 1$, as studied originally by William Shockley.

Another major application of the RD law is the development of high current thermionic cathode used in vacuum electronics, in particular, at high frequency and high voltage regime for high power microwave (HPM) generation [7, 8] in which solid state electronics is unable to meet the performance requirement. For HPM application, a thermionic cathode is required to operate at high current regime, where the emission law becomes the space-charge-limited current (SCLC) as governed by the Child–Langmuir (CL) law. The CL law [9, 10] describes the maximum steady state current density that can be transported in a vacuum gap of spacing d, and with applied voltage V, given by

$$J_{\mathrm{CL}} = C \times V^{3/2} / d^2, \tag{11.4}$$

where $C = \varepsilon_0 \times 4/9 \times \sqrt{2e/m_\mathrm{e}}$ is a constant. The studies of CL law were intensively revised in past 30 years including its extension into multidimensional models, quantum regime, discrete pulse limit, and other effects as reported in a recent review paper [11]. The transition regime from the RD law to CL law is important operating regime for high current thermionic cathode. The transition was formulated in an empirical form called Miram curves, which was also recently rederived by using statistical mechanics [12]. A nonuniform model for the transition from RD law to CL law has also been developed [13].

11.3 Field Emission

Compared to hot emission (or thermionic emission), electrons can be emitted without heat, but through quantum tunneling. Historically, it was either named as *cold emission* or *field emission*. Field emission has a long and complicated history. Here we will focus on the basic field-emission model known as Fowler–Nordheim law first developed by Ralph H. Fowler and Lothar Wolfgang Nordheim in 1928 [14], and a few key extensions well-known to the field-emission community.

According to the FN law, electron scan tunnel through the potential barrier of an emitting surface when a high electric field, F (e.g., a few GV/m for a metal surface with a work function Φ_B of a few eV), is applied to the surface. In the original derivation, the potential barrier was assumed to be a simple triangular shape, which was later revised to include the image charge potential. The simplest and most commonly used form of the FN law can be written as

$$J_{\mathrm{FN}} = \frac{aF^2}{\Phi_{\mathrm{B}}}\exp\left(-\frac{b\Phi_{\mathrm{B}}^{3/2}}{F}\right). \tag{11.5}$$

Here, a and b are constants and F is the surface electrostatic field. Using the typical units for work function Φ_{B} [eV], J_{FN} [A/m^2], and F [V/m], we have a = 1.541 × 10^{-6} A eV V^{-2} and b = 6.831 × 10^9 eV$^{-3/2}$ V m^{-1}. Including the image charge and the finite temperature effects, the FN law was modified into the Murphy-Good expression [15],

$$J_{\mathrm{MGT}}(V,T) \approx \frac{aF^2}{\Phi_{\mathrm{B}} t_0^2} \frac{\pi k_{\mathrm{B}} T / d_{\mathrm{F}}}{\sin(\pi k_{\mathrm{B}} T / d_{\mathrm{F}})} \exp\left(-\frac{b\Phi_{\mathrm{B}}^{3/2}}{F} v_0\right), \tag{11.6}$$

where d_{F} is an energy-like parameter, with $1/d_{\mathrm{F}} = (3t_0 b/2)\,\Phi_{\mathrm{B}}^{1/2}/F$, v_0 and t_0 are image-force correction factors which can be well-approximate, respectively, as

$$v_0 \approx 1 - y_0^2 + \frac{y_0^2 \ln(y_0)}{3} \quad \text{and} \quad t_0 \approx 1 + \frac{y_0^2}{9} - \frac{y_0^2 \ln(y_0)}{9},$$

where $y_0 \equiv cF^{1/2}/\Phi_{\mathrm{B}}$ [16].

In using the FN law to describe the field emissions from a sharp metallic tip, it is customary to assume that the surface electric field is enhanced by a factor β over the applied DC field F_o with the relation of $F = \beta F_o$, where $F_o = V/d$, V is the applied voltage over a distance of d. With sufficient large β, one can reduce the field F_o required for field-emission to MV/m (instead of GV/m). Instead of J–F relationship, the FN law can also be rewritten into the current–voltage (I–V) characteristics given by $I_{\mathrm{FN}} = \left(\frac{a'}{\Phi}\right) V^2 \exp(-b'\Phi^{3/2}/V)$ with modified constants a' and b'. Nevertheless, the plotting of $\ln(J/F^2)$ vs $1/F$ or $\ln(I/V^2)$ vs $1/V$ (commonly known as the *FN plot*) remains practically a straight line. The FN plot thus serves as a signature of field-induced electron tunneling from a metal surface. Importantly, the fitting of FN plotting with experimental results allows a rough estimation of the work function Φ_{B} and the field enhancement factor β by fitting to the experimental data.

It is important to note that FN law is a simple one-dimensional (1D) model that does not include the geometrical effects of sharp tips or protrusive structures consistently in the formulation. To our best knowledge, a consistent multidimensional and analytical FN law remains an open question thus far. The original FN law for cold field emission was reformulated [17] and other improvements on using FN law to characterize the experimental results were also studied [18]. A general model that links thermionic and field emission over a wide range of parameters was formulated by Jensen and Cahay [19], which offers a general theoretical framework for electron emission from bulk metals. More recently, a new formulation of FN law using fractional calculus to account for the field enhancement of a rough emitting surface has also been developed [20]. Beyond the conventional FN law, a field-emission model was developed for CNTs [21]. It should be noted that due to the exceptional mechanical and electrical properties as well as the presence of exposed atomically sharp tips [22], CNT in various forms, such as individual CNTs [23], array [24, 25], composite film [26], and fibers [27, 28], have been actively explored as a novel nanomaterial for high-performance field-emitter applications. The recent demonstration of thermionic electron emitter array using super-aligned CNT film further establishes the promising role of CNTs as on-chip integrated electron sources [29].

11.4 Thermionic Emission Models for 2D Materials

11.4.1 Motivation

From the traditional RD and the FN laws mentioned above, they follow a remarkable symmetry. Both are composed of a material-dependent pre-factor multiplied by a universal exponential factor of $\exp(-\Phi_B/kT)$ for thermionic emission and $\exp(-\Phi_B^{3/2}/F)$ for field emission. Over a wide range of operating temperature and electric field, the plotting of $\ln(J/T^\beta)$ as a function of $1/T$ and $\ln(J/F^\beta)$ as a function of $1/F$, both provide a constant of $\beta = 2$, which is material independent (for bulk 3D materials).

However, it is important to know that the assumptions used in the derivation of the traditional electron emission models for 3D materials are not valid for 2D materials. For example, the parabolic energy dispersion of the bulk materials is often not found in 2D materials. Unlike 3D crystals, the energy dispersion in 2D materials is only well-defined within its 2D plane while exhibiting discrete energy level quantization due to the quantum confinement of electrons within the 2D plane. The direction of the electron emission in 2D materials can also significantly affect the electron emission characteristics, as illustrated in Fig. 11.2 for two different emission configurations: *vertical* (surface) emission and *lateral* (edge) emission.

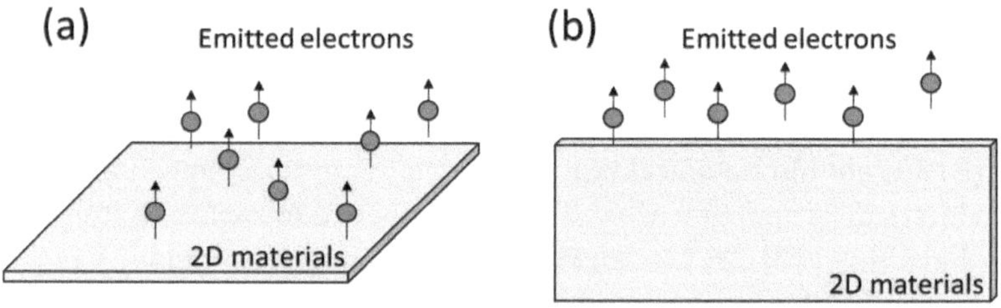

Figure 11.2 Thermionic emitter geometry for 2D materials. (a) Vertical emission geometry in which electrons are emitted from the 2D planar surface; and (b) lateral emission geometry in which electrons are emitted from the 1D edge.

In the following sections, we will focus specifically on the theory of *thermionic emission* in 2D materials, which has attracted the research spotlight in recent years due to its key importance in compact thermionic emitters, and solid state metal/semiconductor interfaces which are omnipresent in modern electronic devices. We will present a universal electron thermionic emission model [30] for 2D materials, using graphene as an example, under both the vertical (Fig. 11.2a) and the lateral (Fig. 11.2b) emission geometries, with and without electron momentum conservation during the emission process.

11.4.2 General Formalism of Electron Emission in 2D Materials

For 2D materials, the traditional electron field-emission models developed for bulk materials should be revised because of the following reasons:

1. Electronic band structure of 2D material is quantized into multiple 2D energy subbands that disperse only in the 2D

crystal plane due to the quantum confinement of electrons within the 2D plane of atomic-scale thickness [31]. The energy–momentum dispersion relation of the i-th 2D subband can be written as a combination of the parallel and vertical components, i.e.,

$$\varepsilon_i(\boldsymbol{k}) = \varepsilon_{\parallel,i}(\boldsymbol{k}_{\parallel,i}) + \varepsilon_{\perp,i}(k_{\perp,i}). \tag{11.7}$$

2. While the parabolic energy–momentum dispersion represents the archetypical band structure of many commonly studied bulk metals, *nonparabolic* energy–momentum relations are rather common in 2D materials. One particularly important example is graphene in which the electrons around the Fermi level is described by the linear Dirac equation,

$$\varepsilon_{\parallel}(\boldsymbol{k}_{\parallel}) = \hbar v_F k_{\parallel}, \tag{11.8}$$

where $v_F = 10^6$ m/s is the Fermi velocity of the linear Dirac conic band structure.

3. Because of the absence of energy dispersion along the vertical direction, which is perpendicular to the 2D crystal plane, the conservation of the lateral electron momentum component (parallel to the 2D plane) plays a nontrivial role in the electron emission characteristics. When this lateral electron momentum $k_{\parallel}$ is conserved, the tunneling probability, in WKB approximation, is:

$$D(k_{\perp,i}) \propto \exp\left(\pm \int_0^x \sqrt{\frac{2m}{\hbar^2}\left[V(x') - \varepsilon_{\perp}(k_{\perp,i})\right]} dx' \right). \tag{11.9}$$

In contrast, in the absence of $k_{\parallel}$-conservation, the $k_{\parallel}$ and the $k_{\perp}$ components of the 2D electron wavevectors are coupled. Under this condition, the transmission probability is a function of the total wavevector, $\boldsymbol{k}$, given by

$$D(\boldsymbol{k}_{\parallel}, k_{\perp,i}) \propto \exp\left(\pm \int_0^x \sqrt{\frac{2m}{\hbar^2}\left[V(x') - \varepsilon_{\parallel,i}(\boldsymbol{k}_{\parallel,i}) + \varepsilon_{\perp,i}(k_{\perp,i})\right]} dx' \right). \tag{11.10}$$

We now illustrate the *vertical* electron emission from a 2D electronic system with N subbands. The emitted current density can be written as

$$J_{2D}^{(C)} = \frac{ge}{(2\pi)^2} \sum_{i=1}^{N} \tau_{\perp,i}^{-1} \int_{B.Z.} d^2 \boldsymbol{k}_{\|,i} D\left(k_{\perp,i}\right) f(\varepsilon_{\|,i}, \varepsilon_{F,i}, T), \quad (11.11)$$

and

$$J_{2D}^{(NC)} = \frac{ge}{(2\pi)^2} \sum_{i=1}^{N} \tau_{\perp,i}^{-1} \int_{B.Z.} d^2 \boldsymbol{k}_{\|,i} D\left(\boldsymbol{k}_{\|}, k_{\perp,i}\right) f(\varepsilon_{\|,i}, \varepsilon_{F,i}, T), \quad (11.12)$$

for $k_{\|}$-conserving and $k_{\|}$-nonconserving emission, respectively. Here, $f(\varepsilon_{\|}, i, \varepsilon_{F,i}, T)$ is the Fermi–Dirac distribution and $\varepsilon_{F,i}$ is the Fermi level of the i-th subband of the 2D electronic system. Using the relation of the electronic density of states (DOS) for an isotropic 2D electron gas,

$$\rho_{2D}\left(\varepsilon_{\|}\right) d\varepsilon_{\|} = \frac{g}{2\pi} k_{\|} dk_{\|}. \quad (11.13)$$

Equations (11.11) and (11.12) become

$$J_{2D}^{(C)} = e \sum_{i=1}^{N} \tau_{\perp,i}^{-1} \int d\varepsilon_{\|} \rho_{2D}\left(\varepsilon_{\|}\right) D\left(\varepsilon_{\perp,i}\right) f(\varepsilon_{\|,i}, \varepsilon_{F,i}, T), \quad (11.14)$$

$$J_{2D}^{(NC)} = e \sum_{i=1}^{N} \tau_{\perp,i}^{-1} \int d\varepsilon_{\|} \rho_{2D}\left(\varepsilon_{\|}\right) D\left(\varepsilon_{\|}, \varepsilon_{\perp,i}\right) f(\varepsilon_{\|,i}, \varepsilon_{F,i}, T), \quad (11.15)$$

where we have rewritten the transmission probability in terms of carrier energies, i.e., $D(k_{\perp,i}) \rightarrow D(\varepsilon_{\perp,i})$ and $D(\boldsymbol{k}_{\|\_i}, k_{\perp,i}) \rightarrow D(\varepsilon_{\|}, \varepsilon_{\perp,i})$. In the following, we focus on graphene and derive the electron thermionic emission current with and without $k_{\|}$-conservation.

11.4.3 Dirac Cone Model of Graphene

We first briefly describe the electronic band structure of graphene [32]. Graphene is composed of carbon atoms arranged in honeycomb lattices. The low-energy electrons around the Fermi level in graphene mimic relativistic massless Dirac fermions, which is described by the 2×2 Weyl Hamiltonian:

$$H\left(\boldsymbol{k}_{\|}\right) = \hbar v_F \begin{pmatrix} 0 & k_x - ik_y \\ k_x + ik_y & 0 \end{pmatrix}, \quad (11.16)$$

expressed in the basis of (ψ_A, ψ_B) where $\psi_{A,B}$ represents the two inequivalent sublattices A and B of the graphene unit cell. The energy–momentum dispersion follows a *linear* relationship:

$$\varepsilon_{||}\left(\boldsymbol{k}_{||}\right)=\hbar v_{\mathrm{F}}\left|\boldsymbol{k}_{||}\right|, \qquad (11.17)$$

where $\boldsymbol{k}_{||} = (k_x, k_y)$ and $v_F = 10^6$ m/s is the Fermi velocity. The corresponding band structure is often denoted as the Dirac cone. Correspondingly, the electronic density of states can be obtained from DOS relation in Eq. (11.13) as:

$$\rho_{\mathrm{Gr}}\left(\varepsilon_{||}\right)=\frac{2\varepsilon_{||}}{\pi\hbar^2 v_{\mathrm{F}}^2}, \qquad (11.18)$$

where both spin and valley degeneracy are included in Eq. (11.18).

The DOS of graphene is thus vanishingly small around the Dirac point ($\varepsilon_{||} = 0$) and it increases linearly with the carrier energy $\varepsilon_{||}$. This is in stark contrast to the 2D electron gas having parabolic energy band, such as the electron gas confined in a quantum well structure, which exhibits a constant DOS, i.e., $\rho_{\mathrm{2DEF}} \propto$ const. Because of the vanishing DOS and the atomic thickness of graphene, the Fermi level can be sensitively tuned by applied voltage, which renders graphene as an exceptional material with tunable electronic and optical properties. The 2D sheet carrier density of graphene (considering only the conduction band) with Fermi level of ε_{F0} is

$$n_{\mathrm{Gr},0}=\int_0^{\varepsilon_{\mathrm{F0}}}\rho_{\mathrm{Gr}}\left(\varepsilon_{||}\right)d\varepsilon_{||}=\frac{\varepsilon_{\mathrm{F0}}^2}{\pi\hbar^2 v_{\mathrm{F}}^2}, \qquad (11.19)$$

and ε_{F0} is related to $n_{\mathrm{Gr},0}$ by

$$\varepsilon_{\mathrm{F0}}=\sqrt{\pi\hbar^2 v_{\mathrm{F}}^2 n_{\mathrm{Gr},0}}. \qquad (11.20)$$

Consider a simple parallel-plate setting [33], the induced carrier density on the graphene is

$$n_{\mathrm{ind}}(V_{\mathrm{G}})=\frac{\varepsilon_{\mathrm{ox}}V_{\mathrm{G}}}{t_{\mathrm{ox}}e}, \qquad (11.21)$$

where t_{ox} is the thickness of oxide layer. The total carrier density is thus

$$n_{\mathrm{Gr,total}}\left(V_{\mathrm{G}}\right)=n_{\mathrm{Gr},0}+n_{\mathrm{ind}}\left(V_{\mathrm{G}}\right)=n_{\mathrm{Gr},0}+\frac{\varepsilon_{\mathrm{ox}}V_{\mathrm{G}}}{t_{\mathrm{ox}}e}. \qquad (11.22)$$

The new Fermi level of the gated graphene is $\varepsilon_{\mathrm{F}}\left(V_{\mathrm{G}}\right)=\sqrt{\pi\hbar^2 v_F^2 n_{\mathrm{Gr,total}}\left(V_{\mathrm{G}}\right)}$, and the change of the Fermi level due to electrostatic gating V_{G} is

$$\Delta\varepsilon_{\mathrm{F}}(V_{\mathrm{G}}) = \hbar v_{\mathrm{F}}\sqrt{\pi}\left(\sqrt{n_{\mathrm{Gr},0} + \frac{\varepsilon_{\mathrm{ox}}V_{\mathrm{G}}}{t_{\mathrm{ox}}e}} - \sqrt{n_{\mathrm{Gr},0}}\right). \tag{11.23}$$

The gate-tunable Femi level of graphene has been demonstrated in experiment and it has been actively explored in electronic and optoelectronic device applications.

A new concept of barristor ("barrier + transistor") has also been proposed [34]. The barristor is essentially a graphene/semiconductor Schottky contact (Fig. 11.3a,b). By applying a gate voltage to tune the Fermi level of graphene, the Schottky barrier across the graphene/semiconductor interface can be modulated. This allows the thermionic injected current density to exhibit exceptionally high on/off ratio of 10^5 at room temperature (Fig. 11.3c).

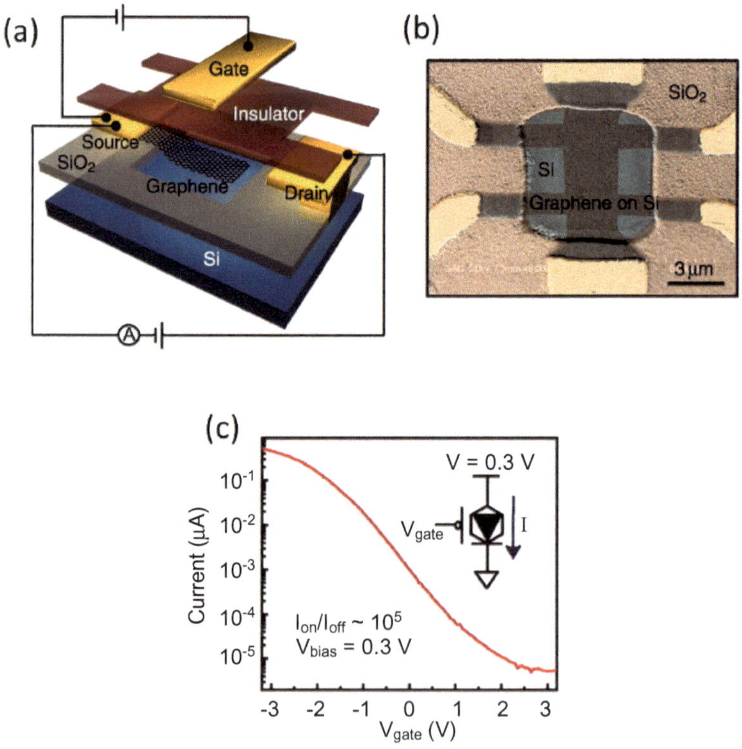

Figure 11.3 (a) Schematic drawing of a graphene barristor. (b) Scanning electron microscopy image of the graphene barristor. (c) Current on/off switching characteristics of the graphene barristor. Copyright American Association for the Advancement of Science [34].

11.4.4 Graphene Vertical Thermionic Emission: $k_{||}$-Conserving Model

We first consider the ideal case where interfacial scatterings are absent in the electron emitter surface. In this case, the $k_{||}$-component of the graphene thermionic emission is strictly conserved, and there is no coupling between $k_{||}$ and $k_{\perp,i}$ in the transmission probability. The transmission probability is only a function of $\varepsilon_{\perp,i}$ (or equivalently $k_{\perp,i}$), which can be captured by the following expression in the case of thermionic emission:

$$D\left(\varepsilon_{\perp,i}\right)=\Theta\left(\varepsilon_{\perp,i}-\Phi_{B0}\right), \tag{11.24}$$

where $\Theta(x)$ is the Heaviside step function and Φ_{B0} is the interfacial potential barrier height measured from the zero energy level. Equation (11.24) requires that only electrons with higher energy $\varepsilon_{\perp,i} > \Phi_{B0}$ to be thermionically emitted from graphene. Furthermore, because of such "over-barrier" electrons are no longer confined by the 2D crystal plane, $\varepsilon_{\perp,i}$ is no longer quantized, which is described by a continuous dispersion of $\varepsilon_{\perp,i} \rightarrow \varepsilon_{\perp}\left(k_{\perp}\right)=\hbar^2 k_{\perp}^2/2m$. In this case, the total energy of graphene follows the partition relation, $\varepsilon(\boldsymbol{k}) = \varepsilon_{||}(k_{||}) + \varepsilon_{\perp}(k_{\perp})$. Equation (11.14) can thus be rewritten as

$$J_{\mathrm{Gr}}^{(C)}=e\int_0^\infty v_{\perp}(k_{\perp})dk_{\perp}D[\varepsilon_{\perp}(k_{\perp})]\int_0^\infty d\varepsilon_{||}\rho_{\mathrm{Gr}}\left(\varepsilon_{||}\right)f(\varepsilon_{||}+\varepsilon_k,\varepsilon_F,T), \tag{11.25}$$

where we have converted the discrete summation in Eq. (11.14) as

$$\sum_{i=1}^{N}\tau_{\perp,i}^{-1}=\sum_{i=1}^{N}\frac{v_{\perp,i}}{L_{\perp}}\rightarrow\int_0^\infty v_{\perp}(k_{\perp})dk_{\perp}. \tag{11.26}$$

Here the injection time is given by the semiclassical relation of $\tau_{\perp,i} = L_{\perp}/v_{\perp,i}$, where $L_{\perp}$ is the thickness of the effective 2D material and $v_{\perp,i}$ is the carrier group velocity component along the emission direction vertical to the 2D plane. In the continuous dispersion limit, we have

$$v_{\perp,i}\rightarrow v_{\perp}(k_{\perp})=\hbar^{-1}\frac{\partial\varepsilon(\boldsymbol{k})}{\partial k_{\perp}}=\hbar^{-1}\frac{\partial\varepsilon_{\perp}(k_{\perp})}{\partial k_{\perp}}. \tag{11.27}$$

Combining Eqs. (11.18), (11.24), (11.25), and (11.27), the $k_\perp$-conserving electron thermionic emission current density of graphene becomes

$$J_{\mathrm{Gr}}^{(C)} = \frac{2e}{\pi\hbar^3 v_{\mathrm{F}}^2} \int_{\Phi_{\mathrm{B0}}}^{\infty} d\varepsilon_\perp \int_0^{\infty} \varepsilon_{||} \exp\left(-\frac{\varepsilon_{||} + \varepsilon_\perp - \varepsilon_{\mathrm{F}}}{k_{\mathrm{B}}T}\right) d\varepsilon_{||} . \quad (11.28)$$

The $\varepsilon_{||}$- and $\varepsilon_\perp$-integral can be straightforwardly solved as

$$\int_0^{\infty} \varepsilon_{||} \exp\left(-\frac{\varepsilon_{||}}{k_{\mathrm{B}}T}\right) d\varepsilon_{||} = (k_{\mathrm{B}}T)^2 , \quad (11.29)$$

$$\int_{\Phi_{\mathrm{B0}}}^{\infty} \exp\left(-\frac{\varepsilon_\perp - \varepsilon_{\mathrm{F}}}{k_{\mathrm{B}}T}\right) d\varepsilon_\perp = k_{\mathrm{B}}T \exp\left(-\frac{\Phi_{\mathrm{B0}} - \varepsilon_{\mathrm{F}}}{k_{\mathrm{B}}T}\right). \quad (11.30)$$

Equation (11.28) can thus be solved as

$$J_{\mathrm{Gr}}^{(C)} = \frac{2ek_{\mathrm{B}}^3}{\pi\hbar^3 v_{\mathrm{F}}^2} T^3 \exp\left(-\frac{\Phi_{\mathrm{B0}} - \varepsilon_{\mathrm{F}}}{k_{\mathrm{B}}T}\right). \quad (11.31)$$

This $k_{||}$-conserving thermionic emission model shown by Eq. (11.31) was first developed by Liang and Ang [35], which exhibits a good agreement with the experimental thermionic emission measurement of suspended graphene. Based on this model, thermionic energy converter and electronic refrigerator based on graphene/vacuum and graphene/2D-material interfaces are theoretically predicted to exhibit exceptional conversion efficiency, reaching a coefficient of performance (COP) of 18.5 [36, 37]. A recent experiment of graphene/WSe_2/gold heterostructures [38] has confirmed the improved performance of graphene thermionic-based energy converter (Fig. 11.4).

11.4.5 Graphene Vertical Thermionic Emission: $k_{||}$-Nonconserving Model

We now consider the case where interfacial scatterings are present to allow $k_{||}$ momentum nonconservation. Such scattering can arise from intrinsic defects, phonons, and the thermal and electrical coupling between graphene and the substrate. The thermionic emission current density with $k_{||}$-nonconservation can be obtained by using the transmission probability in the form of Eq. (11.10), i.e.,

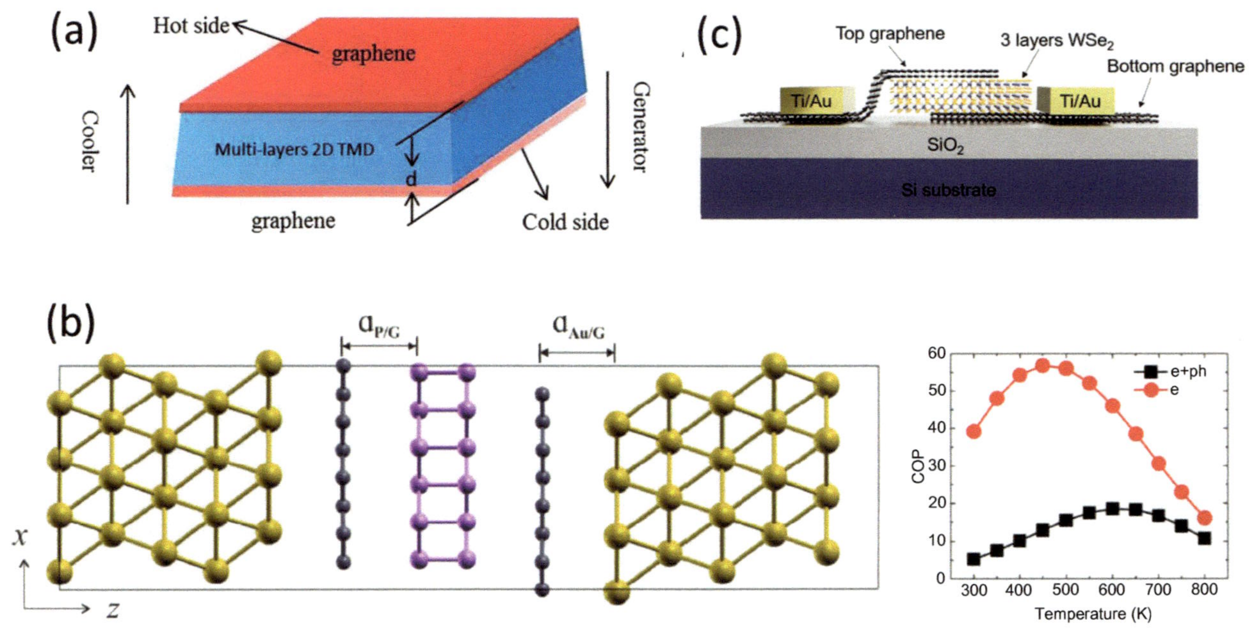

Figure 11.4 (a) Schematic drawing of a graphene/2D-transition-metal-dichalcogenide/graphene thermionic energy converter. Copyright Springer Nature [36]. (b) Schematic drawing of a graphene and phosphorene van der Waals heterostructure thermionic energy converter. The COP reaches 18.5 at 600 K, which is a substantial improvement for nanoscale devices. Copyright, Royal Society of Chemistry [37]. (c) Schematic drawing of a graphene/WSe_2/graphene heterostructure energy converter. Copyright American Association for the Advancement of Science [38].

$$D(\varepsilon_{\perp,i},\varepsilon_{\parallel})=\Theta(\varepsilon_{\perp,i}+\varepsilon_{\parallel}-\Phi_{B0}). \tag{11.32}$$

Following the same procedures in obtaining Eq. (11.30), the $k_{\parallel}$-nonconserving vertical thermionic emission current density in Eq. (11.15) becomes

$$J_{\mathrm{Gr}}^{(\mathrm{NC})}=\frac{2ek_{\mathrm{B}}^{2}}{\pi\hbar^{2}v_{\mathrm{F}}^{2}\tau_{\perp}}T^{2}\left(1+\frac{\Phi_{\mathrm{B0}}}{k_{\mathrm{B}}T}\right)\exp\left(-\frac{\Phi_{\mathrm{B0}}-\varepsilon_{\mathrm{F}}}{k_{\mathrm{B}}T}\right) \tag{11.33}$$

where $\tau_{\perp}^{-1}$ is the vertical carrier injection rate.

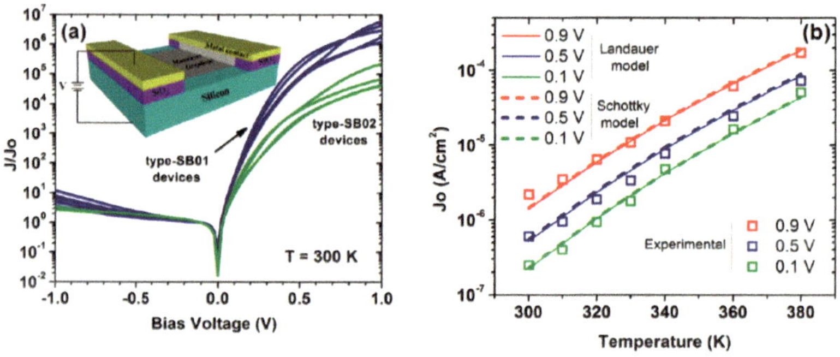

Figure 11.5 (a) Current–voltage characteristics of a graphene/silicon Schottky contact. (b) Fitting of the experimental data with the $k_{\parallel}$-nonconserving thermionic emission model (denoted as Landauer model in the figure). Although the conventional Richardson–Dushman model (denoted as Schottky model in the figure) exhibits a similarly good agreement with the experimental data, the extracted Richardson–Dushman constant is more consistent with that predicted by the $k_{\parallel}$-nonconserving thermionic emission model. Copyright American Chemical Society [39].

This $k_{\parallel}$-nonconserving vertical thermionic emission given by Eq. (11.33) was first developed by Sinha and Lee [39]. A good agreement with the thermionic carrier transport measurement in graphene/semiconductor Schottky diodes is reported (Fig. 11.5). The carrier injection rate $\tau_{\perp}^{-1}$ is, however, determined by fitting the experimental data with the thermionic emission model in Eq. (11.33). The value of $\tau_{\perp}$ varies between 10^{-11} and 10^{-13} s [39–42], which is dependent on the semiconductor in contact with the graphene and the quality of the fabricated contact.

11.4.6 Graphene Vertical Thermionic Emission at High-Energy Regime

In the case of high-temperature thermionic emission across a potential barrier of $\Phi_{B0} > 1$ eV, the emission current density is dominantly contributed by high-energy electrons residing far away from the Dirac point. In this limit, the simple Dirac cone approximation of the energy dispersion used in Eq. (11.17) and the linear electronic DOS in Eq. (11.18) are no longer applicable. Instead, a full-band tight-binding model (Fig. 11.6) is required for a more accurate description of the electronic properties of these energetic electrons and should be incorporated in the formulation of the thermionic emission current density. The high-energy vertical thermionic emission current density from graphene surface is reformulated as [43]:

$$J_{\mathrm{Gr}} = \frac{4e}{\tau_{\perp}\pi^2 t_0^2}\left[\int_{\Phi_{B0}}^{t_0} \frac{\varepsilon_{\parallel} d\varepsilon_{\parallel}}{\sqrt{\Gamma(\varepsilon_{\parallel}/t_0)}} K\left(\frac{4\varepsilon_{\parallel}/t_0}{\Gamma(\varepsilon_{\parallel}/t_0)}\right)\exp\left(-\frac{\varepsilon_{\parallel}-\varepsilon_{\mathrm{F}}}{k_{\mathrm{B}}T}\right) + \int_{t_0}^{3t_0} \frac{\varepsilon_{\parallel} d\varepsilon_{\parallel}}{\sqrt{4\varepsilon_{\parallel}/t_0}} K\left(\frac{\Gamma(\varepsilon_{\parallel}/t_0)}{4\varepsilon_{\parallel}/t_0}\right)\exp\left(-\frac{\varepsilon_{\parallel}-\varepsilon_{\mathrm{F}}}{k_{\mathrm{B}}T}\right)\right], \tag{11.34}$$

where $t_0 = 2.8$ eV, $\Gamma(x) = (1 + x)^2 - (x^2 - 1)^2/4$, and

$$K(x) = \int_0^1 \left[\left(1 - x^2\right)\left(1 - mx^2\right)\right]^{-1/2}.$$

Comparing both approaches, it is found that the Dirac cone approximation severely overestimates that emitted electrical and heat currents by over 50%, thus revealing the necessity of using the full-band thermionic emission model in Eq. (11.34) for high-energy and high-temperature electron emission from graphene, such as the thermionic emission in graphene/vacuum and the photo-thermionic emission in graphene/semiconductor [44].

11.4.7 Universal Thermionic Emission Model

Based on a generalized density of states model [30], the $k_{\parallel}$-conserving and $k_{\parallel}$-nonconserving models for vertical thermionic emission current densities from a wide array of 2D systems—including graphene and its few-layer, transition metal dichalcogenide, thin films of topological insulators, and 2D Rashba spintronic system—are found to follow the universal current–temperature scaling relation:

$$J_{2D} = AT^{\beta} \exp\left(-\frac{\Phi_{B0} - \varepsilon_F}{k_B T}\right), \tag{11.35}$$

where β varies for vertical thermionic emission with $k_{||}$-conservation, $\beta = 1$ for vertical thermionic emission without $k_{||}$-conservation, and $\beta = 3/2$ for lateral thermionic emission (Fig. 11.7). Importantly, the $\beta = 3/2$ for the lateral thermionic emission is in excellent agreement with the thermionic carrier injection measured in 2D/2D lateral Schottky contacts [45–47]. The values of β at different contact settings are given in Table 11.1.

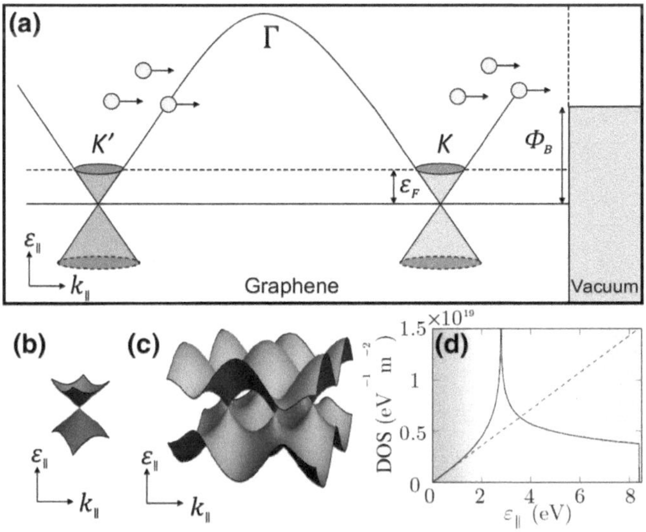

Figure 11.6 (a) Schematic drawing of the thermionic emission of high-energy electrons in graphene. (b) The energy–momentum dispersion relation at low energy can be well described by the Dirac cone approximation. (c) Full-band tight-binding energy–momentum relation of graphene. (d) Electronic density of states of the Dirac cone approximation (dashed line) and the full-band model (solid curve). For electron with energy >1 eV, the Dirac cone approximation is no longer valid, and the full-band model is required to more accurately capture the electronic properties of graphene. Copyright American Physical Society [43].

Table 11.1 Summary of the scaling exponent, β, for various 2D/2D and 2D/3D Schottky contacts

Metal	Semiconductor	Contact geometry	Momentum conservation	β
2D	3D	Vertical	Yes	Varies
2D	3D	Vertical	No	1
2D	2D	Vertical	Yes	3
2D	2D	Vertical	No	1
2D	2D	Lateral	Yes	3/2
3D	2D (not metalized)	Vertical	Yes	3
3D	2D (metalized)	Vertical	Yes	3/2
3D	2D	Lateral	Yes	3/2

Correspondingly, the electrical transport current across a 2D-material-based Schottky contact can be succinctly written as

$$J_{2D,\text{Schottky}}(V,T) = AT^{\beta} \exp\left(-\frac{\Phi_{B0} - \varepsilon_F}{k_B T}\right)\left[\exp\left(\frac{eV}{\eta k_B T}\right) - 1\right], \tag{11.36}$$

where η is the ideality factor of the Schottky contact and V is the bias voltage across the Schottky contact. Equation (11.36) can be regarded as the generalization of Eq. (11.3): *generalized Shockley diode equation* for 2D-material-based Schottky contacts.

In Table 11.1, the β scaling exponent for different 2D/2D and 2D/3D interface with various types of contact configurations are listed. It should be noted that in the case of 2D semiconductor contacted by 3D metal, the thermionic emission is dependent on whether the 2D semiconductor hybridizes with the 3D metal. In the case where 2D is hybridized and metalized by the 3D metal, the charge injection occurs in a side-way fashion from the metalized 2D material directly under the 3D metal into the "bare" 2D semiconductor (Fig. 11.7a). In this case, the scaling exponent follows $\beta = 3/2$. In contrast, when the 2D semiconductor contacted by a 3D metal does not exhibit substantial metalization, charge injection occurs in a vertical fashion, thus following $\beta = 1$ in the case of $k_{||}$-nonconserving thermionic charge injection (Fig. 11.7b).

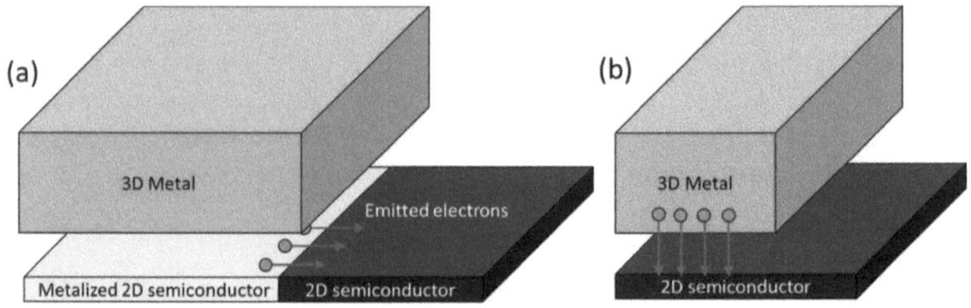

Figure 11.7 (a) Lateral thermionic electron emission in a 3D-metal/2D-semiconductor Schottky contact with metalization in the contacted 2D semiconductor. Electrons are injected sideway from the metalized 2D semiconductor into the adjacent uncontacted 2D semiconductor. (b) Vertical thermionic electron emission in a 3D-metal/2D-semiconductor Schottky contact without metalization in the contacted 2D semiconductor.

11.5 Conclusion and Outlooks

In summary, we have presented some recent models of electron thermionic emission from 2D materials. For 2D materials, the current–temperature scaling relation in the thermionic emission regime is sensitively dependent on the emission direction (i.e., lateral versus vertical) and whether the electron momentum is conserved during the emission process. The Richardson–Dushman thermionic emission is generalized into a universal form [30] [i.e., Eq. (11.35)] in which the scaling exponent, β, takes various values as listed in Table 11.1.

Finally, we discuss several models of 2D material thermionic emission not covered above before closing this chapter. The models presented above involve a charge injection time parameter, $\tau_\perp$, whose physical meaning remains debated. Recently, Trushin proposed a Heisenberg uncertainty principle model [48, 49] that interprets $\tau_\perp$ as a carrier lifetime originating from the energy–time uncertainty relation. Furthermore, a generalized thermionic emission model based on Kane's narrow band gap semiconductor model has been developed, which reveals a continuous transition between $\beta = 2$ and $\beta = 3$ as a function of the band nonparabolicity [50]. Beyond the *intrinsic* thermionic emission model discussed above, extrinsic mechanisms such as light illumination can also generate thermionic emission from 2D materials and the theoretical modeling of such *extrinsic* thermionic emission has been examined recently [51]. Phonon-assisted thermionic emission model has also been proposed and experimentally observed [52].

Despite the intensive development in uncovering the physics of thermionic emission and thermionic charge injection in vacuum and solid state graphene-based systems, the theories of electron field emission [53] and photoemission from 2D materials remain largely incomplete thus far. Recent experiments have reported the observations of field emission from graphene [54], WS_2 [55], MoS_2 [56], and the internal photoemission from graphene [57]. The development of field- and photoemission models of 2D materials to combine with the thermionic model [30] presented in this chapter will allow us to have a unified model, akin to the one formulated for bulk materials [19]. The future work in this direction will provide a comprehensive picture critically important for unearthing the unusual physics of electron emission and charge injection for 2D-material-based systems, and for the development of advanced nanoscale electron emitters, such as microemitter arrays [58], vacuum transistors [59], on-chip electron sources [29], and energy converters [60, 61].

Acknowledgments

This project is funded by a Singapore MOE Tier 2 Grant (2018-T2-1-007), A∗STAR IRG Grant (A1783c0011), U.S. Air Force Office of Scientific Research (AFOSR) through the Asian Office of Aerospace Research and Development (AOARD) under Grant No. FA2386-17-1-4020, and Office of Navy Research Global under Grant No. N62909-19-1-2047.

References

1. Jensen, K. L. (2018). *Introduction to the Physics of Electron Emission*, 1st Ed. (John Wiley & Sons Ltd).
2. Saito, Y. (2010). *Carbon Nanotube and Related Field Emitters*, 1st Ed. (Wiley-VCH).
3. Liang, S. D. (2014). *Quantum Tunneling and Field Electron Emission Theories*, 1st Ed. (World Scientific Press).

4. Richardson, O. W. (1901). On the negative radiation from hot platinum. *Proc. Camb. Philos. Soc.* **11**, pp. 286–295.
5. Dushman, S. (1923). Electron emission from metals as a function of temperature. *Phys. Rev.* **21**, pp. 623–636.
6. Shockley, W. (1949). The theory of p-n junctions in semiconductors and p-n junction transistors. *The Bell Syst. Technical J.* **28** (3), pp. 435–489.
7. Barker, R. J., and Schamiloglu, E. (2001). *High-Power Microwave Sources and Technologies*, 1st Ed. (Wiley-IEEE Press).
8. Nusinovich, G. S., Barker, R. J., Luhmann, N. C, and Booske, J. H. (2005). *Modern Microwave and Millimeter-Wave Power Electronics*, 1st Ed. (Wiley-IEEE Press).
9. Child, C. D. (1911). Discharge of hot CaO. *Phys. Rev. Ser. I* **32**, pp. 492–511.
10. Langmuir, I. (1913). The effect of space charge and residual gases on thermionic currents in high vacuum. *Phys. Rev.* **2**, pp. 450–485.
11. Zhang, P., Valfells, A., Ang, L. K., Luginsland, J. L., and Lau, Y. Y. (2017). 100 years of the physics of diodes. *Appl. Phys. Rev.* **4**, 011304.
12. Longo, R. T. (2019). A statistical mechanics derivation of the empirical asymptotic thermionic emission equation. *AIP Adv.* **9**, 085227.
13. Chernin, D., Lau, Y. Y., Petillo, P. P., Ovtchinnikov, S., Chen, D. Z., Jassem, A., Jacobs, R., Morgan, D., and Booske, J. H. (2020). Effect of nonuniform emission on Miram curves. *IEEE Trans. Plasma Sci.* **48**, pp. 146–153.
14. Fowler, R. H., and Nordheim, L. (1928). Electron emission in intense electric fields. *Proc. R. Soc. London Ser. A* **119**, pp. 173–181.
15. Murphy, E. L., and Good, R. H. (1956). Thermionic emission, field emission, and the transition region. *Phys Rev.* **102**, pp. 1464–1473.
16. Forbes, R. G. (2006). Simple good approximations for the special elliptic functions in standard Fowler-Nordheim tunneling theory for a Schottky-Nordheim barrier. *Appl. Phys. Lett.* **89**, 113122.
17. Forbes, R. G., and Deane, J. (2007). Reformulation of the standard theory of Fowler–Nordheim tunnelling and cold field electron emission. *Proc. R. Soc. London Ser. A* **463**, pp. 2907–2927.
18. Forbes, R. G. (2019). Comments on the continuing widespread and unnecessary use of a defective emission equation in field emission related literature. *J. Appl. Phys.* **126**, 210901.
19. Jensen, K. L., and Cahay, M. (2006). General thermal-field emission equation. *Appl. Phys. Lett.* **88**, 154105.

20. Zubair, M., Ang, Y. S., and Ang, L. K. (2018). Fractional Fowler-Nordheim law for field emission from rough surface with nonparabolic energy dispersion. *IEEE Trans. Electron Device* **65**, pp. 2089–2095.

21. Liang, S. D., and Chen, L. (2008). Generalized Fowler-Nordheim theory of field emission of carbon nanotubes. *Phys. Rev. Lett.* **101**, 027602.

22. Nojeh, A. (2014). Carbon nanotube electron sources: From electron beams to energy conversion and optophononics. *Int. Sch. Res. Notices* **2014**, 879827.

23. deHeer, W. A., Chatelain, A., and Ugarte, D. (1995). A carbon nanotube field-emission electron source. *Science* **270**, pp. 1179–1180.

24. Minoux, E. et al. (2005). Achieving high-current carbon nanotube emitters. *Nano Lett.* **5**, pp. 2135–2138.

25. Fairchild, S. B., Zhang, P. Park, J., Back, T. C., Marincel, D., Huang, Z., and Pasquali, M. (2019). Carbon nanotube fiber field emission array cathodes. *IEEE Trans. Plasma Sci.* **47**, pp. 2032–2038.

26. Shin, D. H., Yun, K. N., Jeon, S.-G., Kim, J.-I. Saito, Y., Milne, W. I., and Lee, C. J. (2015). High performance field emission of carbon nanotube film emitters with a triangular shape. *Carbon* **89**, pp. 404–410.

27. Behabtu, N. et al. (2013). Strong, light, multifunctional fibers of carbon nanotubes with ultrahigh conductivity. *Science* **339**, pp. 182–186.

28. Zhang, P., Fairchild, S. B., Back, T. C., and Luo, Yi. (2017). Field emission from carbon nanotube fibers in varying anode-cathode gap with the consideration of contact resistance. *AIP Adv.* **7**, 125203.

29. Wang, Y., Xiang, L., Yang, W., Li, Z., Fang, Z., Zhou, D., Liu, P., and Wei, X. (2020). High-performance on-chip thermionic electron micro-emitter arrays based on super-aligned carbon nanotube films. *Adv. Funct. Mater.* **40**, 1907814.

30. Ang, Y. S., Yang, H. Y., and Ang, L. K. (2018). Universal scaling laws in Schottky heterostructures based on two-dimensional materials. *Phys. Rev. Lett.* **121**, 056802.

31. Wei, X., Chen, Q., and Peng, L. (2013). Electron emission from a two-dimensional crystal with atomic thickness. *AIP Adv.* **3**, 042130.

32. Geim, A. K., and Novoselov, K. S. (2017). The rise of graphene. *Nat. Mater.* **6**, pp. 183–191.

33. Tongay, S. et al. (2012). Rectification at graphene-semiconductor interfaces: Zero-gap semiconductor-based diodes. *Phys. Rev. X* **2**, 011002.

34. Yang, H. et al. (2012). Graphene barristor, a triode device with a gate-controlled Schottky barrier. *Science* **336**, pp. 1140–1143.

35. Liang, S. J., and Ang, L. K. (2015). Electron thermionic emission from graphene and a thermionic energy converter. *Phys. Rev. Appl.* **3**, 014002.

36. Liang, S.-J., Liu, B., Hu, W., Zhou, K., and Ang, L. K. (2017). Thermionic energy conversion based on graphene van der Waals heterostructures. *Sci. Rep.* **7**, 46211.

37. Wang, X., Zebarjadi, M., and Esfarjani, K. (2016). First principles calculations of solid-state thermionic transport in layered van der Waals heterostructures. *Nanoscale* **8**, pp. 14695–14704.

38. Rosul, M. G. et al. (2019). Thermionic transport across gold-graphene-WSe2 van der Waals heterostructures. *Sci. Adv.* **5**, eaax7827.

39. Sinha, D., and Lee, J. U. (2014). Ideal graphene/silicon Schottky junction diodes. *Nano Lett.* **14**, pp. 4660–4664.

40. Massicotte, M. et al. (2016). Photo-thermionic effect in vertical graphene heterostructures. *Nat. Commun.* **7**, 12174.

41. Song, Y. et al. (2015). Role of interfacial oxide in high-efficiency graphene–silicon Schottky barrier solar cells. *Nano Lett.* **15**, pp. 2104–2110.

42. Xia, F., Perebeinos, V., Lin, Y., Wu, Y., and Avouris, P. (2011). The origins and limits of metal–graphene junction resistance. *Nat. Nanotechnol.* **6**, pp. 179–184.

43. Ang, Y. S., Chen, Y., Tan, C., and Ang, L. K. (2019). Generalized high-energy thermionic electron injection at graphene interface. *Phys. Rev. Appl.* **12**, 014057.

44. Schwede, J. W. et al. (2010). Photon-enhanced thermionic emission for solar concentrator systems. *Nat. Mater.* **9**, pp. 762–767.

45. Behranginia, A. et al. (2017). Direct growth of high mobility and low-noise lateral MoS_2 graphene heterostructure electronics. *Small* **13**, 1604301.

46. Guimarães, M. H. D. et al. (2016). Atomically thin Ohmic edge contacts between two-dimensional materials. *ACS Nano* **10**, pp. 6392–6399.

47. Ling, X. et al. (2016). Parallel stitching of 2D materials. *Adv. Mater.* **28**, pp. 2322–2329.

48. Trushin, M. (2018). Theory of thermionic emission from a two-dimensional conductor and its application to a graphene-semiconductor Schottky junction. *Appl. Phys. Lett.* **112**, 171109.

49. Ang, Y. S., and Ang, L. K. (2016). Current-temperature scaling for a Schottky interface with non-parabolic energy dispersion. *Phys. Rev. Appl.* **6**, 034013.

50. Javadi, M., Noroozi, A., Mazaheri, A., and Abdi, Y. (2019). Sequentially assembled graphene layers on silicon, the role of uncertainty principles in graphene–silicon Schottky junctions. *Adv. Opt. Mater.* **7**, 1900470.

51. Trushin, M. (2018). Theory of photoexcited and thermionic emission across a two-dimensional graphene-semiconductor Schottky junction. *Phys. Rev. B* **97**, 195447.

52. Wei, X., Bando, Y., and Golberg, D. (2012). Electron emission from individual graphene nanoribbons driven by internal electric field. *ACS Nano* **6**, pp. 705–711.

53. Ang, Y. S., Liang, S.-J., and Ang, L. K. (2017). Theoretical modeling of electron emission from graphene. *MRS Bull.* **42**, pp. 505–510.

54. Di Bartolomeo, A. et al. (2016). Leakage and field emission in side-gate graphene field effect transistors. *Appl. Phys. Lett.* **109**, 023510.

55. Suryawanshi, S. R., Kolhe, P. S., Rout, C. S., Late, D. J., and More, M. A. (2015). Spectral analysis of the emission current noise exhibited by few layer WS2 nanosheets emitter. *Ultramicroscopy* **149**, pp. 51–57.

56. Urban, F., Passacantando, M., Giubileo, F., Iemmo, L., and Di Bartolomeo, A. (2018). Transport and field emission properties of MoS2 bilayers. *Nanomaterials* **8**, pp. 151–160.

57. Heide, C. et al. (2020). Attosecond-fast internal photoemission. *Nat. Photonics* **14**, pp. 219–222.

58. Wu, G., Wei, X., Gao, S., Chen, Q., and Peng, L. (2016). Tunable graphene micro-emitters with fast temporal response and controllable electron emission. *Nat. Commun.* **7**, 11513.

59. Wu, G., Wei, X., Zhang, Z., Chen, Q., and Peng, L. (2015). A graphene-based vacuum transistor with a high ON/OFF current ratio. *Adv. Funct. Mater.* **25**, pp. 5972–5978.

60. Yaghoobi, P., Moghaddam, M. V., and Nojeh, A. (2011). "Heat trap": Light-induced localized heating and thermionic electron emission from carbon nanotube arrays. *Solid State Commun.* **151**, pp. 1105–1108.

61. Chang, M., Fan, H. D. E., Chowdhury, M. M., Sawatzky, G. A., and Nojeh, A. (2018). Heat localization through reduced dimensionality. *Phys. Rev. B* **98**, 155422.

Chapter 12

Direct Grown Vertically Full Aligned Carbon Nanotube Electron Emitters for X-Ray and UV Devices

Kyu Chang Park and Joonwon Lim
Department of Information Display,
Kyung Hee University, Seoul 02447, Republic of Korea
kyupark@khu.ac.kr, joonwon.lim@khu.ac.kr

Carbon nanotubes (CNTs) have been considered as a promising material for achieving enhanced electron field emission due to its fascinating properties including high electrical conductivity, high aspect ratio, thermal stability, and nanoscale tip size. Especially, vertically aligned CNT arrays are an ideal configuration for high-performance field emitters. In this chapter, we introduce vertically aligned CNT arrays, directly grown on the heavily doped silicon water. Field-emission properties of the novel high aspect ratio CNT are discussed with operating devices in diode and triode electrode configuration. As a representative application of the novel CNT field emitters, X-ray, and ultraviolet lighting for medical and environmental applications are introduced.

Nanostructured Carbon Electron Emitters and Their Applications
Edited by Yahachi Saito

ISBN 978-981-4877-62-6 (Hardcover), 978-1-003-14199-0 (eBook)
www.jennystanford.com

12.1 Introduction

A focused electron beam is used in a wide spectrum of applications, such as chemical and crystallography analyses, atomic resolution imaging, nanotechnology, and cancer treatment. Recently, the demands for high-performance field-emission (FE) electron sources are ever-increasing to reduce the overall size of the final FE devices and to extend the use of the focused electron beam into diverse applications [1–4]. FE electron sources enable improved FE efficiency as well as simple design of FE devices without additional cooling and heating units, which are necessary for conventional thermionic electron sources. This progress in the field of focused electron beam has been expanded into medical devices, germicidal irradiation, environmental applications, and so on [5–8].

To realize the efficient FE cathode, the kind of materials and the shape of cathode are highly important. Low work function materials are beneficial to extract electrons from cathode materials with small external energy, compared to high work function materials [9]. The cathode having high aspect ratio effectively intensifies the electric field at the tip of cathode materials and leads to highly efficient FE of electrons. With considering those requirements for efficient FE cathode, carbon nanotube (CNT) is one of promising candidate materials [10]. CNTs possess unprecedented electrical and physical properties as well as geometrical advantages for electron FE: (1) a high aspect ratio with a small radius curvature of the tip, which is useful to induce a high-intensity electric field for efficient FE; (2) high thermal end electrical conductivity; (3) outstanding mechanical strength; and (4) high thermal and chemical stability [12]. In addition, CNTs has moderate work function of 5.0 ± 0.1 eV. Therefore, after discoveries of CNTs and FE ability of CNT, it has been attracting much of attention for promising FE cathodes with high performance and long-term stability, enabling future FE devices such as lamp, medical devices, display devices, environmental devices, sensors, and so on [11, 13–16].

12.2 Synthesis of Vertically Aligned Carbon Nanotube Arrays

Vertically aligned CNT arrays were prepared by plasma enhanced chemical vapor deposition (PECVD) with triode configuration [17,

18]. Figure 12.1 shows a schematic diagram of the triode–PECVD system incorporating the mesh electrode. The mesh hole size and pitch were 2 and 1 mm, respectively. The growth of CNTs was carried out with the mesh grid placed 10 mm above the substrate electrode. The sample holder was kept at −600 V, while the top electrode was grounded. The 30-nm-thick Ni catalysts prepared by magnetron sputtering were patterned to yield disks of 5 μm diameters. An ammonia (NH_3) plasma exposure for 2 min was used for granulation of the Ni catalyst film prior to CNT growth. Acetylene (C_2H_2) and NH_3 were used for CNT growth, with the flow rate of 50 and 50 sccm, respectively. The total gas pressure during growth was kept at 2.66 mbar and the CNT growth time was 20–100 min. The mesh electrode was biased from +300 to −300 V and the growth temperature was maintained at 580 °C. In the triode-based PECVD system, the bias of mesh electrode significantly affects the final geometry of grown CNT arrays. The diameter of CNT was changed from 90 to 20 nm with mesh bias of +300 to −300 V, respectively (Fig. 12.2).

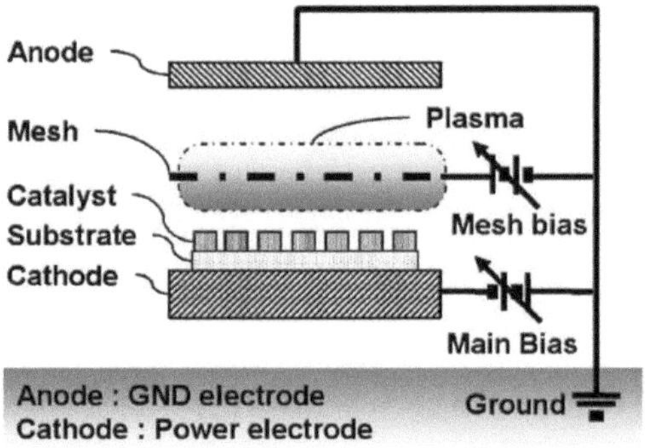

Figure 12.1 A schematic diagram of triode–PECVD system [19]. Copyright 2006, Elsevier.

Figure 12.3 shows various photographs of the plasma shape and extension, depending on the mesh electrode bias voltage. The plasma shape can be controlled by the mesh bias voltage and polarities [19]. Under positive bias, the plasma appears distributed all over inside the reactor and it is difficult to see the sheath region. However, under negative mesh bias, the isolated plasma was clearly observed. The plasma sheath thickness can be controlled by the bias voltage and polarity.

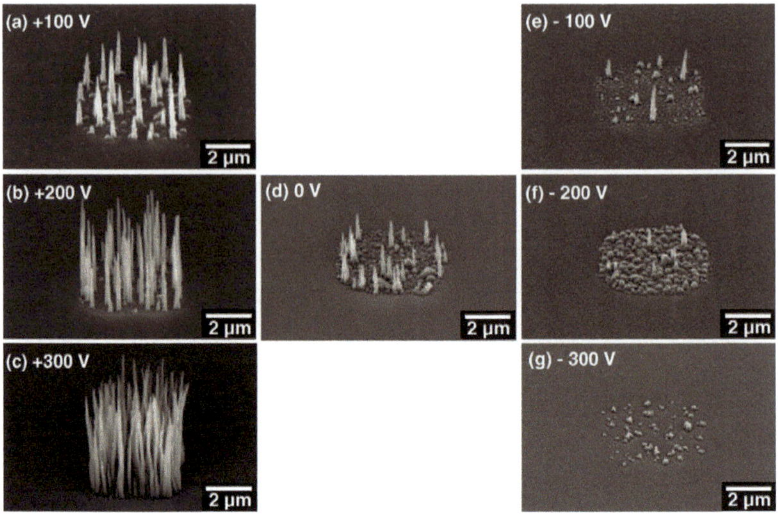

Figure 12.2 SEM images of CNTs grown with mesh bias voltages of +100 V (a), +200 V (b), +300 V (c), 0 V (d), −100 V (e), −200 V (f), and −300 V (g) [19]. Copyright 2006, Elsevier.

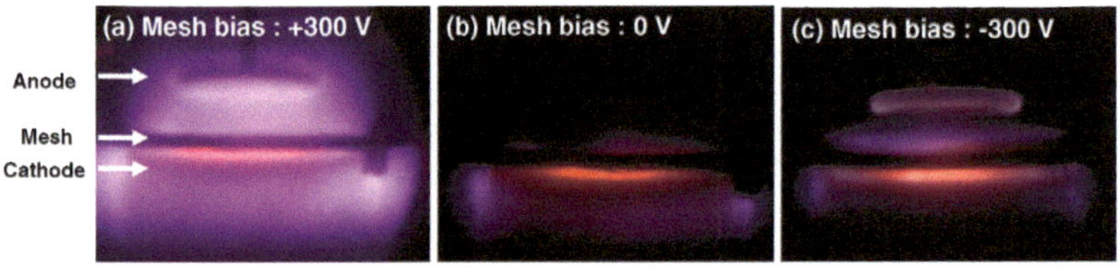

Figure 12.3 Photographs of the plasma shape mesh bias of +300 V (a), 0 V (b), and −300 V (c) [19]. Copyright 2006, Elsevier.

The plasma parameters were analyzed in the space between the sample holder and mesh electrode with the help of a Langmuir probe. The thickness of the plasma sheath can be expressed by $\sigma = 0.47\lambda_d\,(2e|V_p{-}V_d|/kT_e)^{3/4}$, where λ_d, V_p, V_d, and T_e are the Debye length, plasma potential, bias voltage, and electron temperature, respectively. With the parameters measured by a Langmuir probe under +300 V mesh bias, such as n_e = 10^{10} cm^{-3}, T_e = 2.5 eV, and V_p = 20 V, we obtain a sheath thickness of approximately 6.67 mm. Therefore, the field across the sheath at mesh bias of +300 with C_2H_2:NH_3 flow ratio of 50:50 is around 0.18 V/μm. The measured ion flux and electric field with mesh bias is depicted in Fig. 12.4.

The electric fields at mesh bias of +300, 200, 100, and 0 V were 0.18, 0.16, 0.13, and 0.08 V/μm, respectively. The ion flux is decreased by one order of magnitude when the mesh electrode bias voltage is shifted from +300 to −300 V. On the other hand, the electric field is reduced by ~50% when the bias voltage is reduced from 300 to 0 V. From these data, we can understand that the control of the CNT density and length is essentially due to the ion flux on the surface of the sample. It is also concluded that the ion flux and ion bombardment energy has to be optimized in order to obtain uniform and vertically aligned CNTs with controlled density.

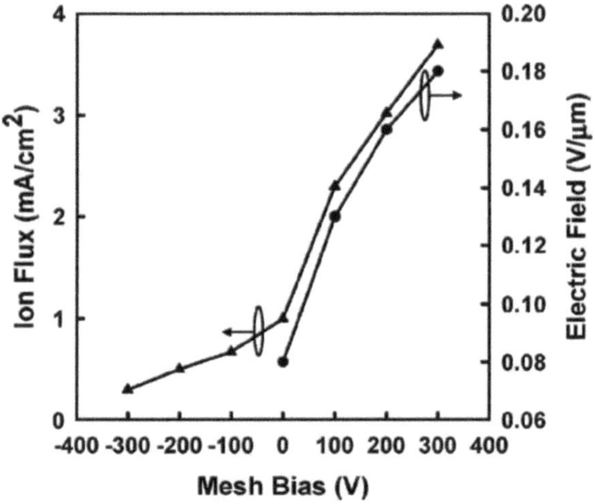

Figure 12.4 The measured ion flux and calculated electric field as a function of the mesh electrode bias voltage [19]. Copyright 2006, Elsevier.

Figure 12.5 shows the effect of mesh bias on the shape of vertically CNT arrays. Figure 12.5a,b summarizes the results concerning the length and density of CNTs grown with various mesh biases. CNTs grown with a positive bias of +300 V on the mesh electrode exhibit a length and a density of ~5 μm and ~1.5 μm^{-2}, respectively. From the above results, it is apparent that the density and length of CNTs depend on the density of positive ion species in the plasma. Negative bias on the mesh electrode can prevent positive ionic species from reaching the substrate, which results in reduced density and length. Also note that the species density in the plasma can depend on the gas flow rate, especially on C_2H_2 flow rate.

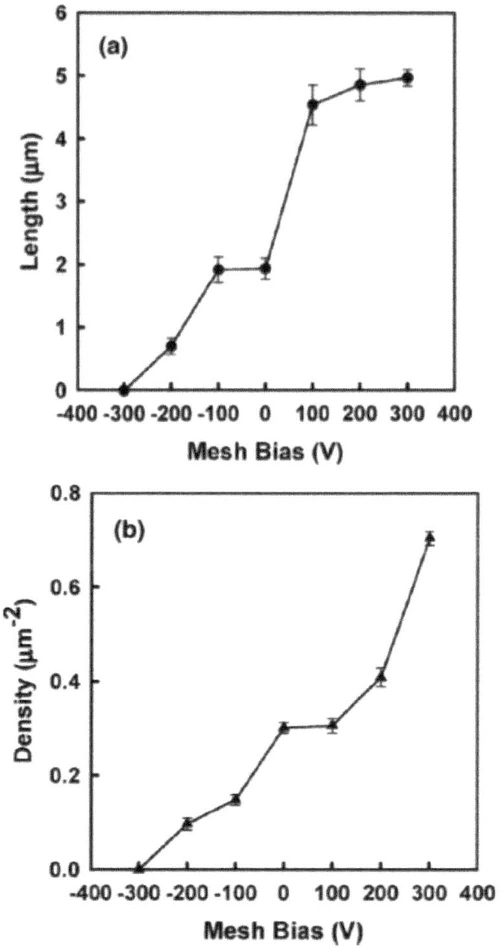

Figure 12.5 The length (a) and density (b) of CNTs grown with various mesh bias voltage [19]. Copyright 2006, Elsevier.

The triode–PECVD system is highly beneficial to adjust the condition of plasma by precise control of mesh electrode and, consequently, enables delicate control of the geometry of the final CNT arrays. The CNT density and length can be easily controlled by changing the bias and polarity of gate electrode (mesh grid). The applied voltage on gate electrode can effectively alter the CNT growth mechanism, modifying the density of ions supplied onto the substrate and the ion bombardment effects. It demonstrated the control of areal density of CNTs without resorting to sophisticated and expensive techniques such as electron beam lithography

12.3 Carbon Nanotube as a Cold Cathode

For the next generation of electron beams for vacuum nanoelectronic devices, many works have focused on the carbon nanotube cold cathode (C-beam) because of its unique properties of fast pulse driving, room temperature operation, selective emission, and low power consumption [20, 21]. To apply the cold cathode for vacuum nanoelectronic devices, structures of the electron beam for an extraction electrode have been devised by many previous works [22–25].

12.3.1 Diode-Based FE Device Structure with CNT Cold Cathodes

Diode configuration for field emission of electrons consists of a pair of anode and cathode, as shown in Fig. 12.6. This simple structure has been used to explain the theory of the Fowler–Nordheim vacuum tunneling effect of FE. This simply designed devices can extract electrons from cathode materials by simple driving, but it is not suitable to realize practical device applications due to the inherent limitation in voltage driving. To properly operate FE devices in a variety of applications, FE current should be precisely controllable, being increased or decreased in a short time period. Unfortunately, diode configuration is difficult to perform this fast driving in FE current or pulse driving. Moreover, short electron trajectory between the cathode and anode electrodes provide insufficient time to control the focus of emitted electron beam from cathode.

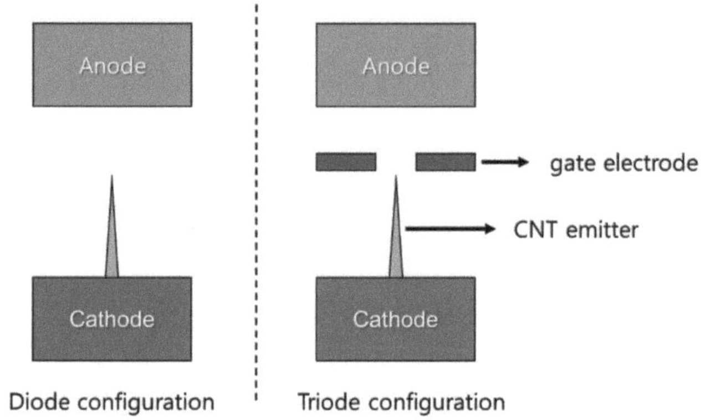

Figure 12.6 Schematic illustration of diode and triode configuration of FE devices.

12.3.2 Triode-Based FE Device Structure with CNT Cold Cathodes

After the discovery of CNTs, a triode structure, consisting of an anode, a cathode, and a gate electrode (Fig. 12.6), was devised by De Heer et al. [25]. Basically, triode-type FE devices allow precise voltage driving compared to diode-type FE devices. Third electrode between an anode and a cathode, i.e., gate electrode, is responsible for delicate and fast voltage driving of FE devices. The electric field at the tip of CNT cold cathode is controlled by adjusting the voltage applied on gate electrode while maintaining the voltage between an anode and a cathode. A film type of this triode structure has been studied for use in FE displays, whose source is equipped with selective electron emission. However, this method was not successful owing to the nonuniformity of the electron emission and the poor insulator properties and stability. To address these limitations, Kang et al. introduced a simple triode structure as an electron beam with a thick grid gate electrode. It has been used for diverse applications, such as an X-ray source [13], FE lamp [15], and handheld UV emitters [7]. With the thick grid mesh, a high emission current, good structural stability, and small focal size of electron beam have been achieved even at high driving voltage. To further improve the beam properties, key factors including electron transmission efficiency and heat degradation of mesh grid are needed to be considered for efficient and long-term stable design of triode-based C-beam devices.

12.3.2.1 The effect of alignment of CNT emitter to gate electrode

Electron transmission efficiency is a crucial parameter to determine the performance triode-based C-beam devices. Electron transmission efficiency can be calculated based on the ratio of the number of electron emitted from cathode to the number of electron reaching to anode passing through gate electrode (mesh grid). High electron transmission efficiency indicates low gate leakage currents, capable of high resolution beam, and long-term stable operation. Achieving high electron transmission efficiency of C-beam devices still remains as challenge with conventional CNT emitters made via electrophoresis, chemical vapor deposition, or screen printing processes due to their vertically nonaligned, rapid degradation during FE accompanying higher electron emission current. In addition, conventional CNT emitters show undesired leakage current of around 30% of electron emitted from cathode through the gate electrode. High current emission characteristics and low leakage current are strongly required for reliable FE operation.

Hong et al. optimized the electron transmission efficiency of carbon nanotube cold cathode electron beam (C-beam) from vertically aligned CNT arrays (Fig. 12.7) [14]. The cold cathode was aligned with the gate mesh electrode and the gap distance between the gate and cathode. The gate mesh electrode was used as photo-mask to predefine the CNT growth area for higher electron transmission efficiency in the electron emission current. The CNT emitters precisely positioned at the center of the hole to reduce the gate leakage current. The electron transmission efficiency drops in proportion with the misalignment. The electron transmission efficiency is decreased 90.2, 82.9, and 74.2% with 0, 35, and 100 μm of misalignment, respectively. It results in a lower electron emission current on the anode. High performance C-beam with more than 90% electron transmission efficiency and more than 5 mA anode current was achieved in a small area with the optimized design. The performance-enhanced C-beam devices enabled to obtain high resolution X-ray images.

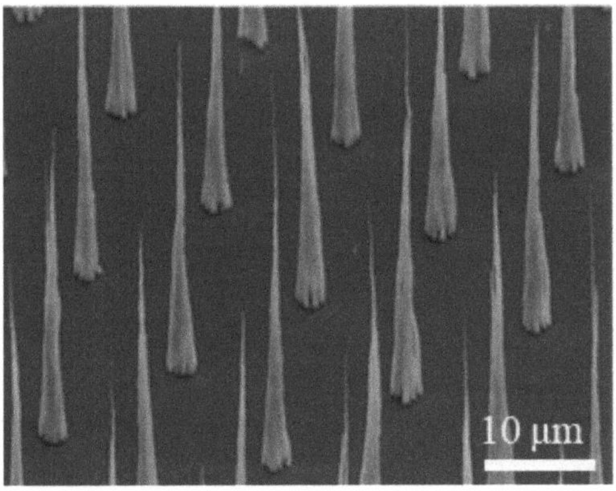

Figure 12.7 SEM images of micro-patterned vertically aligned CNT emitters array and magnified images. The vertically aligned CNTs were grown on the Si substrate by the resist-assisted patterning (RAP) process.

12.3.2.2 The effect of thermal stability of gate electrode

Kang et al. investigated the effect of thermal stability of gate electrode on vertically aligned CNT arrays [26]. Two different gate electrode made of molybdenum (Mo) and SUS304 were tested. Figure 12.8 shows mechanical degradation of Mo and SUS304 gate electrode. Mo gate electrode maintained its initial shape after field-emission test. In contrast, SUS304 gate electrode was deformed after field-emission test, giving rise to change of alignment between CNT emitters and the aperture of gate electrode. The poor thermal stability of SUS304 caused degradation in field-emission efficiency, as shown in Fig. 12.8.

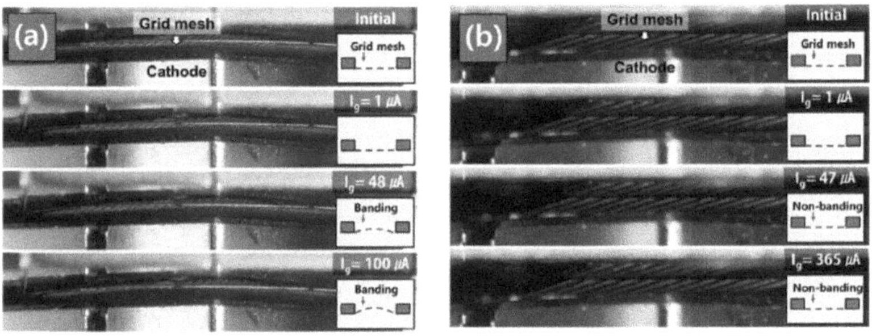

Figure 12.8 Photographs of the mesh grid during electron emission at various emission currents, showing the gap between the gate mesh and the cathode. (a) Bending SUS304 grid mesh with increasing the leakage current and (b) unbending Mo grid mesh [26]. Copyright 2018, AIP Publishing LLC.

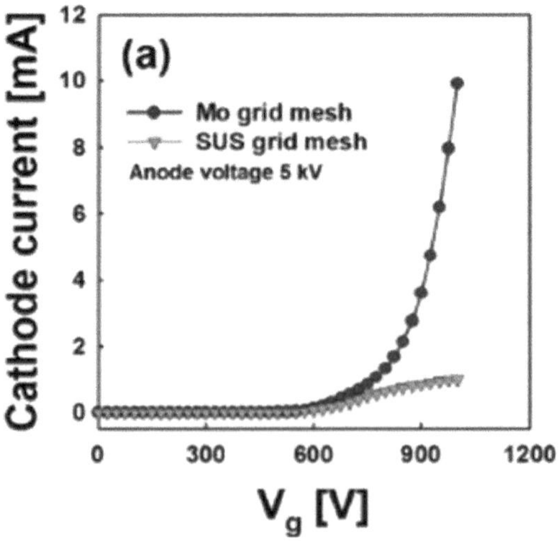

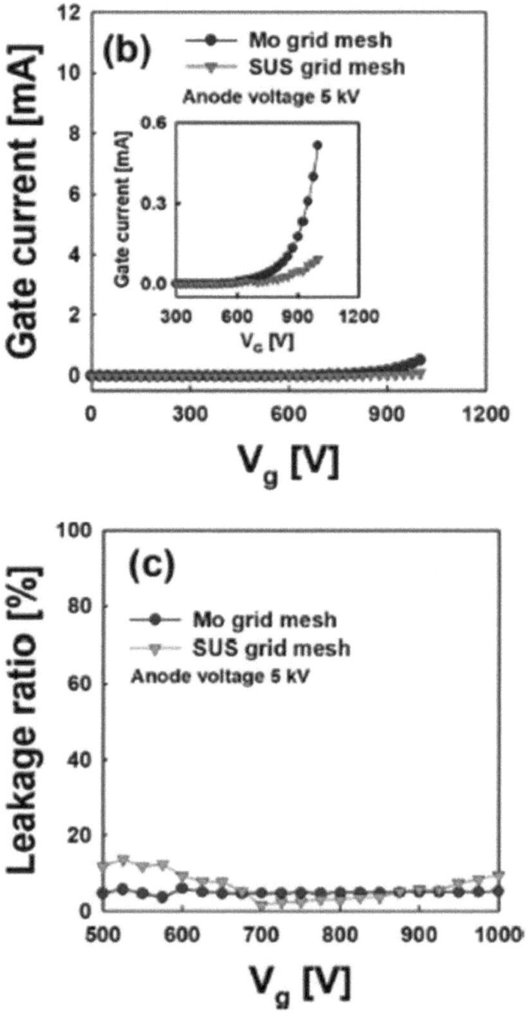

Figure 12.9 Comparison of electron emission characteristics of the Mo and SUS304 gate meshes, showing the (a) total current, (b) leakage current, and (c) leakage ratio [26]. Copyright 2018, AIP Publishing LLC.

Figure 12.9 shows a comparison of the electron emission from identical vertically aligned CNT emitters gating by the SUS and Mo grid meshes with DC driving. The Mo grid mesh improved the cathode current, I_C, by 10 times at a voltage of 5 V, compared to when the SUS grid mesh was used (Fig. 12.9a). In particular, with the SUS304 mesh the I_C and anode current (I_A) were 0.97 mA (J = 23.6 mA/cm^2) and 0.88 mA (J = 21.4 mA/cm^2), respectively, while with the Mo grid mesh the I_C and I_A were 9.9 mA (J = 241.4 mA/cm^2) and 9.38 mA (J = 228.7 mA/cm^2), respectively. The threshold voltage and gate bias at a 10 mA anode current with the Mo mesh in triode mode were only 75 and 210 V higher than that in the diode mode, respectively. With the Mo grid mesh, we obtained very similar emission currents for the triode and diode modes, although a slight difference was found in the threshold voltage (V_{th}) of the two modes. The leakage current as a function of the grid mesh is shown in Fig. 12.9b, where the total leakage current with a Mo grid mesh is higher than the cathode current with the SUS304 grid owing to the tenfold electron emission current exists with the Mo gate mesh. However, though the leakage current with the Mo mesh is higher than that with the SUS304 grid mesh, the Mo mesh exhibits nonsaturation of the anode current. It indicates that the Mo mesh is more robust with thermal load via Joule heating than mesh of SUS304. Figure 12.9c shows the ratio of leakage current ($I_G/I_C \times 100$) as a function of the gate voltage. Mo grid mesh shows lower level of leakage current than that of SUS304 grid mesh. The fluctuation in the ratio of leakage current (standard deviation/average × 100) is small with Mo grid mesh (17%) compared to SUS304 grid mesh (73%). As a result, Mo grid mesh leads to more stable and efficient electron emission from vertically aligned CNT arrays. The results indicate that thermal property of gate electrode should be considered to guarantee uniform and long-term stable C-beam device.

12.4 X-Ray Imaging with C-Beam

Research into the development of alternative X-ray sources has been ongoing since X-ray source technology was first developed more than a century ago, but conventional thermionic emission electron sources are still used in most commercial X-ray sources. CNTs, which have unique structures and characteristics, have been highlighted as potential replacements for conventional X-ray sources [27, 28]. CNTs have been studied by many groups for X-ray source development using a variety of CNT cathode fabrication techniques, including electrophoresis, chemical vapor deposition, and the use of CNT pastes. CNTs, as a promising candidate for next-generation X-ray source material, have been attracting research attention in the field of inspection, security, and medical applications. Conventional thermionic emission electron sources have used metal filaments since the last century. A tungsten filament emits electrons when heated to more than 1000 °C. While thermionic X-ray tubes have certain inherent limitations, CNTs offer advantages including reduced power consumption, faster response times, room temperature driving capability, and narrow electron beam energy dispersion.

Hong et al. reported C-beam-based high-performance X-ray tubes with vertically aligned CNT emitters [13, 14]. The grown CNT emitters were attached to the cathode carrier via Al–Si eutectic bonding, as shown in Fig. 12.10. The gate mesh shown in Fig. 12.10b is formed from rounded square shapes and has an aperture ratio of 67%. The electron beam module was clamped and welded to fix it to the tube for alignment between the CNT field emitter and the gate mesh. The size of the assembled module is suitable for use in a compact system with a diameter of 16 mm and a height of 13 mm. The size is suitable for fitting in a compact glass-sealed X-ray tube with an operating voltage of 70 kV and currents of a few milliamperes.

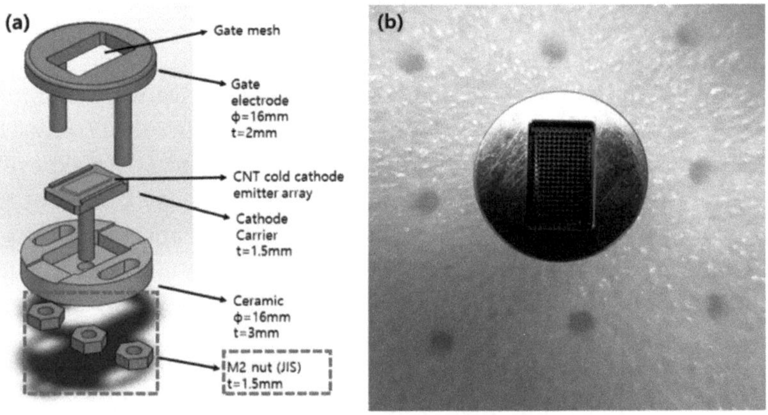

Figure 12.10 Schematic diagram and image of the fabricated compact C-beam module. (a) Schematic of the compact C-beam module comprising a gate, a cathode, and an insulator (ceramic base). (b) Optical image of the gate mesh.

The glass-sealed X-ray tube had a conventional design and comprised a tungsten reflection anode, an evaporation getter and vacuum-sealed glass tubing. As shown in Fig. 12.11, the size of this glass-sealed X-ray tube was reduced to the commercial dental X-ray tube size of 67 × 29 mm (length × diameter). The electron beam module was annealed for degassing purposes before its installation to the X-ray module. The X-ray anode is set at a take-off angle of 9° and a distance of 10 mm from the electron beam module. The tungsten target is inserted into oxygen-free copper.

Figure 12.11 Schematic and image of fabricated glass-sealed X-ray tube. Photograph of the glass-sealed X-ray tube with the C-beam module (left). Cross-sectional view of the X-ray tube. The anode is composed of tungsten-inserted copper with an anode angle of 9° (right) [13]. Copyright 2018, AIP Publishing LLC.

X-ray images of the commercial USB flash drive were obtained from the first glass-sealed X-ray tube, as shown in Fig. 12.12a. The X-ray image was taken with a current of 0.6 mA and a 0.5 s exposure time (0.3 mAs) at 65 kV. Figure 12.12c shows a line profile of the magnified X-ray image shown in Fig. 12.12b. The line profile shows a resolution of 5 lp/mm. We were also able to distinguish a 100 μm metal wire in the USB flash drive. The size of the X-ray sensor is limited (and is normally less than 2.6 × 3.6 cm). In previous X-ray tubes, allowances had to be made for low voltages and prolonged exposure times, but high voltage operation is available in this case and the images can be acquired with short exposure times. Figure 12.13d shows an X-ray image of a human finger. The X-ray irradiation conditions were an anode current of 0.6 mA and an exposure time of 0.5 s at 65 kV. The X-ray absorption image from the low vacuum glass-sealed tube clearly shows the bone structure, the edge and the soft tissue. The X-ray image of the finger shows a

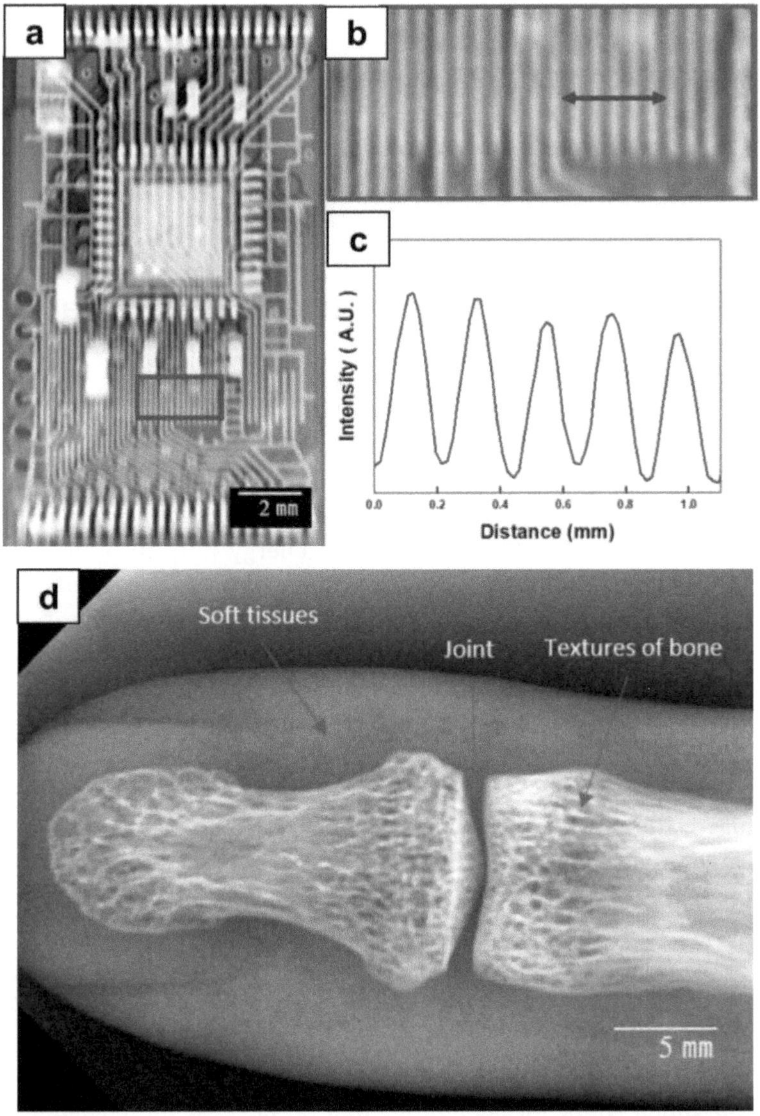

Figure 12.12 (a) X-ray image of the commercial USB flash drive. (b) Magnified image for resolution testing in the square region shown in (a). (c) Line profile at the blue arrow over the distance of 1 mm. (d) X-ray image of the human index finger. The exposure conditions were 0.3 mAs at an anode voltage of 65 kV [13]. Copyright 2018, AIP Publishing LLC.

clear bone structure that can easily be distinguished from the soft-tissue portion of the finger. X-ray images taken from a high-vacuum-pressure (~10^{-8} Torr) closed system using CNT X-ray sources with a beryllium (Be) window show similar images of the bone structure. The resulting X-ray image demonstrates the applicability of the glass sealed X-ray tube with the electron source composed of CNT emitters for use as a bone densitometer.

12.5 UV Irradiative Applications with C-Beam

Ultraviolet (UV) is one component of solar radiation, having a wavelength in the range of 10–400 nm. There are four types of UV radiations according to its wavelengths: UVA (400–320 nm), UVB (320–280 nm), UVC (280–100 nm), EUV (100–10 nm). UVA has been used for chemical reaction including UV curing and organic decomposition due to its low photon energy [29]. UVB has been applied to phototherapy and plant lighting. UVC, which has relatively high photon energy, has been widely used for diverse applications, such as optical cleaning, disinfection, and deodorization. The importance of UV germicidal irradiation (UVGI) is ever-increasing for preventing next generation from the pollution of air and water [30]. Irradiation of UVC effectively prevent reproduction of microorganisms by changing their nucleic acids. The rate of inactivation of microorganisms is dependent on the dose of UVC irradiation defined as the product of the UV irradiation and exposure time.

Conventional mercury-based UV sources (wavelength of 254 nm) for UVGI are harmful to the human body. Accordingly, the conventional UVGI system is currently recommended to use in space without people. According to the Minamata Convention on Mercury, signed in 2013, manufacturing and trading of mercury-containing lamps for general lighting purposes is disallowed after 2020 to reduce and eliminate the adverse effects of mercury on human health and the environment. The alternatives for the harmful UV sources, enabling the use of UVGI without space restriction, have been recently studied [29]. The principle for the harmless UV sources is based on different penetration depth of UVC (wavelength 200–230 nm) into human skins and biological samples. Typically,

UVC in the range of the wavelength is easy to be blocked by human cells of size about 10–25 μm. In contrast, microorganisms less than 1 μm in diameter can be easily penetrated by this light and damaged by UVC.

As a promising candidate for UV sources, EB-pumped UV light sources have been researched and developed over the last 10 years. In 2009, Watanabe et al. explored EB pumping on hexagonal boron nitride powder to obtain far UVC light with a wavelength of 225 nm and a power conversion efficiency (PCE) of 0.6% [31]. This PCE with EB pumping technique is higher than 0.2%, which is the external quantum efficiency (EQE) of UVC-LEDs at wavelengths of 226 and 227 nm. Oto et al. reported an EB pumping technique in 2010 using AlGaN MQWs with an optical output power of 100 mW and a PCE of 40% at a wavelength of 238 nm [32]. In 2016, the Palo Alto Group achieved an optical output power of 230 mW at a wavelength of 246 nm using EB pumping technology and AlGaN MQWs [33]. Yoo et al. also found that Zn_2SiO_4 pumped with C-beam, that is via CNT emitters as electron source, obtained far UVC light at wavelengths of 226 nm and that CNT emitters could be used as excitation sources to generate far UVC light [7].

Figure 12.13a shows the schematic diagram of deep UV generation with C-beam pumping. The anode has a quartz substrate, UVC layer, and an aluminum (Al) electrode. We used a quartz substrate with a length of 40 mm and a width of 20 mm. The UVC layer contains Zn_2SiO_4 as a layer for generating deep UV by electron beam irradiation. The Zn_2SiO_4 was fabricated by spin-coating and annealing ZnO ink solution on a quartz substrate. Zinc acetate dihydrate (ZAD), monoethanolamine, and 2-methoxyethanol were stirred at 70 °C for 5 h to prepare a ZnO ink solution. The ZAD concentration of the ZnO ink solution is 0.3 mol. In the annealing process, nitrogen (N_2) gas was injected into the furnace and the process was performed at 1,000 °C for 1 h. Zn_2SiO_4 was formed after the annealing process and Al electrode was deposited with the Al thermal evaporation technique to form an anode electrode. In order to improve the deep UV light intensity, SiO_x and SiN_x were deposited on the quartz with a thickness of 100 nm using PECVD.

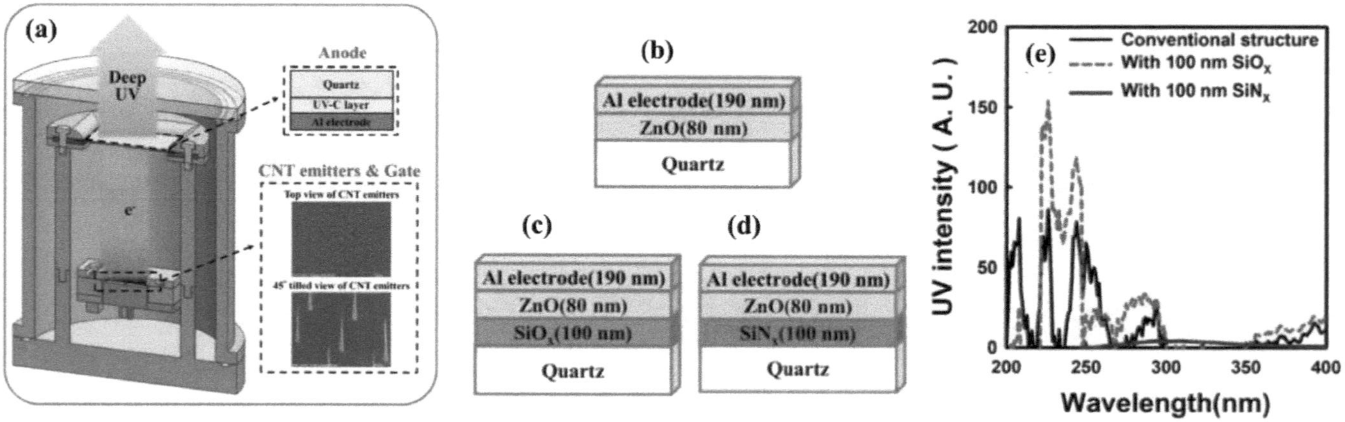

Figure 12.13 (a) Schematic diagram of deep UV generation [Inset: cross-sectional view of anode structure (top) and the SEM image of CNT emitters (bottom)]. UV performance with variable anode structure: (b) conventional structure, (c) with 100 nm SiO_x, and (d) with 100 nm SiN_x. (b), (c), and (d) are structures before annealing at 1000 °C. (e) The UV spectra with various anode structures after annealing process [7].

Figure 12.13b–e shows various anode structures before annealing process and UV spectra depending on the anode structures. The anode structure of Fig. 12.13b is a conventional structure, which is a ZnO ink solution spin-coated onto a quartz substrate. In the case of Fig. 12.13c,d, SiO_x and SiN_x films were deposited on quartz substrate with a thickness of 100 nm and then the ZnO ink solution was spin-coated on the films. For structural modification, annealing was carried out at 1000 °C for 1 h after injection of N_2 gas and maintaining a pressure of 400 mTorr. After annealing, we obtained Zn_2SiO_4 anode film. Figure 12.13e shows the deep UV spectra obtained with C-beam pumping with V_a of 7 kV and I_a of 0.3 mA. The UV spectra depend on the films used on the quartz substrate. When we used the SiO_x layer, a deep UV of 226 and 244 nm wavelength was obtained. The UV intensity at a wavelength of 208 nm is lower than without the SiO_x layer. However, at 226 and 244 nm, the deep UV intensities were increased by 1.8 and 1.5 times, respectively, compared to the conventional structure of Fig. 12.13b. When 100 nm SiN_x was used, the peak wavelength was observed at 308 nm, but the intensity was weak. In addition, UV light of wavelengths of 250 nm or less disappeared. This is related to the bandgap energy. The bandgap energies of crystalline silicon nitride (Si_3N_4) and silicon dioxide (SiO_2) are ~5 and ~9 eV, respectively. Using SiN_x, the generated deep UV is absorbed by Si_3N_4 existing in SiN_x because the bandgap energy is smaller than that of Zn_2SiO_4. However, in the case of SiO_x layer, the bandgap energy is larger than that of Zn_2SiO_4, so deep UV is not absorbed.

12.6 Summary

Hot cathodes based on thermionic emission of electrons have suffered from insufficient efficiency and limitation of miniaturization, arising from inherently limited efficiency of its mechanism and the necessity of additional units for cooling and heating processes. Clearly, vertically aligned CNT-based cold cathode, i.e., C-beam, offers unprecedented opportunities for the high performance vacuum devices based on FE of electrons to expand their applicable field. The physical, electrical, and geometrical features of CNTs addresses remained challenges of conventional cold cathode materials.

Moreover, the progress of CNT growth technique including resist-assisted process enables position-selective synthesis of vertically aligned individual CNT strand. It allows us to obtain the intensified electric field at the tip of individual CNT strand by adjusting the areal density and the aspect ratio of CNT, and consequently leads to highly efficient FE of electrons from cathode materials. In the view of device design, the position-controlled CNT arrays are highly required for high-performance triode-based FE devices. In this triode configuration, alignment between CNT emitters and the aperture of gated electrode is essential factors to determine the FE efficiency as well as reliable operation of FE devices. Triode configuration for FE devices enables fast, precise, and reliable driving of FE devices, addressing limitations of diode-type FE devices. For the further improvement in the performance of FE devices, researches on efficient device design and materials for cathode/gate electrode for triode-type devices should continue.

Addressing all the aforementioned challenges and improving further require global synergistic efforts, both from the CNT community and the FE community to ensure innovative growth in this field. Improving the basic knowledge with the support from the two complementary disciplines can potentially lighten unforeseen possibilities in the interdisciplinary subject of CNT-based FE research in the near future.

References

1. Basu, A., Swanwick, M. E., Fomani, A. A. and Velásquez-García, L. F. (2015). A portable X-ray source with a nanostructured Pt-coated silicon field emission cathode for absorption imaging of low-Z materials, *J. Phys. D: Appl. Phys.*, **48**, 225501.
2. Evans-Nguyen, T., Parker, C. B., Hammock, C., Monica, A. H., Adams, E., Becker, L., Glass, J. T. and Cotter, R. J. (2011). Carbon nanotube electron ionization source for portable mass spectrometry, *Anal. Chem.*, **83**, pp. 6527–6531.
3. Kim, H. J., Ha, J. M., Heo, S. H. and Cho, S. O. (2012). Small-sized flat-tip CNT emitters for miniaturized X-ray tubes, *J. Nanomater.*, **2012**, 854602.
4. Kim, J.-W., Jeong, J.-W., Kang, J.-T., Choi, S., Park, S., Shin, M.-S., Ahn, S. and Song, Y.-H. (2015). Great improvement in adhesion and uniformity

of carbon nanotube field emitters through reactive nanometer-scale SiC fillers, *Carbon*, **82**, pp. 245–253.

5. Yoo, S. T., Lee, H. I. and Park, K. C. (2019). 363 nm UVA light generation with carbon nanotube electron emitters, *Microelectron. Eng.*, **218**, 111142.
6. Yoo, S. T., Hong, J. H., Kang, J. S. and Park, K. C. (2018). Deep-ultraviolet light source with a carbon nanotube cold-cathode electron beam, *J. Vac. Sci. Technol. B*, **36**, 02C103.
7. Yoo, S. T., Lee, H. I. and Park, K. C. (2019). Optimization of Zn_2SiO_4 anode structure for deep ultraviolet generation with carbon nanotube emitters, *IEEE J. Electron Devices Soc.*, **7**, pp. 735–739.
8. Sugie, H., Tanemura, M., Filip, V., Iwata, K., Takahashi, K. and Okuyama, F. (2001). Carbon nanotubes as electron source in an x-ray tube, *Appl. Phys. Lett.*, **78**, pp. 2578–2580.
9. Ilie, A., Hart, A., Flewitt, A., Robertson, J. and Milne, W. (2000). Effect of work function and surface microstructure on field emission of tetrahedral amorphous carbon, *J. Appl. Phys.*, **88**, pp. 6002–6010.
10. Saito, Y., Hamaguchi, K., Uemura, S., Uchida, K., Tasaka, Y., Ikazaki, F., Yumura, M., Kasuya, A. and Nishina, Y. (1998). Field emission from multi-walled carbon nanotubes and its application to electron tubes, *Appl. Phys. A*, **67**, pp. 95–100.
11. Saito, Y. and Uemura, S. (2000). Field emission from carbon nanotubes and its application to electron sources, *Carbon*, **38**, pp. 169–182.
12. Gröning, O., Küttel, O., Emmenegger, C., Gröning, P. and Schlapbach, L. (2000). Field emission properties of carbon nanotubes, *J. Vac. Sci. Technol. B*, **18**, pp. 665–678.
13. Hong, J. H., Kang, J. S. and Park, K. C. (2018). Fabrication of a compact glass-sealed x-ray tube with carbon nanotube cold cathode for high-resolution imaging, *J. Vac. Sci. Technol. B*, **36**, 02C109.
14. Hong, J. H., Kang, J. S. and Park, K. C. (2017). High electron transmission coefficient on carbon nanotube emitters for X-ray sources, *J. Nanosci. Nanotechnol.*, **17**, pp. 7200–7204.
15. Bae, N. Y., Bae, W. M., Ha, A. N., Nakamoto, M., Jang, J. and Park, K. C. (2011). Low-voltage driven carbon nanotube field emission lamp, *Curr. Appl. Phys.*, **11**, pp. S86–S89.
16. De Heer, W. A., Chatelain, A. and Ugarte, D. (1995). A carbon nanotube field-emission electron source, *Science*, **270**, pp. 1179–1180.
17. Lim, S. H., Yoon, H. S., Moon, J. H., Park, K. C. and Jang, J. (2006). Optical emission spectroscopy study for optimization of carbon nanotubes

growth by a triode plasma chemical vapor deposition, *Appl. Phys. Lett.*, **88**, 033114.

18. Park, K. C., Ryu, J. H., Kim, K. S., Yu, Y. Y. and Jang, J. (2007). Growth of carbon nanotubes with resist-assisted patterning process, *J. Vac. Sci. Technol. B*, **25**, pp. 1261–1264.
19. Lim, S. H., Park, K. C., Moon, J. H., Yoon, H. S., Pribat, D., Bonnassieux, Y. and Jang, J. (2006). Controlled density of vertically aligned carbon nanotubes in a triode plasma chemical vapor deposition system, *Thin Solid Films*, **515**, pp. 1380–1384.
20. Calderón-Colón, X., Geng, H., Gao, B., An, L., Cao, G. and Zhou, O. (2009). A carbon nanotube field emission cathode with high current density and long-term stability, *Nanotechnology*, **20**, 325707.
21. Zhou, O., Shimoda, H., Gao, B., Oh, S., Fleming, L. and Yue, G. (2002). Materials science of carbon nanotubes: fabrication, integration, and properties of macroscopic structures of carbon nanotubes, *Acc. Chem. Res.*, **35**, pp. 1045–1053.
22. Jung, Y., Son, G., Park, J., Kim, Y., Berdinsky, A. S., Yoo, J. and Park, C. (2005). Fabrication and properties of under-gated triode with CNT emitter for flat lamp, *Diam. Relat. Mater.*, **14**, pp. 2109–2112.
23. Zhang, Y., Wu, C., Lin, J., Lin, Z. and Guo, T. (2011). An improved planar-gate triode with CNTs field emitters by electrophoretic deposition, *Appl. Surf. Sci.*, **257**, pp. 3259–3264.
24. Saito, Y. (2010). *Carbon Nanotube and Related Field Emitters: Fundamentals and Applications*, John Wiley & Sons.
25. Baughman, R. H., Zakhidov, A. A. and De Heer, W. A. (2002). Carbon nanotubes—The route toward applications, *Science*, **297**, pp. 787–792.
26. Kang, J. S., Hong, J. H. and Park, K. C. (2018). High-performance carbon-nanotube-based cold cathode electron beam with low-thermal-expansion gate electrode, *J. Vac. Sci. Technol. B*, **36**, 02C104.
27. Kawakita, K., Hata, K., Sato, H. and Saito, Y. (2006). Development of microfocused X-ray source by using carbon nanotube field emitter, *J. Vac. Sci. Technol. B*, **24**, pp. 950–952.
28. Zhou, O., Calderon-Colon, X. and Saito, Y. (2010). *Carbon Nanotube-based Field Emission X-ray Technology, Carbon Nanotube and Related Field Emitters: Fundamentals and Applications*, Wiley Online Library, pp. 417–438.
29. Kneissl, M., Seong, T.-Y., Han, J. and Amano, H. (2019). The emergence and prospects of deep-ultraviolet light-emitting diode technologies, *Nat. Photonics*, **13**, pp. 233–244.

30. Kowalski, W. (2009). *Ultraviolet Germicidal Irradiation Handbook: UVGI for Air and Surface Disinfection*, Springer.

31. Watanabe, K., Taniguchi, T., Niiyama, T., Miya, K. and Taniguchi, M. (2009). Far-ultraviolet plane-emission handheld device based on hexagonal boron nitride, *Nat. Photonics*, **3**, pp. 591–594.

32. Oto, T., Banal, R. G., Kataoka, K., Funato, M. and Kawakami, Y. (2010). 100 mW deep-ultraviolet emission from aluminium-nitride-based quantum wells pumped by an electron beam, *Nat. Photonics*, **4**, pp. 767–770.

33. Northrup, J. E., Chua, C., Kneissl, M., Wunderer, T. and Johnson, N. M. (2016). Ultraviolet light emitting devices having compressively strained light emitting layer for enhanced light extraction, U.S. Patent 9252329B2.

Chapter 13

Development of CNT X-Ray Technology for Medical and Dental Imaging

Christina Inscoe, Yueh Z. Lee, Jianping Lu, and Otto Zhou
Department of Physics & Astronomy, Department of Radiology, University of North Carolina at Chapel Hill, 120 E. Cameron Avenue, Chapel Hill, NC 27599, USA
credmon@email.unc.edu

The excellent field-emission capability of the carbon nanotube (CNT) has enabled a new method of X-ray generation. Through 20 years of research and development, CNT-based field-emission X-ray sources have been successfully developed and deployed in a range of novel X-ray imaging systems. This chapter introduces the CNT X-ray research in our lab at the UNC-Chapel Hill for medical and dental imaging applications.

13.1 Introduction

The remarkable field-emission attributes of the carbon nanotube (CNT) and fabrication techniques for robust cathodes have been

---

Nanostructured Carbon Electron Emitters and Their Applications
Edited by Yahachi Saito

ISBN 978-981-4877-62-6 (Hardcover), 978-1-003-14199-0 (eBook)
www.jennystanford.com

detailed in previous chapters. This chapter introduces our work at the University of North Carolina (UNC) on developing CNT X-ray sources and X-ray imaging devices.

13.1.1 Conventional Thermionic X-Ray

The X-ray was discovered in 1895 by Wilhelm Roentgen during experimentation with a gas discharge tube [39]. The value for medical and dental diagnosis was immediately obvious, with over 1000 related scholarly articles published within the first year [35]. The first milestone in X-ray production was by William Coolidge in 1913, upon fabrication of a purpose-built tube containing a tungsten thermionic filament [10]. When resistively heated, the filament produced a well-regulated electron cloud and eliminated several challenges to tube use [3]. A conventional X-ray tube is a vacuum envelope containing the filament and a positively charged metal target, or anode, which emits X-ray radiation when bombarded with electrons. The size of the bombarded region, or focal spot, dictates the ultimate resolution capability of the source. The focal spot size (FSS) must be precisely controlled: if too large, image resolution will be poor, but if too small, high energy density can melt the anode.

In the past century, X-ray imaging has been extensively used in scientific and industrial applications and been an integral component of medicine and dentistry. Two-dimensional (2D) and three-dimensional (3D) X-ray imaging modalities are used for both screening and diagnosis of a wide variety of pathoses, from identifying dental caries and bone fractures to diagnosing stroke and cancer. 3D imaging, or computed tomography (CT), is performed by collecting a series of 2D images and mathematically constructing slice images through the patient. In the past 50 years, CT hardware has evolved rapidly, driving innovations in X-ray tube fabrication for increased flux, spatial and temporal resolution, and tube reliability.

Implementation of sophisticated electron optics and thermal management techniques has been required to meet the ever-increasing demands of modern diagnostic devices. Attempts have been made to improve upon the thermionic emitter, including variations in geometrical construction and composition [3, 25], but it has remained largely unchanged over the past 100 years.

13.1.2 Field-Emission X-Ray

Field emission has been viewed as an attractive alternative to thermionic emission for electron production in many devices, including displays, electron microscopes, and X-ray tubes [28]. The behavior of these emitters is described by the Fowler–Nordheim equation, for which the material should have a low work function or/and high field enhancement factor to reduce the extraction field [12]. Considered a "cold" cathode, the field emitter requires only application of a moderate voltage to a proximal electrode, rather than high temperature, for electron production. Thermal management of the cathode in thermionic X-ray sources is a fundamental design consideration, and elimination of this issue is one primary benefit of field-emission X-ray.

Although the field emitter consumes much less power than its thermionic counterpart, some thermal energy may be generated. In some configurations, a large percentage of the current can be consumed by other electrodes instead of impinging upon the anode. Overall, most of the energy is dissipated by the heating of the anode, and it remains a limitation in field-emission X-ray, just as it is for thermionic X-ray production.

Another benefit to field-emission X-ray is the ability to instantaneously switch the cathode, therefore the X-ray beam, at speeds limited only by the control electronics. The beam can be pulsed at high frequency and triggered with physiological signals. In contrast, thermionic sources require a finite heating and cooling period and are not as well-suited for those applications.

Investigations into metal tip [1, 17] and Spindt-type field emitters [43] for X-ray generation have been performed, but material interactions, such as sputtering and degradation within the tube, have hindered their utilization in medical imaging [3].

13.1.3 CNT Field-Emission X-Ray

The discovery of the carbon nanotube by Sumio Iijima in 1991 [21] sent researchers around the world into a frenzy to synthesize and find novel applications for this unusual new structure. The small diameter and relatively long length give CNTs a very high aspect ratio and atomically sharp tip. These properties yield a large field enhancement factor when voltage is applied, and make CNTs excellent field emitters [56].

Early reports of field-emission X-ray imaging using CNT cathodes showed promise, though results varied [44, 51, 53, 54]. Some studies revealed low current and short cathode lifetime in which poor adhesion, oxidation, and sputtering of the CNT film limited the performance. Other studies revealed higher current and long cathode lifetime. Twenty years have passed since these first reports of CNT field-emission X-ray, and the CNT cathode is now capable of delivering high current density and lifetime comparable to conventional thermionic emitters. Many academic and industrial research groups have since turned their attention to development of this technology. Various nanomaterials have been fabricated for use in field-emission X-ray sources but at this time none have achieved the current densities, lifetime, and reliability of the CNT emitter.

13.2 CNT X-Ray Devices

13.2.1 CNT Cathode

Development of robust CNT cathodes with high current density for X-ray production was initially the primary objective of our team at UNC [54, 55]. Prototype cathodes were fabricated and multiple deposition methods were explored, such as screen printing, spin coating [4], direct CVD [48], and electrophoretic deposition (EPD) [14, 52]. Although electron emission from a CNT-based cathode could be readily achieved by all of the methods, the performance varied widely. Refinement of the raw materials and fabrication techniques were required to generate cathodes with the performance and stability required for medical imaging [7]. Table 13.1 provides examples of cathode performance using the EPD process at UNC.

The CNT fabrication method, which ultimately determines the structure, single- or multiwall, length, as well as purity, including any catalyst particles, was critical to obtaining robust cathodes. Our team at UNC refined the EPD techniques using few-walled CNTs produced by thermal CVD in a method developed by Jie Liu's team at Duke University [36]. An alcohol-based suspension was made by combining the CNTs with a charger salt, $MgCl_2$, and glass particles which would serve as a binder [34]. Insulating substrates, such as glass or quartz, were patterned with conductive deposition regions and interconnects using photolithography. EPD was used to deposit the CNT ink onto the prepared substrate. The photoresist was removed, leaving the desired cathode pattern, but with low adhesion of the CNT ink to the surface. Substrates were annealed at high temperature causing the glass binder particles to fuse together to form a matrix with the CNTs partially bound at the ends.

Table 13.1 Examples of CNT cathode performance fabricated by the electrophoretic deposition at UNC

Area (mm^2)	Current (mA)	Current density (mA/cm^2)	Mode
2 × 15 mm [56]	1000	3333	Diode
0.08 [7]		~1400	Diode
0.92 [7]		~400	Diode
32.5 [15]	41	~130	Triode (30 kVp)

13.2.2 Single-Beam CNT X-Ray Source

The basic design of a single-beam CNT field-emission X-ray source is similar to that of a conventional thermionic X-ray tube. A typical CNT source comprises the cathode, an extracting grid electrode (heretofore referred to as the gate), a focusing structure, and a tungsten anode in a vacuum housing [54]. Maintaining a high vacuum level is critical to ensure long lifetime of the CNT emitters because they are more susceptible to oxidation compared to the W filament/film used in the conventional X-ray tubes. Active or passive pumping is often employed, depending on the power output of the tube.

To generate X-rays, a positive DC voltage is applied to the anode, in our cases, in the unipolar mode. Electrons are extracted from the cathode via application of a bias voltage, typically several V/um, between the cathode and gate. In a triode configuration, approximately 70% of the electrons are transmitted through the gate electrode toward the focusing structure and anode. The focusing lens shapes the electron beam in preparation for bombardment of the anode, to ensure desired FSS (Fig. 13.1a).

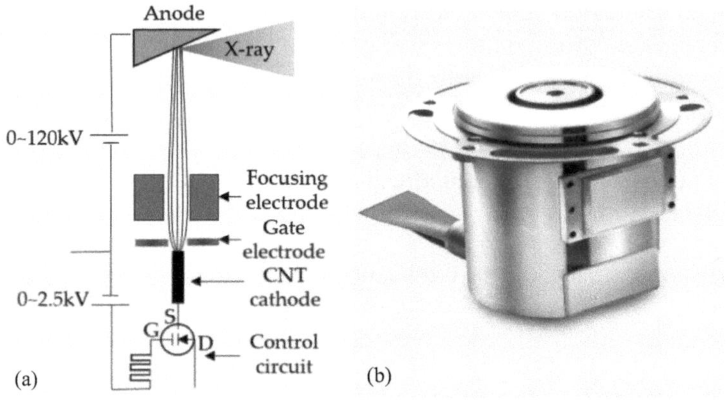

Figure 13.1 (a) Working principle of CNT X-ray source. (b) A commercial CNT X-ray source capable of 110 kVp and 100 mA (MED1101, NuRay Technology Co. [19]).

Compared to a thermionic X-ray tube, one of the advantages of the field-emission X-ray source is that the electron emission, and therefore X-ray exposure, can be readily programmed and synchronized with external triggers. This allows elimination/minimization of image blur caused by, for example, internal organ motion in animal and patient imaging.

After the initial demonstration of X-ray generation, one of the first CNT X-ray sources developed in our lab was a micro-focused CNT X-ray tube optimized for preclinical imaging of small animals for biomedical research [29]. A high-resolution physiologically gated micro-CT scanner was designed based on the X-ray tube [8]. The scanner was successfully utilized for prospective respiratory and cardiac gated micro-CT imaging of different small animal models including mice with human lung tumors [6].

One of the challenges in developing the CNT X-ray source technology is to achieve the output power (current and voltage) and long-term stability required for imaging applications. Early experimental devices often suffered from both low emission current due to the low quality of the CNT cathodes fabricated at the time and high voltage instability due to issues such as arcing. Significant progress has been made over the years to mitigate these issues by (1) optimizing the structure and morphology of the CNT raw materials and the design of the CNT cathodes to achieve both the high emission current density and high total emission current, (2) improving the CNT source design and manufacturing process, and (3) incorporating electronics to compensate and regulate the X-ray flux.

Although there is still room for improvement, the CNT X-ray sources manufactured today can deliver the same power as a conventional thermionic X-ray tube, limited by the thermal management of the anode rather than by the emission current from the cathode. An example of a high-power fixed-anode CNT X-ray source is shown in Fig. 13.1b. The sealed source, without an external active ion pump, can be operated at 110 kVp and 70 mA of tube current with a variable focal spot size (NuRay Technology Co.). This is close to the physical limit for the tungsten anode. Further increasing the output power will require moving from the fixed anode to a rotating anode design.

13.2.3 Spatially Distributed CNT X-Ray Source Array

A conventional thermionic X-ray tube is an analog and single-pixel device in which X-ray radiation is generated from a single focal spot on the X-ray anode. For tomographic imaging such as tomosynthesis and computed tomography, the X-ray source needs to be moved to different viewing angles to collect the projection images needed for 3D image reconstruction. A spatially distributed X-ray source array in which X-ray radiation is generated from multiple and spatially distributed focal spots offers the potential to perform tomographic imaging at improved spatial and temporal resolutions.

The relatively simplistic design and low thermal contribution of the CNT cathode assembly make possible the construction of spatially distributed multisource arrays with variable configurations based on the system requirements. The basic concept and feasibility of a CNT field-emission X-ray source array was first demonstrated in 2005 [53, 54, 55]. Linear [50] and 2D arrays of CNT X-ray emitters were later fabricated [31].

Compared to the single-beam X-ray source, the distributed X-ray source array presents some additional technological challenges, such as consistency in the field-emission characteristics of the individual CNT cathodes, focal spot sizes, and the X-ray outputs from beam to beam. For most imaging applications, the X-ray dose and focal spot size need to be consistent from view to view to ensure the quality of the reconstructed images. This is challenging when 10–10^2 CNT cathodes are used per X-ray source array.

The field-emission current changes exponentially with the bias electrical field. Small variations in the emitter composition, geometry, or morphology can lead to a large variation in the emission behavior between the individual cathodes. Even with continuous improvement in the fabrication technology, some variation in the cathode is inevitable. To mitigate this problem, dedicated control electronics have been developed to compensate in real time the variations in the emission behavior of the individual cathodes and regulate the output X-ray photon flux to achieve the consistency and reliability required for imaging.

Table 13.2 Examples of commercially available CNT X-ray sources and their specifications [19]

No. of focal spots	*V* (kVp)	Anode *I* (mA)	FSS
1	110	70	IEC 0.9–1.9
7	70	7	IEC 0.7
20	42	38	IEC 0.9
45	120	15	IEC 0.8
250	180	30	2.6 mm × 0.5 mm

Today spatially distributed CNT X-ray source arrays with different configurations and performance specifications are manufactured commercially based on the technologies initially developed at UNC (Table 13.2). Figure 13.2 is an illustration of the working mechanism of the source array. To illustrate the capabilities of this technology, a CNT X-ray source array manufactured to our specification by NuRay Technology for a new CT system under development is shown in Fig. 13.2c. The array contains 45 emitters, each with an ~1 mm × 1 mm focal spot size, distributed over ~50 cm, and is designed for operation at 120 kVp and 25 mA of cathode current. Figure 13.2d is a plot of the experimentally measured cathode current from a single CNT cathode operated at 15 mA for 2 ms pulse width. Figure 13.2e is a plot of the cathode current from all emitters at 15 mA, 2 ms pulse width, and 120 kVp anode voltage, demonstrating consistency among emitters. Table 13.2 lists examples of commercially available CNT X-ray sources [19].

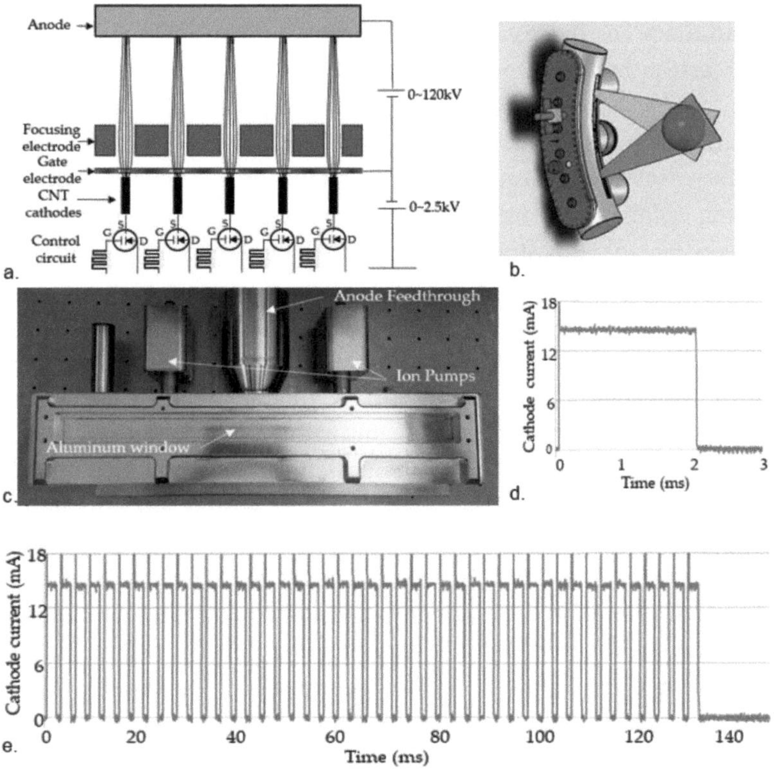

Figure 13.2 (a, b) Working principle of the CNT source array. (c) A CNT X-ray source array (NuRay Technology) containing 45 emitters. The array can be operated up to 120 kVp and 25 mA of cathode current. (d) Pulse profile of single emitter at 15 mA and 2 ms width. (e) Pulse profile of all emitters at 15 mA and 2 ms width.

13.3 Medical and Dental Imaging Applications

13.3.1 Motivation

3D X-ray imaging technologies such as tomosynthesis [11] and CT require acquisition of multiple X-ray projections from different imaging angulations. A traditional limitation and engineering challenge has been acquisition speed, due to the required motion of a single X-ray tube on a rotating gantry. Long scan times can result in patient discomfort and motion, which degrades the images. Stationary (or fixed gantry) tomographic imaging in which all the projection images are collected without mechanical movement of the source, detector, or patient offers the potential to increase the temporal and spatial resolution of 3D imaging and the mechanical robustness of the scanner. The feasibility was first demonstrated using a limited number of X-ray sources in the dynamic spatial reconstructor [38] and electromagnetically scanning electron beam in electron beam CT [5].

The spatially distributed CNT X-ray source array technology provides a new approach to achieve this goal. It allows collection of projection images from different viewing angles by electronically activating the individual X-ray beams without any mechanical motion of hardware (Fig. 13.3). Physiological motion can also be mitigated by prospective triggering of the X-ray pulse with respiratory and/or cardiac signals. Our initial efforts focused on developing stationary tomosynthesis imaging. Prototype devices have been developed and are being evaluated by patient studies for applications including breast cancer detection [27], lung imaging [16], wrist fracture [23], and dental imaging [22]. Here we provide a brief review.

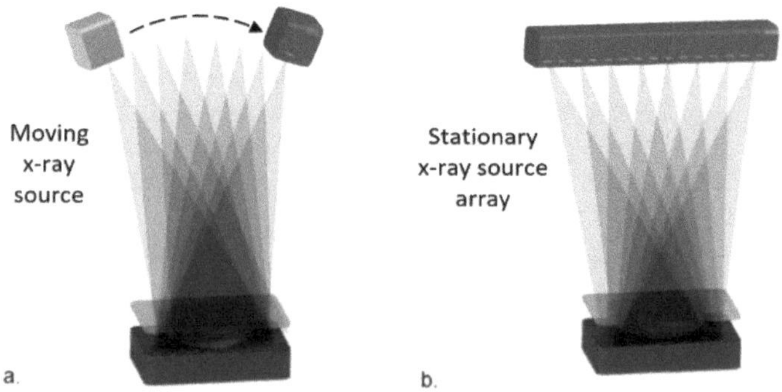

Figure 13.3 (a) Conventional tomosynthesis source motion and (b) stationary tomosynthesis.

13.3.2 CNT X-Ray-Based Stationary Digital Tomosynthesis

Digital tomosynthesis (DT) is a limited angle tomography method that has been proven effective for many applications, most commonly breast, chest, and orthopedic imaging [30, 42]. DT produces quasi-three-dimensional (3D) information at substantially lower dose than CT. Conventional DT utilizes a moving X-ray source to acquire multiple 2D projection images of the subject at varying angular incidence processed to generate a 3D volume. It has several drawbacks due to motion of the X-ray source. Acquisition is performed in either step-and-shoot mode or continuous motion mode, both requiring complex mechanical motion. Step-and-shoot systems can suffer from gantry vibration, and continuous motion systems experience a loss of spatial resolution due to motion of the focal spots. Partially because of the source motion, a DT scan is slower than 2D radiography, which leads to image blur due to patient motion.

CNT source array technology has enabled stationary DT (s-DT) in which the tomosynthesis scan is acquired with no mechanical motion [53]. s-DT scans are performed by sequentially operating individual emitters in the source array. Acquisition can be physiologically gated using respiratory and/or cardiac signals to trigger the X-ray pulses for enhanced resolution.

13.3.2.1 Detection of breast cancer

Digital breast tomosynthesis (DBT) is emerging as the standard of care for cancer screening and diagnostic imaging of some populations [9]. In breast images, cancer presents as a mass and/or microcalcifications. The ability of DBT to remove superimposed dense tissues has increased the conspicuity of masses [24]. Microcalcifications are still better viewed using 2D mammography due to focal spot blurring and reconstruction artifacts in DBT. DBT also requires longer scan time, during which the breast is compressed, leading to patient discomfort and potential motion.

Stationary DBT (s-DBT) has been demonstrated to eliminate focal spot motion and decrease scan time. A clinical prototype was constructed by retrofitting a commercial mammography system (Hologic, Inc., Bedford, MA) with a linear CNT source array (XinRay Systems, Morrisville, NC) (Fig. 13.4a) [37]. The source array contains

31 X-ray emitters, and covers a 30° angular span at the standard breast imaging source-to-detector distance of 700 mm (Fig. 13.4b,c). The average focal spot size of the array is 0.64 mm × 0.61 mm (Fig. 13.4d) [47]. A cathode current of 41 mA, resulting in a tube current of 27 mA, is instantaneously produced in 250 ms pulses using a low voltage trigger for detector synchronization (Fig. 13.4e).

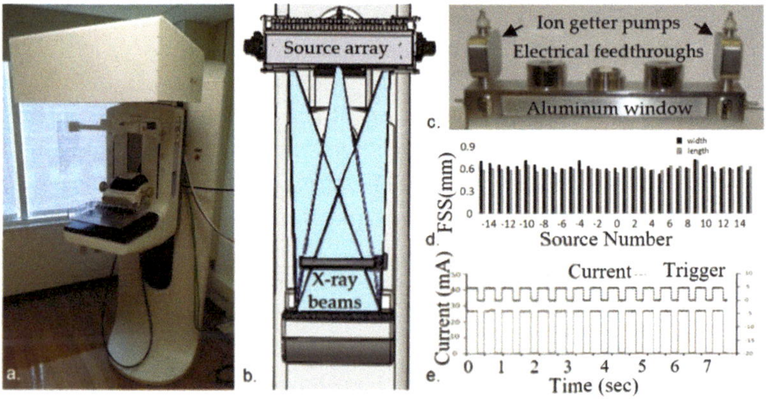

Figure 13.4 (a) The s-DBT prototype in the UNC Mammography clinic. (b) CAD drawing indicating three representative X-ray beams. (c) The CNT source array. (d) Focal spot measurements for the source array. (e) Plot of tube current showing stable output at 27 mA, 250 ms pulses, over 15 emitters.

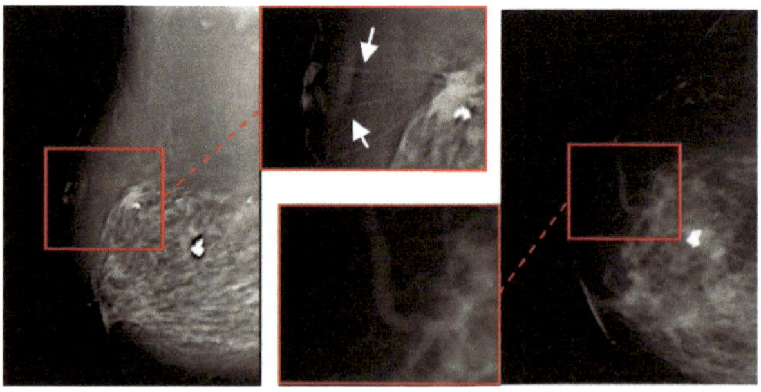

Figure 13.5 (Left) An s-DBT patient image revealing a cancerous lesion (inset) with spiculations extending to the skin surface (white arrows) that are not visualized in the corresponding mammography image (right).

The s-DBT system was evaluated via phantom [46], specimen, and patient imaging studies [27]. Figure 13.5 shows a sample s-DBT patient image from a study comparing s-DBT to mammography, a spiculated mass, later revealed to be invasive lobular carcinoma, is visualized in s-DBT but was occult in the corresponding mammographic image.

13.3.2.2 Chest imaging

Digital chest tomosynthesis (DCT) improves on the conspicuity of lesions as compared to conventional chest radiography, but with about ten-fold reduction in the radiation dose relative to CT. Studies have shown that DCT is capable of visualizing 5 mm+ lung nodules as effectively as CT and has potential to become a screening tool for lung diseases [45]. DCT systems currently in use employ a single, moving source to acquire multiple images.

Our group has recently demonstrated the feasibility of stationary DCT (s-DCT) utilizing CNT X-ray source array [40]. Compared to the regular DCT technology, s-DCT can potentially improve image quality by minimizing image blur caused by source and patient motion with faster data acquisition and physiological gating. The s-DCT prototype system was constructed using a bed-mounted flat panel detector and a CNT source array originally designed for other purposes, with 75 focal spots. Figure 13.6a is a photo of the s-DCT system installed at UNC. Figure 13.6b is a diagram depicting the patient imaging configuration with three representative X-ray beams.

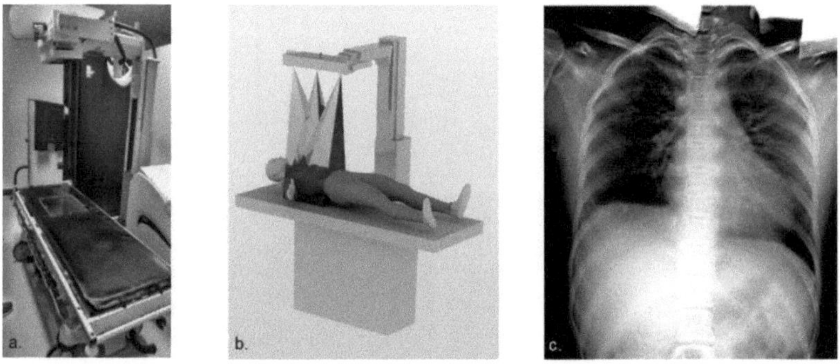

Figure 13.6 (a) The s-DBT prototype at UNC. (b) Diagram of patient imaging with representative X-ray beams. (c) A reconstructed s-DCT lung image of a respiratory-gated CF patient [26].

The s-DCT system has been used to perform patient imaging studies to assess conspicuity of lung nodules and cystic fibrosis (CF) in both adult and pediatric populations [16]. A sample respiratory-gated s-DCT slice image from a CF patient is shown in Fig. 13.6c. Our team is working toward a future version of the s-DCT which will incorporate a longer, purpose-built source array, providing a larger angular span, at higher kVp. Elimination of source motion for shorter scan times, coupled with the ability to perform physiological gating, should improve the image quality and diagnostic accuracy [26].

13.3.2.3 Dental imaging

Dentistry is another field in which radiography is widely used and tools have evolved at a pace similar to medical radiography [20]. Although considered to be an indispensable component of dental care, intraoral radiography has relatively low diagnostic accuracy for many tasks [2]. Cone-beam CT (CBCT) has been incorporated into many dental clinics for diagnostic applications but is not appropriate as a dental screening tool [13]. Tomosynthesis has also been explored using conventional X-ray tube to acquire projection images from different viewing angles and the use of fiducial marker for geometry determination [49]. Although the clinical advantages of this method have been demonstrated, the complexity of moving the X-ray source for data collection was not viable for patient use.

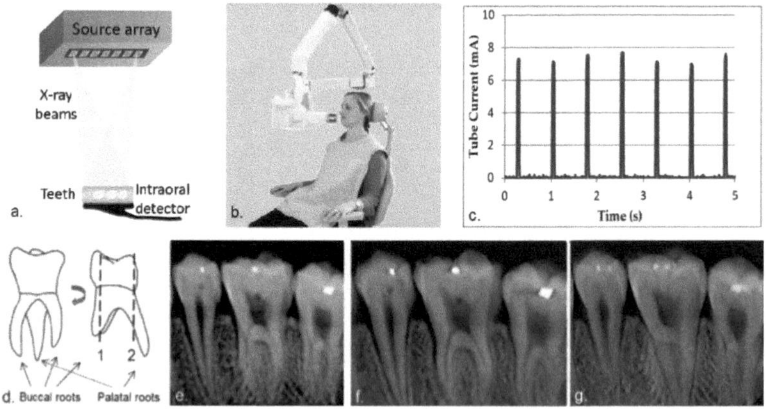

Figure 13.7 (a) Sketch of s-IOT. (b) The s-IOT device shown installed in the UNC ASoD. (c) Plot of tube current for the seven emitters in the source array. (d) A diagram of a molar indicating root orientations for the central molar, obscured in the intraoral image (e) and clearly seen in reconstruction slice images (f) and (g).

Stationary intraoral tomosynthesis (s-IOT) was proposed [57] and initially assessed using a benchtop device with a CNT source array, dental phantoms and extracted teeth, and a standard intraoral detector in collaboration with the UNC Adams School of Dentistry (ASoD) (Fig. 13.7a) [41]. A preliminary reader study revealed a 36% increase in sensitivity for interproximal caries for s-IOT [33]. Based on this finding, XinVivo, Inc. (now Surround Medical Systems) designed and constructed a clinical prototype, installed at the UNC ASoD (Fig. 13.7b). The source array contains seven emitters and is operated at 70 kVp and 7 mA of current per emitter (Fig. 13.7c), similar to a conventional fixed-anode dental tube [22]. Using a dose similar to 2D radiography, s-IOT has been shown to substantially reduce overlap of interproximal contacts, eliminating the need for retakes, which commonly occur in 2D radiographs in the clinic [32]. An example is shown in Fig. 13.7d in which the palatal root is indiscernible in the radiograph (Fig. 13.7e), but all roots are clearly depicted in different reconstruction slice images (Fig. 13.7f—plane 1 and Fig. 13.7g—plane 2). At the time of survey, ~50 patients have been imaged in an Institutional Review Board-approved study conducted to evaluate s-IOT sensitivity for caries detection [18].

13.4 Conclusions

The CNT-based field-emission X-ray technology offers several distinct advantages over the conventional thermionic X-ray tubes because of the field-emission mechanism for electron generation, including the electronic programmability of the X-ray radiation and the flexibility in fabricating distributed X-ray source arrays with different configurations. These capabilities are especially attractive for physiologically gated imaging and for designing tomographic imaging systems with improved resolution, mechanical robustness, and mobility. Although demonstration of X-ray generation using CNT emitters is relatively straightforward, achieving the power, stability, and reliability required for clinical and industrial imaging has been challenging. Through improvement in the quality of the CNT materials, and advances in the design and manufacturing process for the cathode structure and the X-ray sources, the performances of the

CNT X-ray sources have progressed considerably since the original reports 20 years ago. The CNT cathodes have shown the ability to generate the current and current density needed for even the high-power medical CT tubes. Today the power of the commercially manufactured CNT X-ray sources is not limited by the ability of the CNT cathodes, but by the thermal management of the anode, the same as the conventional thermionic X-ray tubes.

Utilizing the ability of the distributed X-ray source array to emit radiation from multiple viewing angles without any mechanical movement, several new tomosynthesis imaging devices have been proposed and demonstrated in patient imaging studies for clinical applications including detection of breast cancer, diagnosis of various lung diseases, imaging of extremities, and dental examination. A CNT X-ray powered mobile digital radiography device has received FDA approval in the United States and is now clinically used. Several CNT X-ray source array based tomographic imaging systems are in the pipeline for regulatory approval. With continuing advances in the CNT X-ray source technology, new applications for imaging and potentially radiotherapy can be expected.

Acknowledgments

The research described in this chapter is largely the result of work performed at the University of North Carolina by several generations of students and postdoctoral fellows and in collaboration with colleagues at the UNC School of Medicine and Adams School of Dentistry. The research was supported largely by grants from the National Institute of Health and by the Department of Defense.

Disclosures

The authors of this chapter are inventors of multiple patents related to the CNT X-ray technologies which have been licensed to Xintek, Inc. Xintek has equity ownership in NuRay and Surround Medical Systems. Zhou serves on the board of directors of Xintek, NuRay, and Surround Medical Systems. Zhou and Lu have equity ownership in Xintek. All activities have been approved by the institutional COI committee.

References

1. Ardran, G., et al. (1974). Field emission 350kVp chest x-ray system, *Radiography*, **XL** (480), pp. 277–284.
2. Bagramian, R., et al. (2009). The global increase in dental caries. A pending public health crisis, *Am. J. Dent.* **21** (1), pp. 3–8.
3. Behling, R. (2015). *Modern Diagnostic X-ray Sources*, CRC Press, Boca Raton, FL.
4. Bower, C., et al. (2001). Method for fabrication of patterned nanotube films, U.S. Patent 6,277,318
5. Braun, J., Oldendorf, M., Moshage, W., et al. (1996). Electron beam computed tomography in the evaluation of cardiac calcifications in chronic dialysis patients, *AJKD* **27** (3), pp. 394–401.
6. Burk, L., et al. (2011). Carbon nanotube based respiratory gated micro-CT imaging of a murine model of lung tumors with optical imaging correlation, *Proc. SPIE 7965, Medical Imaging 2011*: 79651L.
7. Calderon-Colon, X., et al. (2009). A carbon nanotube field emission cathode with high current density and long-term stability, *Nanotechnology* **20**, p. 325707
8. Cao, G., et al. (2009). A dynamic micro-CT scanner based on a carbon nanotube field emission x-ray source, *Phys. Med. Biol.* **54** (8), pp. 2323–2340.
9. Chong, A., et al. (2019). Digital breast tomosynthesis: Concepts and clinical practice, *Radiology* **292** (1), pp. 1–14.
10. Coolidge, W. (1913). A powerful Rontgen ray tube with a pure electron discharge, *Phys. Rev.* **6** (2), pp. 409–430.
11. Dobbins III, J. (2009). Tomosynthesis imaging: At a translational crossroads, *Med. Phys.* **36** (6), pp. 1956–1967.
12. Fowler, R., and Nordheim, L. (1928). Electron emission in intense electric fields, *Proc. R. Soc.* **CXIX**, pp. 173–181.
13. Gaalaas, L., et al. (2016). Ex vivo evaluation of new 2D and 3D dental radiographic technology for detecting caries, *DMFR* **45** (3).
14. Gao, B., et al. (2001). Fabrication and electron emission properties of carbon nanotube films by electrophoretic deposition, *Adv. Mater.* **13** (23), pp. 1770–1773
15. Gidcumb, E., et al. (2014). Carbon nanotube electron field emitters for x-ray imaging of human breast cancer, *Nanotechnology* **25** (245704), pp. 1–10.

16. Gunnell, E., et al. (2018). Initial clinical experience of stationary digital chest tomosynthesis in adult patients with cystic fibrosis, *Eur. Rad.* **29**, pp. 1665–1673.

17. Hallenbeck, G. (1975). Clinical evaluation of the 350-kV chest radiography system, *Radiology*, **117**, pp. 1–4.

18. http://www.clinicaltrials.gov/NCT02873585, last accessed 06/11/2020.

19. http://www.nuraytech.com, last accessed 06/11/2020.

20. Iannucci, J., and Howerton, L. (2012). *Dental Radiography: Principles and Techniques*, Elsevier.

21. Iijima, S. (1991). Helical microtubules of graphitic carbon, *Nature* **354**, pp. 56–58.

22. Inscoe, C., et al. (2018). Characterization and preliminary imaging evaluation of a stationary intraoral tomosynthesis system, *Med. Phys.* **45** (11), pp. 5172–5185.

23. Inscoe, C., et al. (2020). Preliminary imaging evaluation of a compact tomosynthesis system for point-of-care extremity imaging, *Proc. SPIE: Med. Imaging 2020*, **11312-04**.

24. Korhonen, K., et al. (2019). Breast cancer conspicuity on simultaneous digital mammographic images versus digital breast tomosynthesis images, *Radiology* **292** (1), pp. 69–76.

25. Langmuir, I. (1923). The electron emission from thoriated tungsten filaments, *Phys. Rev. J.* **22** (357), p. 1923.

26. Lee, K., and Chung, M. (2018). Limitations for detecting lung nodules by using digital chest tomosynthesis, *Radiology* **287** (3), pp. 1028–1029.

27. Lee, Y., et al. (2019). Initial clinical experiencee with stationary digital breast tomosynthesis, *Acad. Radiol.* **26** (10), pp. 1363–1372.

28. Lilienfeld, J. (1925). Art of releasing electrons in vacuum discharge devices, U.S. Patent 1,559,714.

29. Liu, Z., et al. (2006). Carbon nanotube based micro-focus field emission x-ray source for micro-computed tomography, *Appl. Phys. Lett.* **89** (103111), pp. 1–2.

30. Machida, H., Yuhara, T., Tamura, M., et al. (2016). Whole body clinical applications of digital tomosynthesis, *Radio Graphics* **26**, pp. 735–750.

31. Maltz, J., Sprenger, F., Fuerst, J., et al. (2009). Fixed gantry tomosynthesis system for radiation therapy image guidance based on a multiple

source x-ray tube with carbon nanotube cathodes, *Med. Phys.* **36** (5), pp. 1624–1636.

32. Mauriello, S., et al. (2020). The role of stationary intraoral tomosynthesis in reducing proximal overlap in bitewing radiography, *DMFR* 20190504.
33. Mol, A., et al. (2015). *Intraoral Tomosynthesis Using Carbon Nanotube (CNT) X-ray Technology: Primary Caries Detection*, AAOMR, San Francisco.
34. Oh, S., et al. (2004). Liquid-phase fabrication of patterned carbon nanotube field emission cathodes. *Appl. Phys. Lett.* **87** (19), p. 3738.
35. Posner, E. (1970). Reception of Rontgen's discovery in Britain and U.S.A., *Br. Med. J.*, **4**, pp. 357–360.
36. Qian, C., et al. (2006). Fabrication of small diameter few-walled carbon nanotubes with enhanced field emission property, *J. Nanosci. Nanotechnol.* **6**, pp. 1346–1349
37. Qian, X., et al. (2009). Design and characterization of a spatially distributed multibeam field emission x-ray source for stationary digital breast tomosynthesis, *Med. Phys.* **36** (10), pp. 4389–4399.
38. Robb, R. A., et al. (1980). The dynamic spatial reconstructor. *J. Med. Syst.* **4**, pp. 253–288.
39. Roentgen, W. (1896). On a new kind of rays, *Nature*, **53**, p. 274.
40. Shan, J., et al. (2015). Stationary chest tomosynthesis using a carbon nanotube x-ray source array: A feasibility study, *Phys. Med. Biol.* **60** (81), pp. 81–100.
41. Shan, J., et al. (2015). Stationary intraoral digital tomosynthesis using a carbon nanotube x-ray source array, *DMFR* **44**, 2015098.
42. Soderman, C., et al. (2015). Evaluation of accuracy and precision of manual size measurements in chest tomosynthesis using simulated pulmonary nodules, *Acad. Radiol.* **22** (4), pp. 496–504.
43. Spindt, C. (2003). A thin-film field-emission cathode, *J. Appl. Phys.* **39**, p. 3504.
44. Sugie, H., et al. (2001). Carbon nanotubes as electron source in an x-ray tube, *Appl. Phys. Lett.* **78** (17), pp. 2578–2580.
45. Terzi, A., et al. (2015). Lung cancer detection with digital chest tomosynthesis: Baseline results from the observational study SOS, *J. Thor. Oncol.* **8** (6), pp. 685–692.

46. Tucker, A., et al. (2013). Dependence of image quality of system configuration parameters in a stationary digital breast tomosynthesis system, *Med. Phys.* **40** (3), 031917, pp. 1–10.
47. Tucker, A. (2014). Development of a stationary digital breast tomosynthesis system for clinical applications, PhD dissertation, UNC.
48. Wang, P., et al. (2008). Microwave plasma enhanced chemical vapor deposition growth of few-walled carbon nanotubes using catalyst derived from an iron-containing block copolymer precursor, *Nanotechnology* **19** (8), p. 185605.
49. Webber, R, et al. (1997). Tuned-aperture computed tomography (TACT). Theory and application for three dimensional dento-alveolar imaging, *DMFR* **26**, pp. 53–62.
50. Yang, G., et al. (2010). Design and feasibility studies of a stationary digital breast tomosynthesis system, *Nucl. Instr. Meth. A* **648** (Suppl. 1), pp. S220–S223.
51. Yue, G. (2002). Generation of continuous and pulsed diagnostic imaging x-ray radiation using a carbon-nanotube-based field-emission cathode, *Appl. Phys. Lett.* **81** (355), pp. 355–357.
52. Zhang, J., et al. (2004). Efficient fabrication of carbon nanotube point electron sources by dielectrophoresis, *Adv. Mater.* **16** (14), pp. 1219–1221
53. Zhang, J., et al. (2005). Stationary scanning x-ray source based on carbon nanotube field emitters, *Appl. Phys. Lett.* **86**, pp. 184104-1–184104-3
54. Zhou, O., and Lu, J. (2003). X-ray generating mechanism using electron field emission cathode, U.S. Patent 6,553,096.
55. Zhou, O., and Lu, J. (2005). X-ray generating mechanism using electron field emission cathode, U.S. Patent 6,850,595.
56. Zhou, O., and Calderon-Colon, X. (2010). *Carbon Nanotube X-ray Field Emission Technology, in Carbon Nanotube and Related Field Emitters: Fundamentals and Applications*, John Wiley & Sons, pp. 418–421.
57. Zhou, O., et al. (2015). Digital tomosynthesis systems, methods, and computer readable media for intraoral dental tomosynthesis imaging, U.S. Patent 20160317107A1.

Chapter 14

Graphene Cold Field-Emission Sources for Electron Microscopy Applications

Anjam Khursheed and Xiuyuan Shao
Department of Electrical and Computer Engineering,
National University of Singapore, Singapore 117576, Singapore
eleka@nus.edu.sg, eleshao@nus.edu.sg

The theoretical and experimental emission characteristics of the room-temperature field-emission source are reviewed as they relate to scanning electron microscopy, transmission electron microscopy, and scanning transmission electron microscopy applications. A brief history of the development of the cold-field electron source is given. Work function, energy distribution, source electron optics, and current fluctuations are the main source characteristics considered. The general focus of this chapter is on the newly developed single-crystal graphene coated nickel cold field-emission sources.

14.1 Introduction

The emission of electrons being emitted from a sharply pointed cold metal surface under the presence of a strong electric field, a

Nanostructured Carbon Electron Emitters and Their Applications
Edited by Yahachi Saito

ISBN 978-981-4877-62-6 (Hardcover), 978-1-003-14199-0 (eBook)
www.jennystanford.com

process called "field emission (FE)," was first reported by Robert Williams Wood, in 1897 [1]. A potential barrier is established by the potential step at the metal surface confining the electrons to the solid, which is to a first approximation, triangular in shape. The height of the potential barrier is given by the work function of the solid. The first clear picture of the FE mechanism in terms of the newly developed theory of quantum mechanics was developed by Fowler and Nordheim in 1928 [2]. Under the influence of a high electric field, the width of the potential barriers becomes thinner and the electrons close to the Fermi level have a finite probability to tunnel through the barrier and escape. In the early 1960s, many of the successful investigations of field-emission phenomena came from three groups led by Professors Erwin Muller (Pennsylvania State University), Robert Gomer (University of Chicago), and Walter Dyke (Linfield College and Linfield Research Institute) [3–5]. Dyke and coworkers proposed a variety of applications of a cold field-emission (CFE) source and formed the first company to produce CFE-based commercial products. The successful application of CFE cathodes to electron microscopy, including both scanning electron microscopy (SEM) and scanning transmission electron microscopy (STEM), was achieved by a group led by Professor Albert Crewe at the Enrico Fermi Institute in Chicago in the late 1960s [6, 7].

A locally high electric field needs to be created for FE to take place, which requires the field emitting tip to have large field enhancement factors. For the past half a century, a great variety of refractory metals, semiconductors, nanowires, single-atom tips, and a variety of nanotips have been investigated for their FE properties [8–10]. The investigations and developments of CFE electron sources turned out to be quite discouraging by the fact that CFE electron sources are found to be not very stable and reliable. For over 40 years, the most common type of CFE electron source used in high-resolution microscopy utilizes a single crystal (310)-oriented tungsten-sharpened wire as the cathode, whose tip radius is in the 100–200 nm range with a work function (ϕ) of 4.4 eV [11]. Typically, electron emission from the cathode occurs when the field strength at the cathode tip exceeds values around 4 V/nm for single-crystal tungsten, allowing electrons to escape from the cathode surface by quantum tunneling. However, single-crystal tungsten CFE sources suffer from some practical engineering problems that have limited their widespread use. One difficulty comes from the native oxide material that forms on the cathode surface and subsequently quenches electron emission, requiring the use of ultrahigh vacuum (UHV) levels in the gun unit ($<10^{-10}$ Torr). Even at such stringent vacuum conditions, the intensity of electron emission falls in 1 to 2 h of continuous operation and the cathode needs to be warmed or rapidly heated (flashed to ~2000 K) regularly to blow off material. One disadvantage of this flashing process is that it will interrupt the continuous operation of the electron beam instruments. Recently, Hitachi has developed a "Mild flashing" technology for their latest generation of CFE source SEMs, which is executed automatically in the background to gently clean the FE source at regular intervals while maintaining a high voltage [12]. This technology enables stable and continuous operation of the new CFE source. However, the feasibility of using this technology requires extreme-stringent UHV conditions ($\sim 10^{-12}$ Torr), and this will be a challenge for many practical electron beam applications.

Since the discovery of two-dimensional (2D) graphene in 2004, graphene has attracted much attention as a potential candidate for CFE source emitters, due to its high aspect ratio (the lateral size to the thickness) and excellent thermal, mechanical, and electrical properties [13]. Excellent field emission from graphene has been demonstrated from a wide variety of different microfabricated structures, and most of them consist of densely packed nano-emitters that produce a large array of defocused overlapping electron beams, and therefore cannot be subsequently defocused down to a single nanometer electron probe [14–16]. For an FE electron source to be useful for high-resolution electron optical instruments, it must have its dominant emission direction on the optical axis, be stable over a sufficient long period of time, and be resistant to different types of current fluctuations in practical vacuum environments.

14.2 Work Function

The key parameter that determines the highest achievable brightness with the lowest energy spread of a CFE point-cathode electron source is the low work function of the emitting material. For electron microscopy applications, as electrons are accelerated through the gun unit, the wide-angle ones are filtered out by the first, second anode plate holes and gun apertures, and they exit the gun unit in the form of a rotationally symmetric electron beam that appears to emanate from a single point on the axis. This single point on the axis, known as virtual source, is defined as the spot at which the back-projected trajectories of emitted electrons are focused. The relatively small virtual source size of CFE guns (a few nanometers in diameter) is one of the main reasons for them having the highest brightness of all electron sources used in focused electron beam instruments. In fact, only a small portion of the beam current contained in a small solid angle centered at the emitter axis is useful. Therefore, the state-of-the-art single-crystal tungsten-sharpened wire needs to be etched in a specific plane (such as the <310> plane) to generate the lowest work function on the apex of the tip, in order to have sufficient emission for a small acceptance aperture centered on the emitter axis.

Graphene is well known for possessing several desirable properties: (1) it has excellent thermal, mechanical, and electrical characteristics [17, 18]; (2) its work function can be significantly lowered both by direct contact with metals (effectively doping it) and through the application of intense electric fields [19, 20]; and (3) it is extremely flexible. Recent progress shows that the characteristic electronic structure of graphene can be significantly altered by chemisorption on metal substrates, resulting in doping either with electrons or holes. The work function (ϕ) of graphene can be reduced to as low as 3.66 eV when chemisorbed on Ni (111). Fast development of 2D graphene in recent years has led to a variety of graphene-based FE electron sources, including graphene film emitters, graphene point/edge-cathode emitters, and graphene ring-cathode emitters [21]. However, none of these graphene sources has yet been utilized for commercial high-resolution electron beam instruments.

Recently, the research group at the National University of Singapore, led by A. Khursheed, have presented the development of a completely new type of CFE electron source, a Graphene–Ni (Gr–Ni) point cathode for electron microscopy systems [22]. One of the standout features of the Gr–Ni point cathodes reported in their work are ultralow work function values (~1.1 eV), as evidenced in Fig. 14.1. The work function for the Gr–Ni CFE source is determined by using two Fowler–Nordheim (F–N) plots, one for the bare Ni tip and the other for the graphene-coated Ni tip, and then by taking the ratio of their F–N slopes, eliminating the anode–cathode separation d and field-enhancement factor β. This procedure assumes that the addition of graphene does not change the tip geometry (confirmed by SEM imaging). From the ratio of the two F–N slopes, the effective work function is calculated from:

$$\phi_{\text{Graphene+Ni}} = \phi_{\text{Ni}} \sqrt[1.5]{\left(\frac{m_{\text{Graphene+Ni}}}{m_{\text{Ni}}}\right)}, \tag{14.1}$$

where $\phi_{\text{Graphene+Ni}}$ and ϕ_{Ni} are the work functions of graphene-coated and bare Ni-pointed cathodes, respectively, while $m_{\text{Graphene+Ni}}$ and m_{Ni} are the slopes of their respective F–N plots.

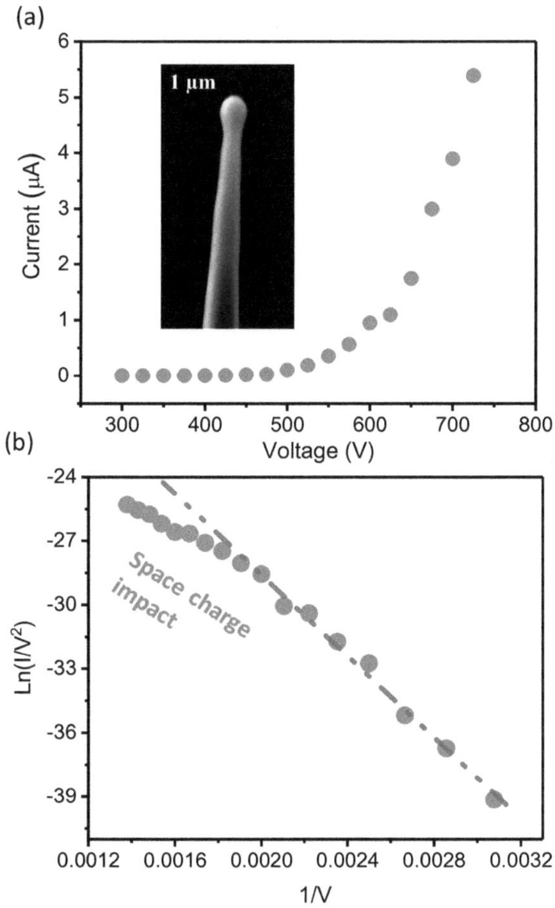

Figure 14.1 Field electron emission characteristics from a Ni tip with a tip radius of 500 nm after coating with graphene, with an anode–cathode spacing of 0.5 mm.

14.3 Energy Distribution

14.3.1 Theoretical Background

The energy spread (ΔE), through the chromatic aberration of the objective lens, degrades the final spatial resolution of an electron optical system, particularly at high relative beam energy spreads. Ideally, the energy distribution of the emitted electron beam should be as narrow as possible while maintaining sufficient brightness. The energy distribution is a convolution of an intrinsic contribution, determined solely by the properties of the material of the emitting surface and extrinsic contributions. An analytical expression for the intrinsic energy distribution (known as total energy distribution (TED)) of electron emission in the thermal field regime was first derived by Young [23] based on the free-electron model as:

$$P(E) = \frac{J_{\text{FN}}}{d}\left[\frac{\exp\left(\dfrac{E-E_f}{d}\right)}{1+\exp\left(\dfrac{E-E_f}{kT}\right)}\right], \tag{14.2}$$

where $kT = 25.9$ meV at 300K, J_{FN} is the F–N emission current density formula developed by Murphy and Good (later modernized), and d is the tunneling parameter (in eV) given by:

$$d = (2/3b)\, t^{-1}(y)\, \phi^{-1/2} F$$

$$(2/3b) = 9.76 \times 10^{-11} \quad [\text{eV}^{3/2}\,\text{V}^{-1}\,\text{m}] \tag{14.3}$$

where F and ϕ are the electric field strength (in V/m) and the work function (in eV), b is the 2nd F–N constant, and $t(y)$ is a slowly varying function of $y = 3.79 \times 10^{-5} F^{1/2}/\phi$ and can be approximated by the formula $t(y) = 1 + 0.1107y^{1.33}$. The analytical formula is valid provided $kT/d < 0.7$ and $y < 1$ [24]. Figure 14.2 shows the room temperature TED for a typical W(310) cold field emitter at ϕ = 4.32 eV and F = 4.34 V/nm as well as for a Gr–Ni tip at ϕ = 1.10 eV and F = 0.781 V/nm. From these plots, the intrinsic full-width at half-maximum (FWHM) energy spreads are calculated to be 0.23 and 0.14 eV, respectively. The smaller predicted intrinsic FWHM of the Gr–Ni cathode compared to the typical W(310) CFE source comes mainly from its smaller field strength requirement.

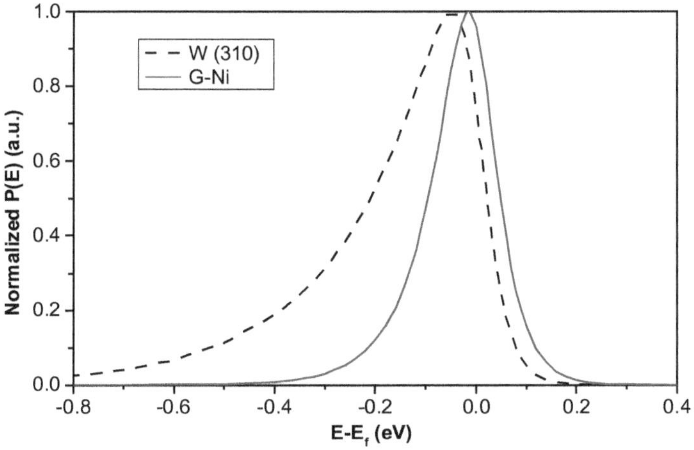

Figure 14.2 Room temperature intrinsic energy distributions for a typical W(310) cold field emitter (ϕ = 4.32 eV, F = 4.34 V/nm) and a Gr–Ni emitter (ϕ = 1.10 eV, F = 0.781 V/nm). The FWHM energy spreads are 0.23 and 0.14 eV, respectively.

14.3.2 Statistical Coulomb Interactions

Charged particle optics is made possible by the fact that trajectories of charged particles are shaped and influenced by electrostatic and magnetic field distributions, this makes it possible to design and make electron lenses and deflectors. However, it also means that the trajectories of electron beams are also affected by statistical coulomb effects, caused by the mutual electrostatic repulsion force between neighboring electrons in the beam. These statistical coulomb interactions manifest as energy broadening or more commonly known as Boersch effect (due to longitudinal interactions) and the trajectory displacement effect (due to lateral interactions), and thus degrades the quality of an electron beam's optical properties [25]. The former adds to the intrinsic source energy spread while the latter causes an additional enlargement of the virtual source size which effectively reduces the source brightness. These two statistical coulomb effects can be estimated in an approximate way by Monte Carlo numerical simulations, and also by simple analytical approximations [26, 27].

The energy broadening, representing the change in axial velocity, was firstly found and investigated by Boersch using a thermionic cathode in 1954 [25]. It has been further experimentally verified for point-cathode FE sources by Bell and Swanson in 1979 [28]. The Boersch effects can be approximated by analytical equations for different regimes of the beam, including Gaussian, Holtsmark, Lorentzian, and the pencil beam regime [27]. For the gun section, the electrons are assumed to be accelerated to V_{ext} in the whole gun segment, the appropriate regime is the Holtsmark regime, and the contribution to the apparent source size is (given by Knauer) [26]:

$$\Delta E_{\text{Boersch}} = 15.9\frac{(I')^{2/3}}{r_{\text{tip}}^{1/3} V_{\text{ext}}^{1/3}}, \tag{14.4}$$

where I' is the emitted angular current density, r is the emitter radius, and V is the beam voltage.

Usually, to a first approximation, the energy broadening term is added in quadrature with the intrinsic emission energy distribution width term to obtain the total beam energy width:

$$\Delta E_{\text{total}} = \left(\Delta E_{\text{intrinsic}}^{2} + \Delta E_{\text{Boersch}}^{2}\right)^{1/2}. \tag{14.5}$$

Predictions using these equations for a conventional W(310) electron source and different diameter tip Gr–Ni electron sources are shown in Table 14.1.

For the smaller tip sizes, the smaller energy spreads expected for the Gr–Ni cathode (using typical parameters from Gr–Ni) compared to conventional tungsten CFE sources (of comparable tip size) based upon the TED distribution will be approximately offset by the Boersch effect. It is interesting to note that since both the TED distribution and Boersch effect on energy spread decrease with increasing tip radius, a significantly smaller energy spread is predicted for the 800 nm radius Gr–Ni tip (a factor of 2 smaller than that of the 170 nm radius tip). This would ordinarily not be possible for conventional large field emitters (tip diameters over 1 μm), such as the Schottky emitter, since the Schottky field emitter only functions by heating the tip up to 1800 K, enlarging the energy spread by thermal effects to around 0.5 eV.

Table 14.1 Energy spread predicted by semianalytical calculations

Source	r_{tip} (nm)	V_{ext} (V)	I' (μA/sr)	d_v (nm)	B_r (A/m^{2}srV)	$\Delta E_{intrinsic}$ (eV)	$\Delta E_{Boersch}$ (eV)	ΔE_{total} (eV)
W(310)	160	4255	62	2.98	2.09×10^9	0.232	0.283	0.366
Gr–Ni	170	975	40.7	4.55	2.51×10^9	0.144	0.342	0.371
Gr–Ni	400	1300	45.5	8.27	7.09×10^8	0.140	0.252	0.288
Gr–Ni	800	1450	11.4	11.79	7.23×10^7	0.136	0.077	0.156

The trajectory displacement effect (Radial broadening), first investigated by Loeffler in 1964, represents the lateral shift in a particle's position and a change in the velocity component perpendicular to the beam axis [29]. A "slice method" can be used to calculate the trajectory displacement, in which the region to be calculated is divided into small segments, where the beam voltage and size is assumed to remain constant within each segment [30]. The trajectory displacement is calculated by applying the slice method to an analytical approximation in the gun region.

$$d_{\text{blur}} = \int \frac{l(z)}{M(z)} \phi_{\text{J}}(r(z), V(z), I)\, dz\,, \tag{14.6}$$

where $l(z) = r(z)/a(z)$ and $M(z)$ is the magnification $\left(= \left(\frac{\alpha_{ref}}{\alpha(z)} \right) \sqrt{\left(\frac{V_{ref}}{V(z)} \right)} \right)$.

The slice method requires the angular displacement per meter, given by:

$$\phi_{\text{J}} = \left[\frac{T_1 D_\lambda^{18/7} D_{\text{r}}^6 I^{18/7} r^{6/7} V^{-15/7}}{T_4 + T_2^{1/7} D_{\text{r}}^6 D_\lambda^2 I^2 r^2 V^{-1}} \right]^{7/6}, \tag{14.7}$$

where $D_\lambda = m^{1/2} / (\pi 2^{7/2} \epsilon_0 e^{1/2})$ and $D_{\text{r}} = \left(\frac{2\epsilon_0 \pi}{e} \right)^{1/3}$. The constants are $T_1 = 4.618 \times 10^{-2}$, $T_2 = 2.041 \times 10^5$, and $T_4 = 6.25 \times 10^{-2}$. $V(z)$ and $r(z)$ are determined using numerical field solving and ray-tracing simulations. One example of this is shown in Fig. 14.3.

The integration is done with Simpson's 1/3 rule applied to unequal intervals. Radial broadening has the effect of increasing the intrinsic virtual source size d_{v}, which in turn will lower the brightness estimate by a correction factor K given by:

$$K = \frac{d_v^2}{d_v^2 + d_{\text{blur}}^2}\,, \tag{14.8}$$

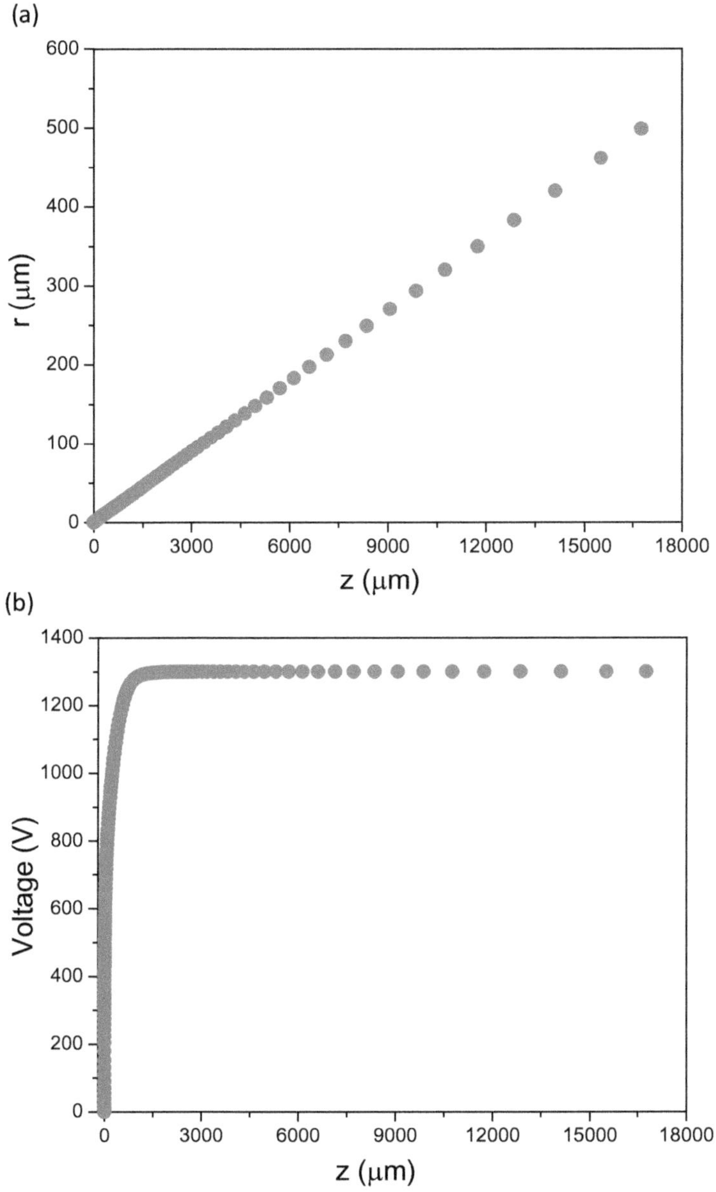

Figure 14.3 (a) Beam radius versus axial distance and (b) potential versus axial distance for trajectory displacement calculation.

14.3.3 Measured Values of the FWHM Energy Spread for the Graphene Cold Field Electron

The measured energy distribution is a convolution of several different effects. They include: (i) the intrinsic energy distribution, (ii) the Boersch effect, and (iii) the finite energy resolution of the analyzer. If the cathode current density is maintained to be relatively low and the resolution of the analyzer is high, the Boersch effect is minimized and the electron beam contains only the energy distribution from the emission process. Figure 14.4 shows the measured beam intensity with respect to electron kinetic energy from the Gr–Ni CFE source at different pass energies for a tip diameter of 200 nm. An upper bound to the energy spread is approximately given by the lowest pass energy value, since the energy resolution of the energy analyzer varies linearly with pass energy value. The overall energy spread values are lower than or comparable with state-of-the-art conventional CFE sources, which is ordinarily not possible for conventional large-size FE electron sources, such as the Schottky emitter, since the Schottky FE source only functions by heating the tip up to 1800 K, enlarging the energy spread by thermal effects to around 0.5 eV. The low energy spread is of particular importance since chromatic aberration of the objective lens is the main factor limiting spatial resolution.

For electron optical analysis, measurement of the beam energy distribution does not have any practical meaning without the simultaneous measurement of the cathode current density J_{FN}. There is a relationship between J_{FN} and angular current density I' with apex radius r, and is useful, since I' is more easily measured. The relationship is given by [31]:

$$I' = J_{FN}\left(\frac{r}{m_{\alpha}}\right)^2, \tag{14.9}$$

where m_{α} is the angular magnification. Trajectory ray-tracing simulation needs to be used to compute the exact value of angular magnification $m_{\alpha} = \alpha/\theta$, where α is the final extraction angle (limited by a small acceptance aperture) and θ is the initial emission angle. The simulated m_{α} is found to lie between 0.5 and 0.7, depending on details of the emitter shape. Alternatively, the cathode current density can be directly calculated by simulating the cathode emission area S_p. By using a Faraday cup with a small acceptance aperture (restricting the semiangle entry to say 30 mrad), the emission current can be traced back to a small cathode emission area S_p. For tip radii in the range of 130–800 nm, S_p was estimated to be in the range of 130–3965 nm^2. The detailed relationship between electron energy distribution and angular current density I' for Gr–Ni electron sources will be discussed in the next section.

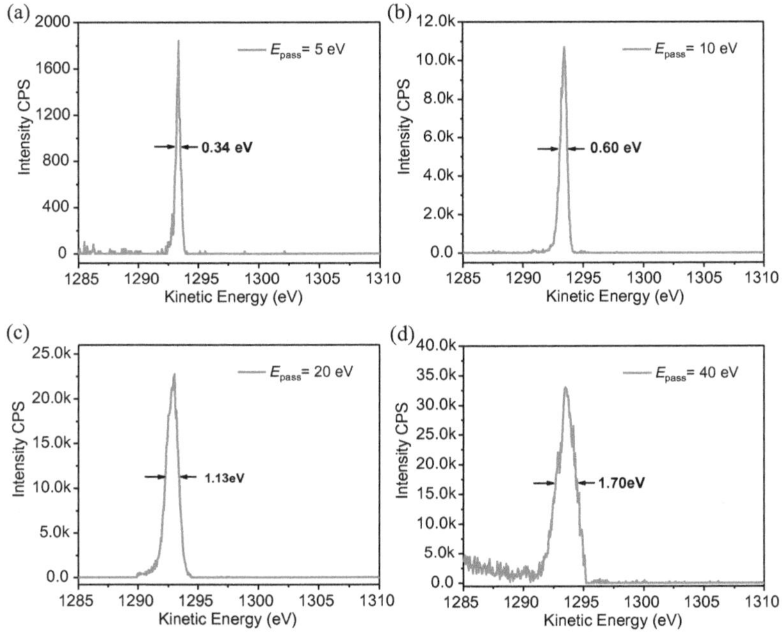

Figure 14.4 Measured FWHM values with different pass energies (5–40 eV) for a tip diameter of 200 nm measured by a cylindrical sector analyzer (CSA 200).

14.4 Source Electron Optics

For successful electron microscopy applications, a high source brightness (B_r) and a low energy spread (ΔE) are always needed to provide a high-resolution focused electron beam. The calculation of B_r requires knowledge of the virtual source size (d_v), defined to be the waist of the crossover formed by the back projection towards the emitter from tangents to the exit electron trajectories. It is given by the following expression [32]:

$$B_r = \frac{4I'}{\pi d_v^2 V_{ext}}, \tag{14.10}$$

where I' is the angular current density, d_v is the virtual source size, and V_{ext} is the extraction voltage. The virtual source size can be calculated using the formula:

$$d_v = 1.67\frac{r_{tip}}{m_\alpha}\sqrt{\frac{<E_t>}{eV_{ext}}}, \tag{14.11}$$

where r_{tip} is the tip radius and $<E_t>$ is defined as $<E_t> = e\hbar F/\sqrt{(8m\phi)}$, with F the local electric field, ϕ the work function, and $\hbar$ the reduced Planck constant. After simplification, the source-reduced brightness for CFE electron sources is given by:

$$B_r = 1.44\frac{eI'}{\pi <E_t>}\left(\frac{m_\alpha}{r_{tip}}\right)^2. \tag{14.12}$$

For high brightness electron beam applications, statistical Coulomb interactions lead to radial broadening of the virtual source size due to lateral electron–electron interactions, which in turn lowers the practical reduced brightness. It should be noted that the analytical method just outlined avoids the need to measure the virtual source diameter d_v experimentally, and this is important, because direct measurements of the virtual source diameter are not trivial.

Figure 14.5a presents semianalytically calculated B_r values for graphene-coated point cathodes versus tip radii, together with state-of-the-art FE electron sources. The data in the plot can be broadly grouped into three categories: high brightness tungsten CFE electron sources with very small tip radii (<200 nm) [11, 33], Schottky emitters having relatively low brightness with large tip radii (~1000 nm) [33, 34], and Gr–Ni CFE electron sources having tip dimensions that roughly fall in the middle region. The estimated B_r value of 2.51 × 10^9 A/m^{2}srV of a graphene-coated cathode of tip radius 170 nm is very close to the highest reported B_r value of 3.98 × 10^9 A/m^{2}srV for a conventional tungsten cold field W(310) emitter with a tip radius of 160 nm. When corrected with the trajectory displacement effect (via Eq. (14.1.8)), the brightness will be lowered by Coulomb interaction effect significantly for the smaller Gr–Ni cathode tip radius (reduction by around 30% for the 170 nm tip radius) and is not expected to significantly lower the brightness estimates for the larger Gr–Ni cathode tip radii. The I' can be easily obtained by dividing the beam current over the solid angle defined by the beam acceptance aperture. With the simultaneous measurement of the electron energy distribution, the reduced brightness versus energy spread is plotted in Fig. 14.5b. The overall energy spread values are lower than or comparable with state-of-the-art conventional CFE sources.

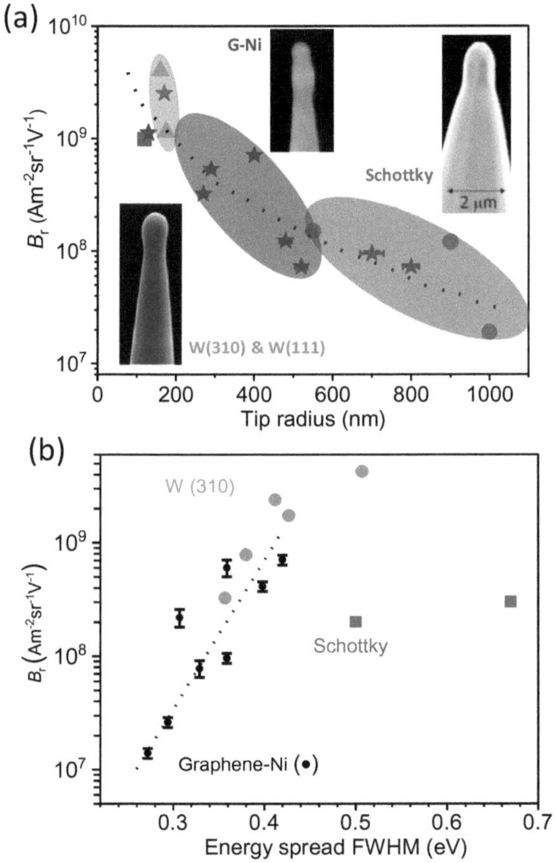

Figure 14.5 (a) Comparison of reduced brightness to the state-of-the-art FE electron sources (Yellow triangle: W(310) [11, 33]; Green square: W(111) [33]; Red circle: Schottky sources [33, 34]; and Blue star: Gr–Ni sources). (b) Reduced brightness versus energy spread and state-of-the-art electron sources.

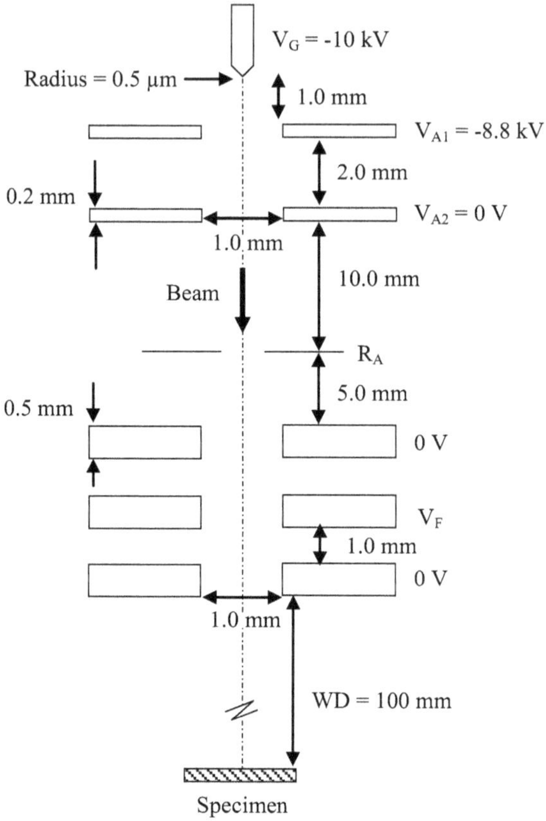

Figure 14.6 Schematic diagram for a 10-kV test column designed to measure the virtual size experimentally.

Figure 14.6 shows the schematic diagram of a 10-kV test column that can be used to determine the practical virtual source size d_v experimentally. Simulation has been done to obtain the optimum values for the final aperture radius R_A for a 100-mm working distance with an accelerating Einzel lens for a high (10-kV) beam voltage. The optimal aperture radius here gives the probe radius that is dominated by the virtual source size contribution. In the test column, the total beam size in the image plane is limited by spherical aberration d_s, diffraction d_l, and virtual source size d_v. Chromatic aberration is negligible (e.g., $C_C(\Delta E/E)\alpha$ = 6 nm for ΔE = 0.2 eV at R_A = 15 μm) and hence not shown. The probe radius–aperture radius relation (shown in Fig. 14.7) is calculated for the 10-kV beam voltage using the Barth-Kruit formula with C_S = 6249.415 m and C_C = 1.540 m. For the 10-kV test column, at the optimum R_A = 15 μm, the middle Einzel lens voltage is V_F = 6625.4 V, giving a magnification factor of M = 4.5. The virtual source radius is 15 nm (F = 0.6 V/nm and m_α = 0.16 predicted and take ϕ = 2 eV), giving a source image radius of 68 nm, while the spherical aberration radius and diffraction radius are minimized to be 11 and 18 nm, respectively.

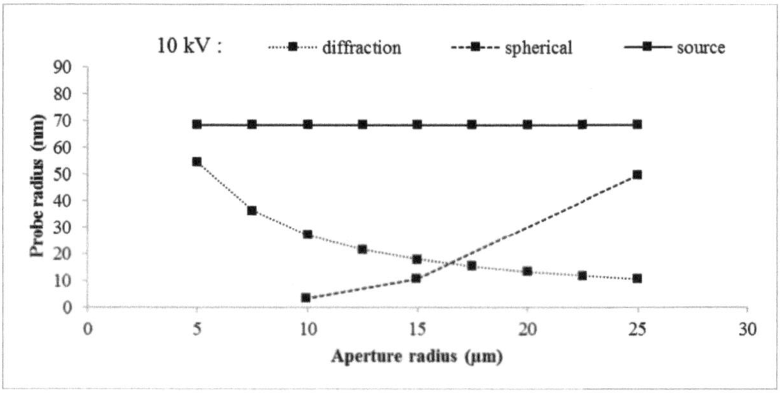

Figure 14.7 Predicted probe diameter for 10-kV beams as a function of final aperture radius. In this plot, spherical aberration is calculated as the RMS radius at the least-confusion plane. Chromatic aberration is ignored since it is negligible.

14.5 Current Fluctuations

14.5.1 Short-Term Current Fluctuations

The stability of the electron beam is of major concern for focused electron beam applications. By ignoring the preexponential term, the Murphy–Good version of F–N theory can be used to show

$$\frac{\delta I}{I} = -\frac{\delta \phi}{d} = -\left[\frac{3b}{2t(y)}\right](\phi^{1/2}\,\delta\phi)(\beta V)^{-1}. \tag{14.13}$$

Therefore, the drift in ϕ and/or β in the emitting area will cause inevitable current fluctuations. Previous studies have shown that conventional CFE electron sources are prone to instability in practical vacuum environments due to [5, 11, 35]:

1. A $1/f$ type noise component, in the presence of adsorbed layers this noise becomes dominant.
2. A generation–recombination process on the cathode surface.
3. Shot noise, caused by uncertainties taking the form of a Poisson distribution, which describes the occurrence of independent random events.
4. Velocity fluctuations caused by the ion oscillations, also known as ion back-bombardment problems.

It has also been shown by L. W. Swanson that current fluctuations arise from the generation of mobile atoms on net plane terraces for single-crystal tungsten emitters [35]. One of the most important advantages of Gr–Ni point cathodes over the conventional metal CFE emitters is the chemical inertness of its carbon surface, which is less likely to adsorb residual gas molecules and is therefore obviously much more stable [36]. In addition, a lower turn-on electric field is desirable for CFE electron sources since it will reduce the kinetic energies up to which the back-bombarding gas-ions are accelerated to when they collide with the cathode surface. On the negative side, there are step jumps and spikes in the short-term beam current, which are always observed at low and moderate emission current densities for CFE electron sources with nanoscale emission area [37, 38]. The magnitude of these step jumps and spikes can sometimes be up to double the size of the preceding current, and the frequency of these step jumps and spikes can be as low as once in every tens of seconds. These step jumps and spikes can be significantly improved by improving the vacuum conditions, but they cannot be eliminated for the moment.

The degree of current instability and damage to the cathode tip depends not only on the chemical inertness of the carbon surface but also on the size of the cathode tip. It is theoretically predicted that the root mean square (RMS) noise ratio ($<\delta I^2>^{1/2}/\bar{I}$) varies inversely with the emission surface area, under constant conditions of temperature and pressure. Since for the same emission angle, a larger radius tip has a greater area of emission, the relatively large diameter graphene-coated cathode tips (in the micrometer range) are therefore expected to have an order of magnitude lower RMS noise ratio values as compared to conventional tungsten cold field emitters. The success of Schottky electron beam sources is primarily due to its large physical emitting surface area.

In electron beam noise measurements, the mean square noise power $<\delta I^2>$ is usually obtained, which is related to the well-known spectral density function $W(f)$ as follows:

$$\delta I^2_{f_1...f_2} = \int_{f_1}^{f_2} W(f)df \,, \tag{14.14}$$

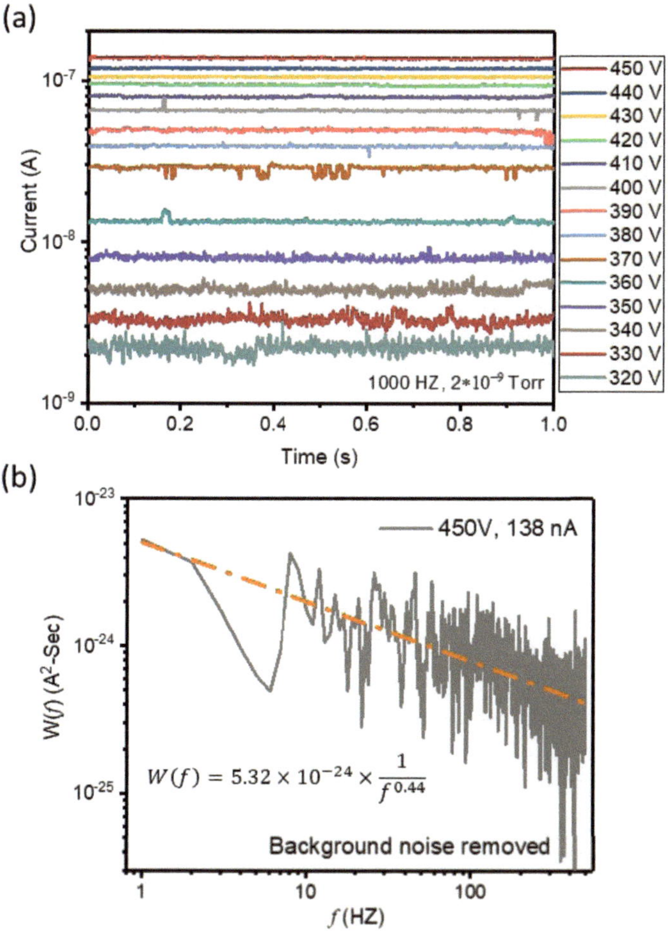

Figure 14.8 (a) Plot of current fluctuations from a graphene-coated Ni point-cathode CFE source at different extraction voltages with a sampling rate of 1000 Hz. (b) Power spectrum analysis of the probe current (138 nA) with a sampling rate of 1 kHz for a time span of 1 s.

Figure 14.8a presents a typical current–time trace from a Gr–Ni CFE source with a sampling rate of 1 kHz (a time span of 1 s), and the frequency dependence of $W(f)$, as defined in Eq. (14.13), is plotted in Fig. 14.8b. No flashing or other tip conditioning procedures were performed. An RMS noise ratio percentage, $<\delta I^2>^{1/2}/I = 0.576\%$, or a signal-to-noise ratio of 173.6 is obtained with a high measured current of 138 nA. The data also shows a $1/f$ roll off for $W(f)$, which was carried out for I = 138 nA at an extraction voltage of 450 V from a Gr–Ni CFE electron source. Due to the $1/f$ nature of the noise spectra, the emission noise is confined to the lower frequencies of the spectrum. The noise components at the high frequency part are entirely due to shot noise. This measured value is lower than that obtained from a conventional tungsten CFE emitter (2–5% at 10^{-10} Torr, T = 25 °C) [39], and is comparable to that obtained from CNT emitters (~0.2% at 6×10^{-8} Torr, T = 500 °C, I = 2.4 nA) [37] and Schottky emitters (~0.23% at 10^{-8} Torr, T = 1527 °C, I = 30 nA) [40] for the same frequency range.

14.5.2 Long-Term Current Drift

The failure to successfully apply many nanotips with promising electron optics performance to electron beam instruments is largely due to limitations in lifetime. The erratic and unstable performance in a UHV environment is caused by the sputtering effect from nanoscale emitting sites. Even for the successful commercial single-crystal tungsten CFE electron sources, the limitation in lifetime still exists. Their total emission current decays significantly (about 70–90% of the initial value) within the first hour of usage. Subsequently, regular cathode flashing (Joule heating at high temperatures) to blow-off the residual contaminant molecules at intervals as short as a few hours is required in order to reach original emission current levels. The following mechanisms given by Dyke's group may help explain the large long-term current drift [5]:

1. Localized changes in cathode work function, caused by the selective adsorption of electronegative gas such as oxygen, result in changes in the emission pattern and a decrease of total current with time.

2. Changes in β caused by deformation of the emitting surface. A surface projection tends to grow outward or extrude when either the temperature is high enough to cause appreciable surface migration or when surface atoms gain energy from ionic bombardment.

Later investigations by other researchers show that the ions formed near the emitter surface, and centered on the emitter axis, will impinge on the actual emitting area. Other ions generated from electron-stimulated desorption at the anode electrode will hardly be able to reach the emitting area. It was suggested that the ion bombardment effect will only dominate when $P \times I_t > 7 \times 10^{-12}$ PaA [41]. However, it should be noted that the energies of those ions impinging on the cathode are only ~10% of the inter-electrode electric potential (V_{ext}) [42]. Excessive ions may gather under a very high emission current around the emission surface and eventually trigger a catastrophic arc, which will end the life for the CFE source.

The long-term performance of the Gr–Ni CFE source has been studied for a sampling rate of 2 Hz in our lab. Before the long-term operation, the tip was carefully investigated under an SEM and a voltage–current measurement was carried out. No obvious current drift or decay was observed for Gr–Ni CFE sources, so thermal flashing is not required. The latest results show that the emission current from a Gr–Ni CFE source in UHV vacuum conditions is stable with no degradation over 6 months, which may be attributed to its ultralow work function and relatively low applied electric field strength at the tip apex. The structural robustness of Gr–Ni CFE sources against ion back-bombardment is confirmed by SEM characterizations.

14.6 Summary

High brightness, low energy spread, and good current stability are critical for CFE electron sources for electron microscopy applications. Although considerable research has been carried out into developing single-crystal tungsten and other nanotip CFE sources, they have as yet faced many practical technological difficulties, such as stringent UHV requirements, poor current stability, and short lifetime. This chapter highlighted some recent promising developments in the category of single-point Gr–Ni CFE electron sources for electron microscopy applications. These current results establish the promising prospect of using them as high brightness high-resolution electron sources for electron microscopy and lithography applications, similar in performance to conventional single-crystal tungsten cathode CFE sources, while at the same time having better emission stability and less stringent vacuum requirements. However, the investigation of graphene-based CFE electron sources is still in an early stage of development, and there is still much room for further improvement. More emission characterization tests need to be performed, ones that can measure the source virtual source size, transverse coherent length, and emitter lifetime. Current fluctuations remain a practical challenge, and the mechanism of step jumps and spikes noises is not yet clear. An electron gun unit that can accommodate promising graphene emitters also needs to be developed in the future, which has its optical axis well aligned to the central axis of the cathode. Beyond that, electron guns based upon the new class of graphene-based field emitters need to be fitted on to electron microscopes, and their performance critically compared to conventional systems, in terms of parameters such as image resolution and signal-to-noise ratio.

References

1. R. W. Wood, "A new form of cathode discharge and the production of X-rays, together with some notes on diffraction. Preliminary communication," *Physical Review (Series I)*, vol. 5, no. 1, pp. 1–10, 1897.
2. R. H. Fowler and L. Nordheim, "Electron emission in intense electric fields," *Proceedings of the Royal Society of London. Series A, Containing Papers of a Mathematical and Physical Character*, vol. 119, no. 781, pp. 173–181, 1928.
3. R. H. Good and E. W. Müller, "Field emission," in *Electron-Emission Gas Discharges I/Elektronen-Emission Gasentladungen I*: Springer, pp. 176–231, 1956.
4. R. Gomer, "Current fluctuations from small regions of adsorbate covered field emitters: A method for determining diffusion coefficients on single crystal planes," *Surface Science*, vol. 38, no. 2, pp. 373–393, 1973.
5. W. Dyke and W. Dolan, "Field emission," in *Advances in Electronics and Electron Physics*, vol. 8: Elsevier, pp. 89–185, 1956.

6. A. Crewe, J. Wall, and L. Welter, "A high-resolution scanning transmission electron microscope," *Journal of Applied Physics*, vol. 39, no. 13, pp. 5861–5868, 1968.

7. A. V. Crewe, M. Isaacson, and D. Johnson, "A simple scanning electron microscope," *Review of Scientific Instruments*, vol. 40, no. 2, pp. 241–246, 1969.

8. H. Zhang et al., "An ultrabright and monochromatic electron point source made of a LaB 6 nanowire," *Nature Nanotechnology*, vol. 11, no. 3, pp. 273–279, 2016.

9. X. Shao, A. Srinivasan, Y. Zhao, and A. Khursheed, "A few-layer graphene ring-cathode field emitter for focused electron/ion beam applications," *Carbon*, vol. 110, pp. 378–383, 2016.

10. H.-S. Kuo, S. Hwang, T.-Y. Fu, Y.-C. Lin, C.-C. Chang, and T. T. Tsong, "Noble metal/W (111) single-atom tips and their field electron and ion emission characteristics," *Japanese Journal of Applied Physics*, vol. 45, no. 11R, pp. 8972–8983, 2006.

11. L. Swanson and G. Schwind, "A review of the cold-field electron cathode," *Advances in Imaging and Electron Physics*, vol. 159, pp. 63–100, 2009.

12. https://www.hitachi hightech.com/global/science/technical/tech/microscopes/cold_fe/ (accessed April 20, 2020).

13. A. K. Geim and K. S. Novoselov, "The rise of graphene," in *Nanoscience and Technology: A Collection of Reviews from Nature Journals*: World Scientific, pp. 11–19, 2010.

14. D. Ye, S. Moussa, J. D. Ferguson, A. A. Baski, and M. S. El-Shall, "Highly efficient electron field emission from graphene oxide sheets supported by nickel nanotip arrays," *Nano Letters*, vol. 12, no. 3, pp. 1265–1268, 2012.

15. G. Eda, H. Emrah Unalan, N. Rupesinghe, G. A. Amaratunga, and M. Chhowalla, "Field emission from graphene based composite thin films," *Applied Physics Letters*, vol. 93, no. 23, pp. 233502(1)–233502(3), 2008.

16. Z. S. Wu et al., "Field emission of single-layer graphene films prepared by electrophoretic deposition," *Advanced Materials*, vol. 21, no. 17, pp. 1756–1760, 2009.

17. K. S. Novoselov et al., "Two-dimensional gas of massless Dirac fermions in graphene," *Nature*, vol. 438, no. 7065, pp. 197–200, 2005.

18. A. A. Balandin et al., "Superior thermal conductivity of single-layer graphene," *Nano Letters*, vol. 8, no. 3, pp. 902–907, 2008.

19. G. Giovannetti, P. Khomyakov, G. Brocks, V. v. Karpan, J. Van den Brink, and P. J. Kelly, "Doping graphene with metal contacts," *Physical Review Letters*, vol. 101, no. 2, pp. 026803(1)–026809(4), 2008.

20. K. C. Kwon, K. S. Choi, B. J. Kim, J.-L. Lee, and S. Y. Kim, "Work-function decrease of graphene sheet using alkali metal carbonates," *The Journal of Physical Chemistry C*, vol. 116, no. 50, pp. 26586–26591, 2012.

21. X. Shao and A. Khursheed, "A review paper on "Graphene Field Emission for Electron Microscopy"," *Applied Sciences*, vol. 8, no. 6, pp. 868–882, 2018.

22. X. Shao, A. Srinivasan, W. K. Ang, and A. Khursheed, "A high-brightness large-diameter graphene coated point cathode field emission electron source," *Nature Communications*, vol. 9, no. 1, pp. 1–8, 2018.

23. R. D. Young, "Theoretical total-energy distribution of field-emitted electrons," *Physical Review*, vol. 113, no. 1, pp. 110–114, 1959.

24. A. Bahm, G. Schwind, and L. Swanson, "Range of validity of field emission equations," *Journal of Vacuum Science & Technology B: Microelectronics and Nanometer Structures Processing, Measurement, and Phenomena*, vol. 26, no. 6, pp. 2080–2084, 2008.

25. H. Boersch, "Experimentelle Bestimmung der Energieverteilung in thermisch ausgelösten Elektronenstrahlen," *Zeitschrift für Physik*, vol. 139, no. 2, pp. 115–146, 1954.

26. W. Knauer, "Energy broadening in field emitted electron and ion beams," *Optik*, 1981, 59, 335–354.

27. P. Kruit and G. H. Jansen, "Space charge and statistical Coulomb effects," in *Handbook of Charged Particle Optics*: CRC Press, pp. 341–389, 2017.

28. A. Bell and L. Swanson, "Total energy distributions of field-emitted electrons at high current density," *Physical Review B*, vol. 19, no. 7, pp. 3353–3364, 1979.

29. K. H. Loeffler, "Ph.D. thesis," University of Berlin, 1964.

30. B. Cook, T. Verduin, C. Hagen, and P. Kruit, "Brightness limitations of cold field emitters caused by Coulomb interactions," *Journal of Vacuum Science & Technology B, Nanotechnology and Microelectronics: Materials, Processing, Measurement, and Phenomena*, vol. 28, no. 6, pp. C6C74–C6C79, 2010.

31. J. Wiesner and T. Everhart, "Point-cathode electron sources-Electron optics of the initial diode region: Errata and addendum," *Journal of Applied Physics*, vol. 45, no. 6, pp. 2797–2798, 1974.

32. M. Bronsgeest, J. Barth, L. Swanson, and P. Kruit, "Probe current, probe size, and the practical brightness for probe forming systems," *Journal*

of Vacuum Science & Technology B: Microelectronics and Nanometer Structures Processing, Measurement, and Phenomena, vol. 26, no. 3, pp. 949–955, 2008.

33. G. Schwind, G. Magera, and L. Swanson, "Comparison of parameters for Schottky and cold field emission sources," *Journal of Vacuum Science & Technology B: Microelectronics and Nanometer Structures Processing, Measurement, and Phenomena*, vol. 24, no. 6, pp. 2897–2901, 2006.
34. L. Swanson and N. Martin, "Field electron cathode stability studies: Zirconium/tungsten thermal-field cathode," *Journal of Applied Physics*, vol. 46, no. 5, pp. 2029–2050, 1975.
35. L. Swanson, "Current fluctuations from various crystal faces of a clean tungsten field emitter," *Surface Science*, vol. 70, no. 1, pp. 165–180, 1978.
36. P. Blake et al., "Graphene-based liquid crystal device," *Nano Letters*, vol. 8, no. 6, pp. 1704–1708, 2008.
37. N. de Jonge, M. Allioux, J. T. Oostveen, K. B. Teo, and W. I. Milne, "Low noise and stable emission from carbon nanotube electron sources," *Applied Physics Letters*, vol. 87, no. 13, pp. 133118(1)–133118(3), 2005.
38. K. Yeong and J. Thong, "Field-emission properties of ultrathin 5 nm tungsten nanowire," *Journal of Applied Physics*, vol. 100, no. 11, p. 114325, 2006.
39. S. Yamamoto, S. Fukuhara, N. Saito, and H. Okano, "Stabilization of field emission current," *Surface Science*, vol. 61, no. 2, pp. 535–549, 1976.
40. D. Tuggle, L. Swanson, and J. Orloff, "Application of a thermal field emission source for high resolution, high current e-beam microprobes," *Journal of Vacuum Science and Technology*, vol. 16, no. 6, pp. 1699–1703, 1979.
41. N. Saitou, "Trajectory analysis of ions formed in the field emitter inter-electrode region," *Surface Science*, vol. 66, no. 1, pp. 346–356, 1977.
42. H. Todokoro, N. Saitou, and S. Yamamoto, "Role of ion bombardment in field emission current instability," *Japanese Journal of Applied Physics*, vol. 21, no. 10R, pp. 1513–1516, 1982.

Chapter 15

CNT Field-Emission Cathode for Space Applications

Yasushi Ohkawa
Research and Development Directorate, Japan Aerospace Exploration Agency, 7-44-1, Jindaiji-Higashi-machi, Chofu, Tokyo 182-8522, Japan
okawa.yasushi@jaxa.jp

A CNT field-emission cathode is a promising candidate for small electron emitters needed in space for various applications. This chapter describes the details of the CNT cathodes that have been tested in low Earth orbit in 2017.

15.1 Introduction

There is a demand for electron emission devices for various applications in space. The major application is electric propulsion [1], which in recent years has been employed in many satellites and probes. In electric propulsion, electron emission devices are required for plasma production and neutralization of the positive ion beam.

An electrodynamic tether (EDT) [2] also needs an electron emitter. The EDT is a propulsion system that uses no consumable

Nanostructured Carbon Electron Emitters and Their Applications
Edited by Yahachi Saito

ISBN 978-981-4877-62-6 (Hardcover), 978-1-003-14199-0 (eBook)
www.jennystanford.com

propellant but instead utilizes the interaction between the geomagnetic field and electric current flowing through the long conductive tether. An active cathode to emit electrons to the ionospheric plasma is required for a certain level of current to flow through the tether. There are also other applications for electron emission devices, such as for suppressing discharge problems caused by excessive charging of spacecraft in orbit, and for various scientific experiments.

There are various types of electron emission devices, depending on the current levels required, lifetime requirements, and compatibility with the on-orbit systems. A hollow cathode using a combination of thermionic electron emission and plasma discharge is a typical example of cathodes for space. However, it is difficult to apply complicated cathodes using gas and heaters to small electric propulsion such as miniaturized ion engines, electrospray thrusters, and EDTs, which are expected to be used for orbital debris deorbit, because of the high system impact. For this reason, a small and simple cathode system is required.

A field-emission cathode featuring a simple structure that only applies voltage between the electrodes is a promising candidate for an electron emitter to meet such a demand. In particular, carbon nanotube (CNT)-based field-emission cathodes (CNT cathodes) are considered to be promising, and extensive related research and development have been conducted such as the CNT-based emitter arrays developed at the Georgia Institute of Technology for the ALICE mission [3], BUSEK's CNT cathodes for the LISA-Pathfinder mission [4], and the miniaturized CNT cathodes developed at TU Dresden for the UWE-4 satellite [5].

This chapter describes the technologies and results of an in-space experiment of the CNT cathode developed for the EDT demonstration experiment called the Kounotori Integrated Tether Experiment (KITE) [6] as an example of the CNT cathodes for space.

15.2 Overview of KITE

The increase in orbital debris (or space debris) is becoming a major concern for human space activities [7]. One of the key technologies to resolve this problem is a propulsion system for the efficient orbital transfer of space debris. The EDT is one of the candidates for this propulsion system, and the KITE project was carried out to demonstrate its key technologies and raise the technological readiness levels.

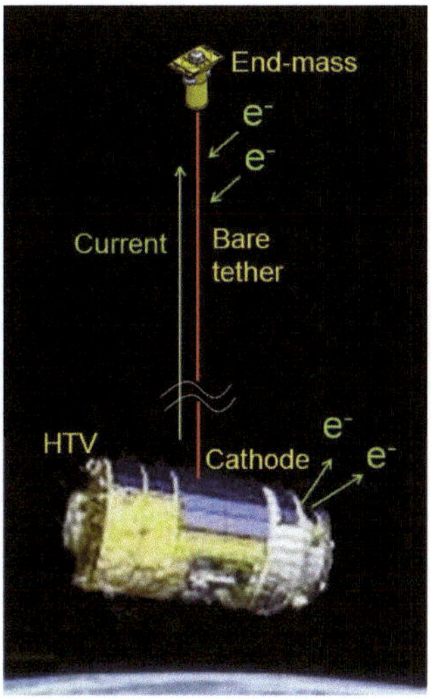

Figure 15.1 Planned outline image of KITE [6].

Figure 15.1 shows the planned outline image of KITE. In this experiment, the H-II Transfer Vehicle 6 (HTV-6 or Kounotori-6) was used as the experimental platform [8]. The HTV is Japan's unmanned cargo transfer spacecraft (measuring about 4 m in diameter and about 10 m in length), and its main mission is to transport supplies to the International Space Station (ISS). In KITE, it was planned to deploy a bare conductive tether (720 m long) from the HTV, and electrical current of about 10 mA was expected to flow through the tether via electron emission from the CNT cathodes. The 1-week mission was scheduled to begin after the HTV left the ISS.

KITE was started on January 28, 2017 (GMT) at an altitude of approximately 370 km and an inclination angle of about 52°. Unfortunately, the EDT system demonstration was not possible as the end-mass was not released due to a mechanical malfunction, but the CNT cathode performed well throughout the 8-day mission period, including an additional day.

15.3 CNT Cathode for KITE

15.3.1 CNT Cathode Module

Figure 15.2 shows the CNT cathode module consisting of the cathodes and related components mounted on the HTV. The left photo in this figure was taken from the ISS shortly after the release of the HTV.

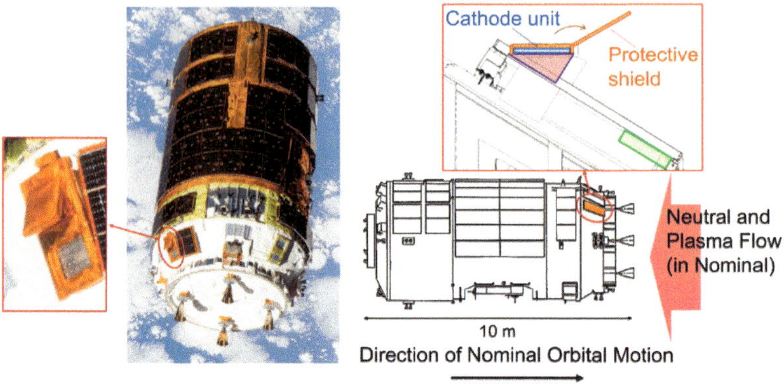

Figure 15.2 CNT cathode module mounted on HTV [6, 9].

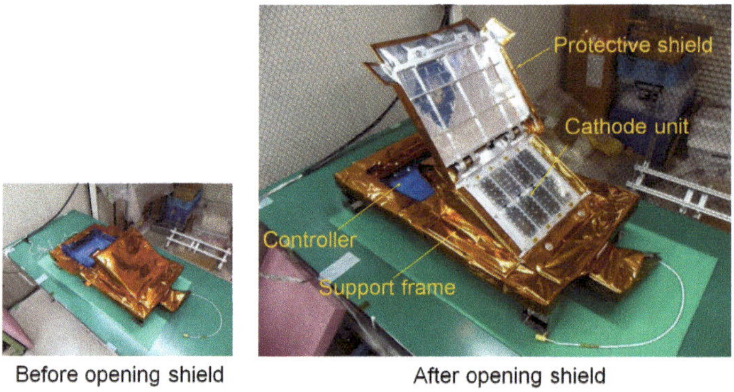

Figure 15.3 CNT cathode module [6].

Figure 15.3 shows the appearance of the CNT cathode module. This module consists of a cathode unit, controller, protective shield, and support frame. The cathode unit consists of eight independent CNT cathodes. The controller includes the high-voltage power supplies and associated electronics to drive the cathodes. The protective shield is designed to protect the CNT cathodes from small-flake contamination and the atomic oxygen/plasma on orbit, and was opened just prior to initiating CNT cathode operation. After opening, the shield acts as a screen for neutrals and plasma flow.

The support frame includes a triangular base for ensuring that the electron emission surface of the CNT cathodes is parallel to the HTV's nominal orbital motion (as shown in Fig. 15.2). The gold-colored film that covers the entire module is multilayered thermal insulation.

15.3.2 Structure and Characteristics of CNT Cathode

Figure 15.4 shows the appearance of the CNT cathode, a schematic drawing of the electrode structure, and a microscopic image of the CNT emitter. Multiwalled carbon nanotube made by an arc-discharge process was used as the emitter material. The CNT emitter was formed by screen-printed CNT paste on the molybdenum substrate. CNTs are randomly distributed on the substrate as shown in Fig. 15.4. A mask electrode made of a thin molybdenum plate is located on the CNT emitter to reduce electron impingement to the gate electrode [10]. A spacer made of polyimide is inserted between the mask and gate electrode to maintain the appropriate resistance between them. The slot-shaped open area of both electrodes is aligned and the open area fraction is approximately 80%. The footprint of the cathode is $80 \times 40\ mm^2$.

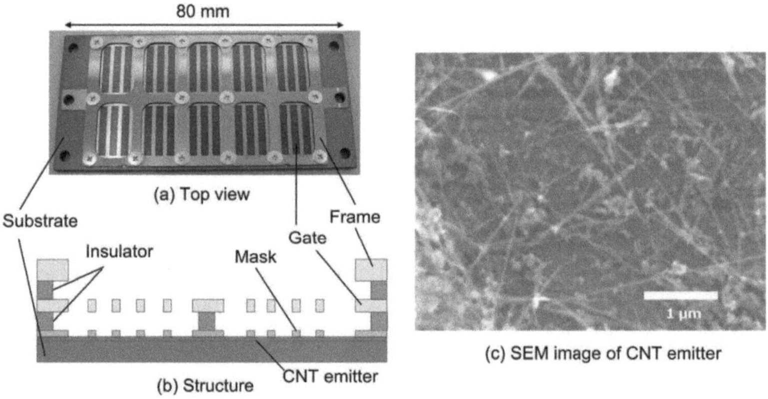

Figure 15.4 CNT cathode [9].

Figure 15.5 shows the typical current–voltage (I–V) characteristics of the CNT cathode in a vacuum chamber test [11]. Here, the emitter current refers to the all-electron current emitted from the CNT emitter, and the gate current refers to the loss current that flows into the gate electrode. In this experiment, an anode plate for electron collection was placed 200 m above the CNT cathode, and its biasing voltage was 200 V relative to the emitter potential. The maximum emitter current of a single cathode was set at 2.2 mA by considering the durability and output current limit of the onboard high-voltage converter. This figure shows that about 20% of the emitter current flows into the gate electrode, that is, the electron extraction efficiency in this case is approximately 80%.

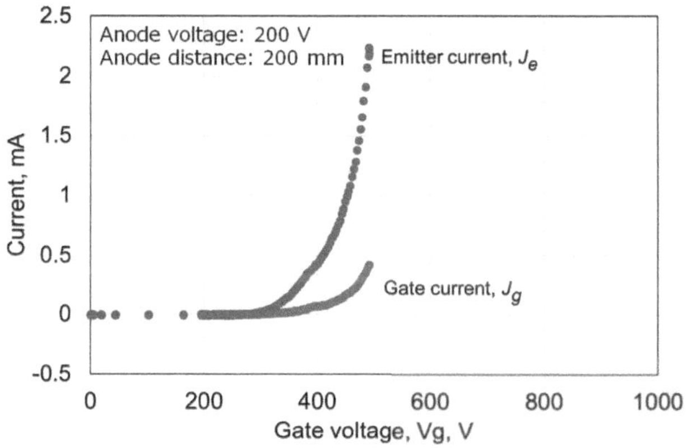

Figure 15.5 Typical I–V characteristics of CNT cathode [11].

15.3.3 Electrical Circuit and Operation of CNT Cathode

Figure 15.6 shows a simplified schematic diagram of the electrical circuit (installed in the controller shown in Fig. 15.3) [11] for driving the CNT cathodes. Eight CNT cathodes (CH1 to CH8) arranged in parallel are driven by eight individual high-voltage converters so that failure in one cathode system does not affect the operation of other cathode channels. Miniaturized high-voltage converters for commercial applications were used in KITE. The emitters of the CNT cathodes were connected to the HTV body, and the gate electrodes were positively biased up to approximately 900 V relative to the HTV body.

The CNT cathodes were operated using the onboard software that includes some operation modes. In I–V measurement mode, the gate voltage is automatically swept twice up to the configurable gate current levels for obtaining the I–V characteristics. In autonomous operation mode, the gate voltage of each cathode is controlled automatically so that the gate current remains constant. This mode was provided so that CNT cathode operation does not become unstable even when the potential difference between the ambient space plasma and the HTV body (corresponding to the anode voltage in the ground test) fluctuates on-orbit. In all operating modes, a single or multiple cathode can be operated optionally.

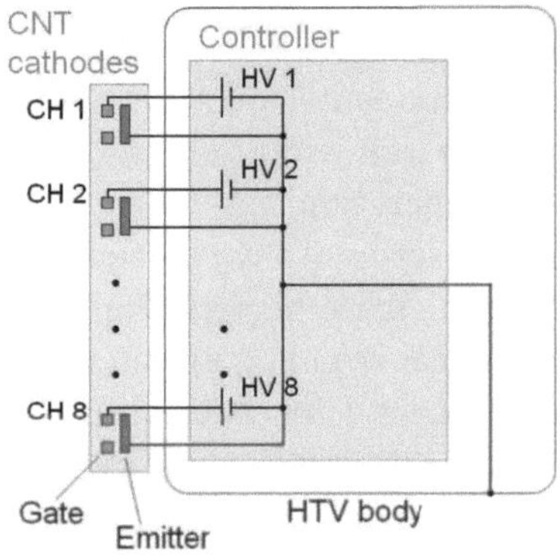

Figure 15.6 Simplified circuit diagram for CNT cathodes [11].

The software also possesses some anomaly detection functions. For example, if high-voltage breakdown or excessive temperature is detected on a specific cathode channel, the operation of that channel is halted automatically.

15.3.4 Unique Treatment for Use in Space

15.3.4.1 Consideration of atomic oxygen in low Earth orbit

Atomic oxygen (AO) is a dominant gas species in low Earth orbit (LEO). AO is chemically active and collides with spacecraft at orbital velocity (e.g., approximately 7.7 km/s at an altitude of 370 km in KITE), causing damage to the exposed materials of the spacecraft. Since AO is expected to deteriorate the CNT emitters, it is necessary to evaluate the effects and prepare countermeasures in advance.

In KITE, the CNT cathode module was arranged so that the electron emission surface was parallel to the direction of orbital motion when the HTV was in the nominal attitude (as shown in Fig. 15.2), because the results of preliminary experiments [12] indicated that the CNT cathodes could not withstand long-term AO irradiation from the front side at the altitude in KITE. The reason why the module was not placed on the back side of the HTV is that the wake region of the plasma where the number density is low is formed on this side, and the electron extraction efficiency from the CNT cathode to the plasma is expected to decrease.

15.3.4.2 Tolerance to mechanical and thermal environments

In general, space components must be able to withstand harsh environments such as the vibration during launch, mechanical impact during stage separation, and thermal cycles on orbit. In the development of KITE, a series of environmental tests were conducted to evaluate the tolerance to such specific environments in space.

As for the mechanical strength, vibration tests were performed for the cathode unit, controller, and entire CNT cathode module after assembly by considering the vibration environment during launch. In addition, an impact test was performed for the assembled CNT cathode module by assuming the environment during stage separation.

In LEO, a spacecraft orbits the Earth in about 90 min, so it passes through alternately sunlit and eclipse in a short time depending on the orbit. The temperature of a spacecraft is dominated by thermal radiation in space, so the spacecraft experiences a wide range of temperature cycles. In particular, the devices exposed to space, such as the CNT cathodes, are subjected to severe thermal cycling due to the limitations of passive and/or active thermal treatment to narrow the temperature range.

During the development of the CNT cathode module for KITE, thermal cycling tests were performed in a vacuum to evaluate tolerance to the space environment. Especially for the cathode unit, the tolerance was carefully confirmed in advance given the concerns about electrical breakdown in the short gap electrode structure due to the repetition of thermal expansion and contraction. The cathode unit was subjected to temperature cycling in the range of approximately −50 to +125 °C with electron emission, resulting in no operation-related problems being confirmed.

15.4 Results and Findings of On-Orbit Operation of CNT Cathodes

15.4.1 Overview of Cathode Operation

Table 15.1 lists the major events in CNT cathode operation [6] in KITE. On the first day (January 28, 2017), a low voltage (~30 V) was applied to the gate electrodes of the eight cathodes to confirm the health of each cathode and the electrical circuit. The protective shield was then opened on Day 3 and the initial I–V characteristics on-orbit were acquired at several current levels. For the next 3 days, autonomous and I–V measurement operations were conducted on a daily basis cycle. On Day 7, the HTV attitude was temporarily changed from the nominal attitude (where the electron emission surface of the CNT cathodes is parallel to the direction of orbital motion) to the attitude where AO impinges the CNT cathodes directly, and its effect was evaluated. The I–V characteristics were finally acquired on Day 8, the last day of the mission.

Table 15.1 Summary of CNT cathode operation events

Day	Event
Day 1	Health check
Day 2	No operation
Day 3	Opening of protective shield/measurement of Initial I–V characteristics
Day 4	Autonomous operation at low current level
Day 5	Measurement of Midterm I–V characteristics
Day 6	Autonomous operation at high current level
Day 7	Autonomous operation at different HTV attitudes
Day 8	Measurement of Final I–V characteristics

Table 15.2 lists the important statistics on CNT cathode operation. The CNT cathodes were exposed to space for about 130 h, of which the electron emission operation was conducted for about 50 h. The maximum total electron current emitted from the CNT

emitter (defined as the emitter current) was approximately 10 mA, of which the maximum total electron current extracted outside the CNT cathodes (defined as the emission current) was approximately 6 mA. These current levels are the largest in-orbit operation of the CNT cathode as of now (i.e., May 2020).

Table 15.2 Major statistics on CNT cathode operation

Accumulated exposure time	130 h
Accumulated operation time	50 h
Maximum emitter current (sum of eight cathodes)	10.2 mA
Maximum emission current (sum of eight cathodes)	5.8 mA

The eight CNT cathodes worked well throughout the 8-day mission period without any malfunction. The I–V measurement, autonomous operation, and overcurrent detection function provided by the onboard software were also confirmed as being effective in a fluctuating space plasma environment.

15.4.2 I–V Characteristics and Tolerance to Atomic Oxygen Environment On-Orbit

In the on-orbit operation, the I–V characteristics of eight CNT cathodes were acquired at three different timings: "Initial" (about 10 h after opening of the protective shield), "Midterm" (about 60 h after said opening), and "Final" (about 125 h after said opening and just before the end of the mission).

Figure 15.7 shows the I–V characteristics (i.e., relationship between emitter current and gate voltage) of the eight CNT cathodes at these three phases and at the final ground test before launch [6]. Each I–V was obtained by twice making ascending and descending sweeps of the gate voltage. The AO fluences (atoms/cm^2) noted in the figure was estimated using the MSIS-E-90 atmospheric model [13]. This figure shows that all eight cathodes operated successfully from the beginning to the end of the mission. The irregular plots observed in CH6 were probably due to output saturation of the high-voltage converter under low-temperature conditions.

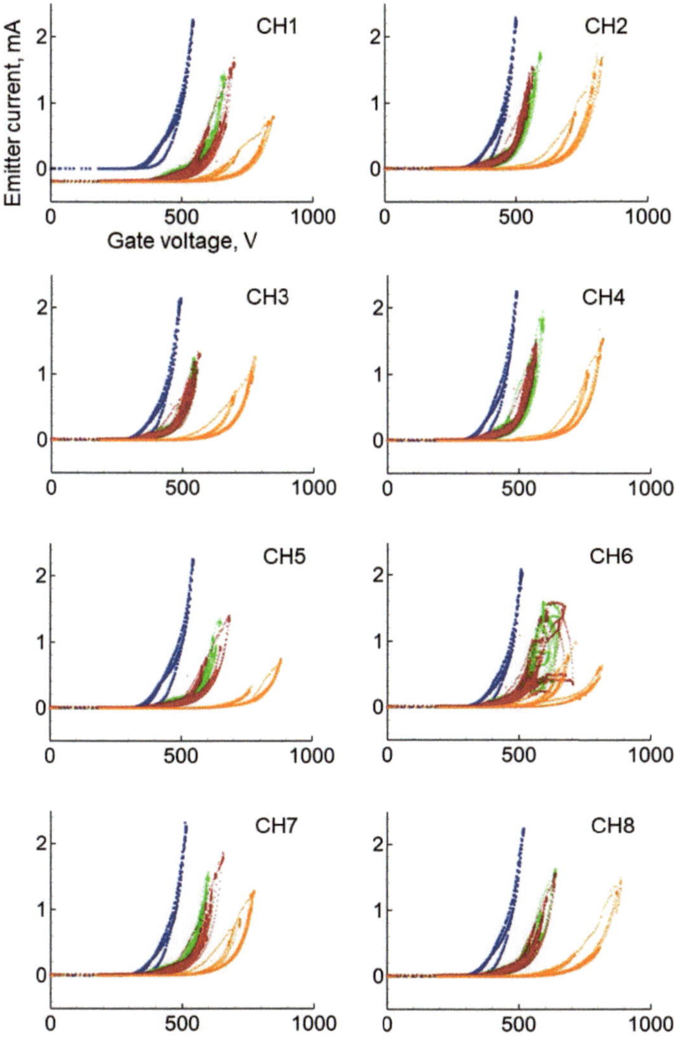

"Before launch" (No AO fluence)
"Initial on-orbit" (After 10-h exposure (parallel to AO flow). AO fluence of 3×10^{18} atoms/cm^2)
"Midterm on-orbit" (After 60-h exposure (parallel to AO flow). AO fluence of 1×10^{19} atoms/cm^2)
"Final on-orbit" (After 125-h exposure (parallel to AO flow) & 5-h exposure (at front side)). AO fluence of 2×10^{19} atoms/cm^2)

Figure 15.7 I–V characteristics of eight CNT cathodes before and during KITE [6]. (AO fluences were estimated using MSIS-E-90 Atmosphere Model [13].)

Three important indications can be seen from Fig. 15.7 as follows:

(i) Comparing the final ground test and "Initial," there was a certain level of performance change between them. This degradation may be attributed to long-term storage (6 months) in the atmosphere before launch, a long waiting period (a month and a half) in LEO inside the protective shield, or short-time exposure (about 10 h) after the shield opening, although the specific cause has yet to be identified. The important point is that the degradation did not proceed as shown below.

(ii) Comparing "Initial" and "Midterm," virtually no performance deterioration was observed. This comparison indicates that the CNT cathode is at least tolerable to AO fluence of 1×10^{19} atoms/cm^2. This result is important in applying the CNT cathode to various LEO missions because the AO flux at the KITE altitude (370 km) is several orders higher than that at an altitude of 600–1400 km, where many LEO satellites are operated. For instance, the total AO fluence at "Midterm" is comparable to that estimated for a 1-year mission at 600-km altitude; therefore, a service life of more than 1 year can be expected for CNT cathodes used in regular LEO missions in terms of AO resistance.

(iii) Comparing "Midterm" and "Final," there were large shifts in the I–V curves, which mean that the electron emission characteristics in "Final" were considerably deteriorated. This performance degradation probably occurred during the 5 h of frontal AO exposure to the CNT cathodes by changing the attitude of the HTV as expected from the preliminary experiments [12].

These results indicate that the CNT cathode can be operated for a long period of time, even in LEO where active AO exists, by avoiding the frontal impingement of AO.

15.4.3 Electron Emission Behavior to Ambient Space Plasma

Estimating how much electrons from the CNT cathodes can reach the ambient space plasma by overcoming the space-charge limitation is important for evaluating CNT cathode capabilities in practical use. Given the difficulty for ground-based experiments to simulate the plasma and magnetic environment in space, on-orbit experiments are indispensable for a quantitative evaluation.

Figure 15.8 shows an example of the relationship between the electron extraction efficiency and the potential difference between the CNT cathode and space plasma obtained in simultaneous operation of all eight cathodes [9]. A result of the final ground experiment conducted with the plate anode located 200 mm above the cathode is also shown in the figure for comparison purposes.

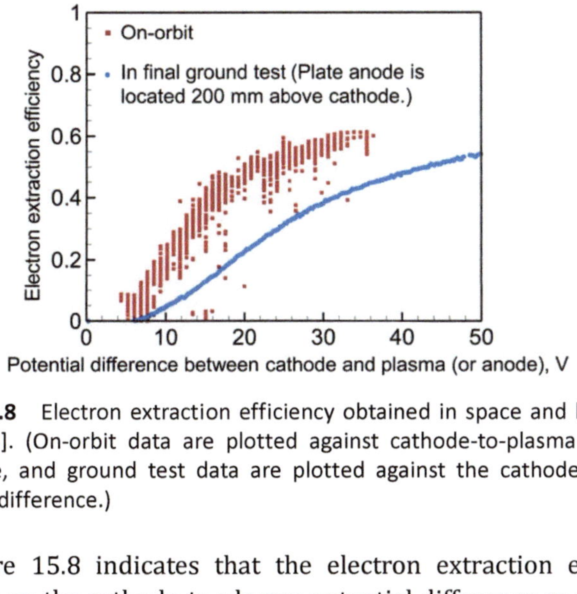

Figure 15.8 Electron extraction efficiency obtained in space and by ground testing [9]. (On-orbit data are plotted against cathode-to-plasma potential difference, and ground test data are plotted against the cathode-to-anode potential difference.)

Figure 15.8 indicates that the electron extraction efficiency depends on the cathode-to-plasma potential difference as expected theoretically and experimentally. An interesting point is that the efficiency on-orbit was better than that in the ground experiment. This is probably attributable to the underestimation of anode current due the electron reflection or secondary electron emission in the ground test. The efficiencies under various plasma conditions were obtained in KITE, and those data will prove important in estimating the on-orbit characteristics when conducting ground experiments in the future.

As another interesting finding related to space plasma interactions, the HTV's electrical charging state relative to the ambient space plasma was observed to change by the CNT cathode operation [11]. For instance, negative charging of the HTV body was mitigated approximately 15 V by the electron emission of about 2 mA from the CNT cathodes. This result suggests that the CNT cathode can be used as a spacecraft charging mitigation device.

15.5 Next Steps

Following the successful on-orbit demonstration of CNT cathode operation in KITE, studies on improving the cathode characteristics are being conducted by using some revised samples as shown in Fig. 15.9.

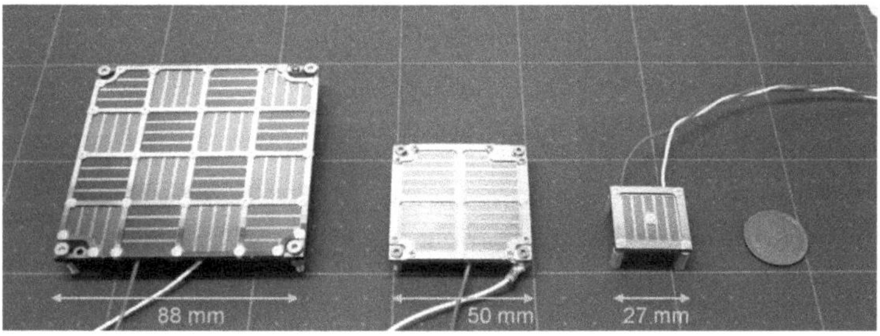

Figure 15.9 CNT cathodes in research and development following KITE.

One of the new promising applications of the CNT cathode using such devices is the neutralizer of a miniaturized throttleable ion engine [14]. As this ion engine features a wide variable thrust range by the high-frequency switching of the ion beam, the electron source for ion beam neutralization should also have a fast response, and in that respect, the field-emission CNT cathode has a high affinity.

Another application example of the CNT cathode is the EDT for post mission disposal (PMD) [15]. PMD means that the spacecraft after its own mission deorbits to prevent itself from becoming orbital debris, and the devices for PMD should be small and lightweight. Given its simplicity, the CNT cathode is suitable for the EDT devices used for PMD.

These examples suggest the continued use of the CNT cathode in space in the future.

15.6 Conclusion

This chapter mainly described CNT cathodes that have been tested in space as the electron source for electrodynamic tether propulsion. As the CNT cathode system is essentially a simple and low-power system compared with other types of electron emitters, it is promising and competitive as an electron emission device for space applications, especially for low-current level options. Although the durability of the CNT cathode in low Earth orbit has been a major concern for its space application, our on-orbit demonstration experiment showed that the CNT cathode can be sufficiently tolerable in low Earth orbit when the direct impact of atomic oxygen is avoided by some geometrical treatment. The use of CNT cathodes in space is thus expected to expand by making further improvements in CNT materials and device structures.

Acknowledgments

The author wishes to thank the KITE and HTV teams for their significant efforts in pursuing the project.

References

1. Jahn, R. G. (1968). *Physics of Electric Propulsion* (McGraw-Hill).
2. Drell, S. D., Foley, H. M. and Ruderman, M. A. (1965). Drag and Propulsion of Large Satellites in the Ionosphere: An Alfven Propulsion Engine in Space, *J. Geophys. Res.*, **70**, pp. 3131–3145.
3. Williams, L. T., Kumsomboone, V. S., Ready, W. J. and Walker, M. L. R. (2010). Lifetime and Failure Mechanisms of an Arrayed Carbon Nanotube Field Emission Cathode, *IEEE Trans. Electron Devices*, **57**, pp. 3163–3168.
4. Ziemer, J, Marrese-Reading, C., Dunn, C., Romero-Wolf, A., Cutler, C., Javidnia, S., Le, T., Li, I., Franklin, G., Barela, P., Hsu, O., Maghami, P., O'Donnell, J., Slutsky, J., Thorpe, J. I., Demmons, N. and Hruby, V. (2017). Colloid Microthruster Flight Performance Results from Space Technology 7 Disturbance Reduction System, *Proc. 35th Int'l Electric Propulsion Conf.*, IEPC-2017-578.
5. Laufer, P., Bock, D. and Tajmar, M. (2018). CNT Cold Electron Emitters for Space Applications, *Proc. 31st Int'l Vac. Nanoelectronics Conf.*, 04-3.

6. Ohkawa, Y., Kawamoto, S., Okumura, T., Iki, K., Okamoto, H., Inoue, K., Uchiyama, T. and Tsujita, D. (2020). Review of KITE: Electrodynamic Tether Experiment on the H-II Transfer Vehicle, *Acta Astronautica*, **177**, pp. 750–758.
7. Kessler, D. J. and Cour-Palais, B. G. (1978). Collision Frequency of Artificial Satellites: The Creation of a Debris Belt, *J. Geophys. Res.*, **83**, pp. 2637–2646.
8. Tsujita, D., Kasai, T., Uematsu, H., Harada, M., Fukatsu, T. and Sasaki, H. (2014). Experiments Plan on the HTV, *Trans. JSASS Aerospace Tech. Jpn.*, **12**, pp. Tg_1–Tg_4.
9. Ohkawa, Y., Okumura, T., Iki, K., Okamoto, H. and Kawamoto, S. (2019). Operation of a Carbon Nanotube Field-Emission Cathode in Low Earth Orbit, *J. Vac. Sci. Tech. B*, **37**, pp. 02203.
10. Ohkawa, Y., Izawa, A., Yamagiwa, Y., Kawamoto, S., Nishida, S. and Kitamura, S. (2010). Research and Development of Carbon Nanotube Cathodes for Electric Propulsion, *Trans. JSASS Aerospace Tech. Jpn.*, **8**, pp. Pb_27–Pb_32.
11. Ohkawa, Y., Okumura, T., Kawamoto, S. and Kobayashi, Y. (2019). Charging Behavior of the H-II Transfer Vehicle by Active Electron Emission, *IEEE Trans. Plasma Sci.*, **47**, pp. 3852–3857.
12. Shimada, A., Tanaka, Y., Ohkawa, Y., Matsumoto, K., Tagawa, M., Matsui, M. and Yamagiwa, Y. (2014). Effect of Atomic Oxygen Irradiation on Field Emission Cathodes in Low Earth Orbit, *Trans. JSASS Aerospace Tech. Jpn.*, **12**, pp. Pb_59–Pb_64.
13. MSIS-E-90 Atmosphere Model (Online). Available: https://ccmc.gsfc.nasa.gov/modelweb/models/msis_vitmo.php.
14. Yamamoto, N., Morita, T., Ohkawa, Y., Nakano, M. and Funaki, I. (2019). Ion Thruster Operation with Carbon Nanotube Field Emission Cathode, *J. Propulsion & Power*, **35**, pp. 490–493.
15. Sato, T., Kawamoto, S., Watanabe, T., Kamachi, K., Ohkawa, Y., Oikawa, Y., Okajima, L. and Okubo, H. (2021). A Study on PMD Device for Microsatellites Using Electrodynamic Tether, *Trans. JSASS Aerospace Tech. Jpn.*, **19**, pp. 61–67.

Chapter 16

Growth of Long Linear Carbon Chains after Serious Field Emission from a CNT Film

Koji Asaka[a] and Yahachi Saito[b]

[a]*Department of Applied Physics, Graduate School of Engineering, Nagoya University, Furo-cho, Chikusa-ku, Nagoya, Aichi 464-8603, Japan*

[b]*Toyota Physical and Chemical Research Institute, Nagakute 480-1192, Japan*

asaka@nagoya-u.jp, ysaito@toyotariken.jp

Carbyne is a long one-dimensional chain of sp-hybridized carbon atoms and is known as one of the allotropes of carbon. Mechanical and electrical properties of carbyne expected from theoretical studies have attracted much interest because of many promising applications to advance nanotechnology and nanoscience, and much effort has been dedicated to the synthesis of carbyne. Recently, we found a novel way to produce a bulk amount of long linear carbon chains encapsulated in carbon nanotubes by field electron emission of single-wall carbon nanotube films. Here, we focus on Raman spectroscopy and electron microscopy characterization of the long linear carbon chains generated by field electron emission of single-wall carbon nanotube films.

Nanostructured Carbon Electron Emitters and Their Applications

Edited by Yahachi Saito

ISBN 978-981-4877-62-6 (Hardcover), 978-1-003-14199-0 (eBook)

www.jennystanford.com

16.1 Introduction

An infinite linear chain consisting of sp-hybridized carbon atoms is known as one of carbon allotropes and called carbyne [1]. Carbyne may theoretically take two possible configurations, i.e., the cumulene form with repeating double bonds throughout the chain and the polyyne form with alternating single and triple bonds. Due to a Peierls distortion, the conductive cumulene form may change into the semi- or nonconductive polyyne one, which leads to emergence of a band gap. According to theoretical studies, carbyne has been expected to possess unique mechanical [2], thermal [3], and electrical properties [4], and therefore attracted significant interest owing to the potential applications such as switching nanodevices [5], spintronic nanodevices [6], and hydrogen storage [7].

Long linear carbon chains encapsulated in multiwall carbon nanotubes (MWCNTs), which were produced by the arc-discharge method in hydrogen gas [8], liquid nitrogen [9, 10], argon gas [11], and helium gas [12, 13], have been reported. The linear carbon chains were also fabricated by high-temperature fusion reaction of molecules such as $C_{10}H_2$ [14] and adamantane [15] within double-wall carbon nanotubes (DWCNTs). Furthermore, the linear carbon chains consisting of more than 6000 atoms were found in high-temperature heat-treated DWCNTs [16]. Inner cavity of these carbon nanotubes is suggested to provide a space for stabilizing the long linear carbon chains.

Recently, we found the presence of a bulk amount of long linear carbon chains in single-wall carbon nanotube (SWCNT) films after field electron emission experiments accompanied with electrical discharge [17, 18]. The SWCNT film was used as an emitter of field emission. The long linear carbon chains were confined within SWCNTs as well as DWCNTs and triple-wall carbon nanotubes (TWCNTs). In this chapter, Raman spectroscopy and electron microscopy of the long linear carbon chains generated by field electron emission of SWCNT films are described.

16.2 Field Electron Emission Accompanied with Electrical Discharge for Single-Wall Carbon Nanotube Films

Figure 16.1 shows an illustration of the field electron emission setup in a vacuum chamber. The distance between an aluminum cathode and an anode grid in parallel-plate configuration is 0.16 cm. In the present field-emission experiments, SWCNT films produced by an enhanced direct injection pyrolytic synthesis method (Meijo Nano Carbon Co., Ltd.) are used as emitters. The film thickness is approximately 40 μm and a nominal diameter of SWCNTs is 1–3 nm. DWCNTs and TWCNTs are originally included as constituents in the as-supplied SWCNT films. The SWCNT film is cut into typically 5 × 5 mm pieces, and a piece of the film is glued on the surface of the cathode using an Ag paste adhesive. The surface of the SWCNT film has homogeneous morphology composed of the entangled SWCNTs, as shown in Fig. 16.2.

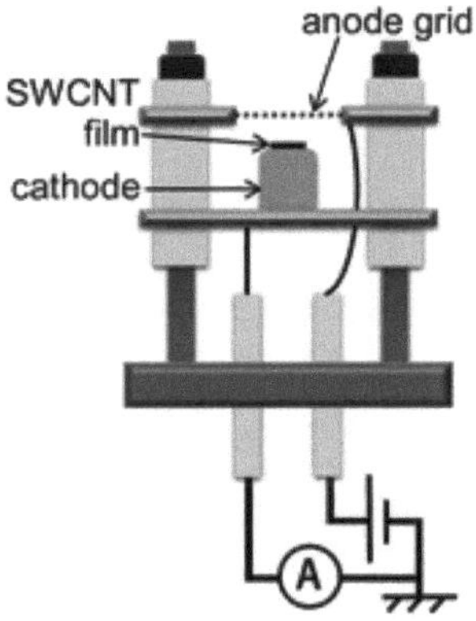

Figure 16.1 Illustration of field electron emission setup with an SWCNT film cathode and an anode grid in parallel-plate configuration in a vacuum chamber.

Field electron emission experiments of an SWCNT film were carried out under a vacuum pressure of 8.3×10^{-6} Pa at room temperature. Figure 16.3 shows field-emission characteristics of the SWCNT film. When the voltage was applied, the emission current increased till an electrical discharge occurred. The process of field-emission followed by electrical discharge was repeated. The emission

Figure 16.2 SEM image of an as-prepared SWCNT film glued on a cathode surface. (Reprinted with permission from Asaka et al. [18], Copyright 2019, Springer Nature.)

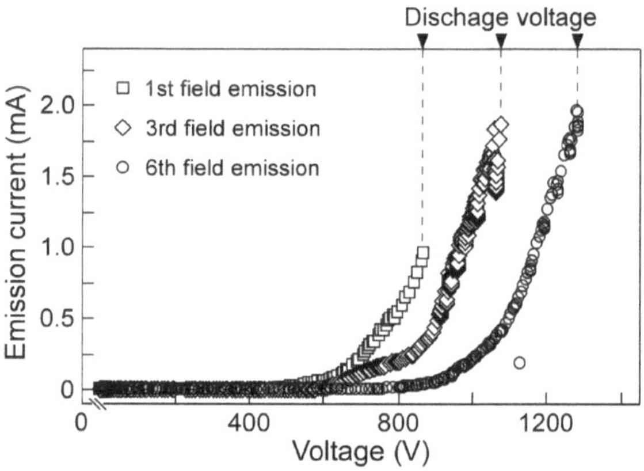

Figure 16.3 Field-emission current–voltage curves measured for an SWCNT film. (Reprinted with permission from Asaka et al. [18], Copyright 2019, Springer Nature.)

current–voltage curves for the 1st, 3rd, and 6th field-emission experiments are shown in Fig. 16.3. The emission currents at the electrical discharge are less than 2.5 mA. The turn-on voltage (here defined as the applied voltage needed to obtain an emission current of 10 nA) and discharge voltage tend to increase as the same SWCNT film is repeatedly used for the field electron emission experiments, i.e., it becomes more difficult to emit electrons under the same applied voltage after electrical discharge. The field-emission process is described by the Fowler–Nordheim (F–N) theory [19]. The theory relates the field-emission current I to the applied voltage V through the equation,

$$\ln\left(\frac{I}{V^2}\right) = -6.49\times10^7 \frac{d\phi^{3/2}}{\beta}\cdot\frac{1}{V} + \left\{\ln\left(1.4\times10^{-6}\frac{A\beta^2}{d^2\phi}\right) + \frac{9.8}{\phi^{1/2}}\right\}, \tag{16.1}$$

where d is the distance between electrodes, ϕ is the work function of CNT, β is the field enhancement factor, which is a constant determined by the geometry of the emitting surface, and A is the emitting area. The graph obtained by plotting $\ln(I/V^2)$ against $1/V$ is called the F–N plot and used to analyze the emission current–voltage curve. From the F–N plot, using $\phi = 4.6$ eV for graphite and $d = 0.16$ cm, the field enhancement factors for the 1st, 3rd, and 6th current–voltage curves shown in Fig. 16.3 are calculated to be 1.5, 1.1, and 0.9×10^4, respectively, which decreases with repetition of the field-emission experiment. The increase of the turn-on voltage and decrease of β suggest that surface morphology of the SWCNT film has changed due to the damage which the SWCNT film undergoes by the electrical discharge.

16.3 Long Linear Carbon Chains in Single-Wall Carbon Nanotube Films after Electrical Discharge

Figure 16.4 shows scanning electron microscopy (SEM) images of the SWCNT film after electrical discharge. An enlarged image of a damaged region indicated by an arrowhead A in Fig. 16.4a is shown in Fig. 16.4b. Some parts of the SWCNT film are damaged and torn up, resulting in the formation of holes, as seen in Fig. 16.4. The electrical discharge in field emission brings about significant morphological changes on the surface of the SWCNT film.

Figure 16.5a,b shows Raman spectra of the damaged and nondamaged regions of the SWCNT film after electrical discharge, measured from the regions indicated by arrowheads A and B in Fig. 16.4a, respectively. In the Raman spectrum of the nondamaged regions (Fig. 16.5b), two peaks are observed at 1343 and 1583 cm^{-1}, which are assigned to the D and G bands for SWCNTs, respectively. On the other hand, in the Raman spectrum measured from the

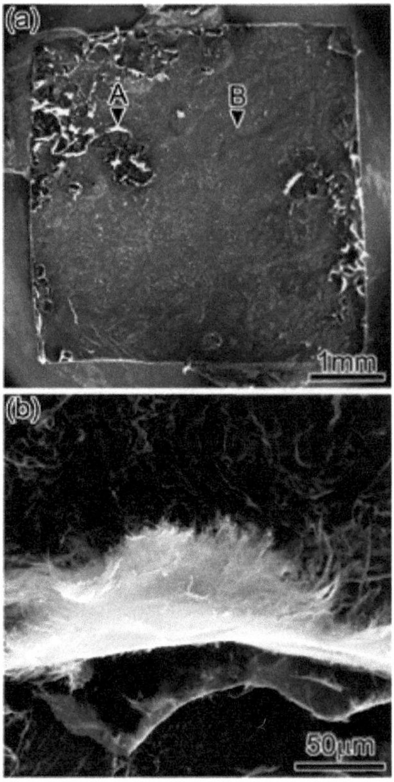

Figure 16.4 SEM image of the SWCNT film after field electron emission accompanied with electrical discharge. The regions indicated by arrowheads A and B are the damaged and nondamaged regions, respectively. (b) A high-magnification image of the damaged region indicated by an arrowhead A in (a). (Reprinted with permission from Toma et al. [17], Copyright 2018, John Wiley & Sons.)

damaged region (Fig. 16.5a), an intense peak appears at 1859 cm^{-1} in addition to the D and G bands. The peak observed at 1859 cm^{-1} does not make an appearance in the nondamaged regions, and is not observed in ordinary SWCNT films. A Raman band appearing in 1780–1870 cm^{-1} range is termed the L band, which is associated with C–C stretching-mode vibrations in a long linear carbon chain [8–11, 13–16]. Figure 16.6a shows a Raman mapping image of the L band for the damaged regions in the SWCNT film after electrical discharge along with the corresponding SEM image (Fig. 16.6b). The regions where the L band signal is detected are colored gray (see Fig. 16.6a).

The L bands are detected exclusively in the damaged regions of the SWCNT films, while no L band is observed in the nondamaged regions, as shown in Fig 16.6. The present results demonstrate that the formation of the long linear carbon chains in the SWCNT films are caused by electrical discharge in the field-emission experiment. The present filling rate of the long linear carbon chains encapsulated inside CNTs in the damage regions of the SWCNT films is estimated to be approximately 36% using the Raman-peak-intensity ratio of the L to G bands [17].

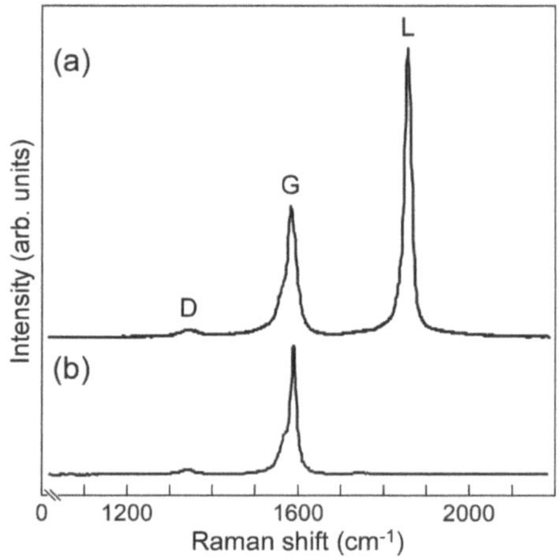

Figure 16.5 Raman spectra of (a) damaged and (b) nondamaged regions of the SWCNT film after electrical discharge, measured from the regions indicated by arrowheads A and B in Fig. 1.4a, respectively. (Reprinted with permission from Toma et al. [17], Copyright 2018, John Wiley & Sons.)

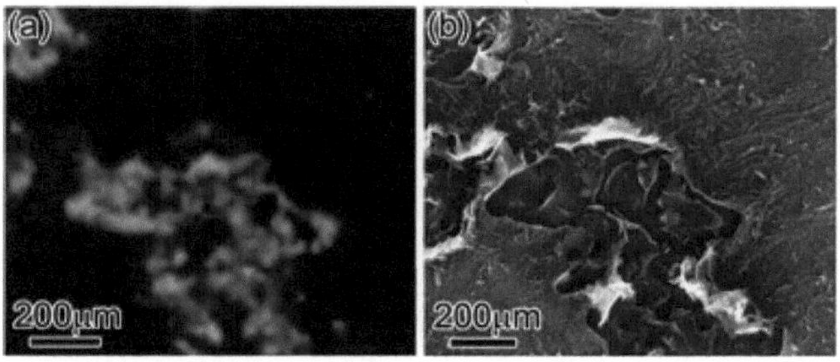

Figure 16.6 (a) and (b) Raman mapping image of the L band for the damaged region of the SWCNT film and the corresponding SEM image. The regions where the L band signal is detected are gray-scaled in (a). (Reprinted with permission from Toma et al. [17], Copyright 2018, John Wiley & Sons.)

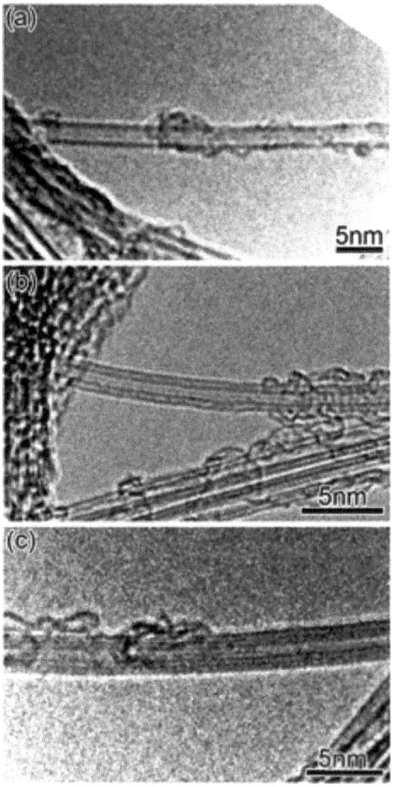

Figure 16.7 TEM images of long linear carbon chains encapsulated in (a) SWCNT, (b) DWCNT, and (c) TWCNT, included in damaged regions of the SWCNT films after electrical discharge. (Reprinted with permission from Toma et al. [17], Copyright 2018, John Wiley & Sons, and Asaka et al. [18], Copyright 2019, Springer Nature.)

The presence of long linear carbon chins in the damaged regions of the SWCNT films after electrical discharge can be recognized by high-resolution transmission electron microscopy (TEM). Figure 16.7a shows a TEM image of a long linear carbon chain encapsulated within an SWCNT. The length of the long linear carbon chain shown in Fig. 16.7a is more than 30 nm, and thus the long linear carbon chain is estimated to contain at least more than 230 contiguous carbon atoms [17, 18]. When the inner diameter of CNTs is larger than about 1.0 nm, some encapsulated carbon chains are located close to the nanotube wall away from the central axis [18]. The off-centered deviation probably results from the electronic

π–π interaction between the carbon chain and CNT wall. The long linear carbon chains encapsulated in DWCNTs and TWCNTs, which are originally included as constituents in the as-supplied SWCNT films, are also observed, as shown in Fig. 16.7b,c. These innermost diameters of the DWCNT and TWCNT, 0.7–0.8 nm, are well agreement with the reported innermost diameters of the DWCNT and MWCNT encapsulating a long linear carbon chain [8, 9, 12, 16]. These long linear chains are aligned along the central axis of the innermost tube, which is approximately at van der Waals distance from the inner wall.

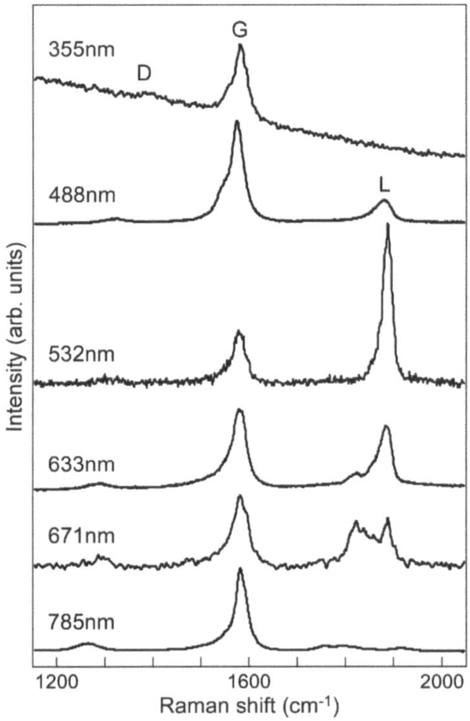

Figure 16.8 Raman spectra of the damaged regions in the SWCNT film after electrical discharge using laser excitation sources with wavelengths of 355, 488, 532, 633, 671, and 785 nm. (Reprinted from [20], licensed under CC BY 4.0.)

The band gaps of the long linear carbon chains confined within the CNTs in the SWCNT films after electrical discharge were investigated by Raman spectroscopy using laser excitation sources with wavelengths from 355 to 785 nm (Fig. 16.8) [20]. Resonant

Raman scattering of the L band for the long linear carbon chain occurs when the excitation energy is close to the band gap of the long linear carbon chain. In Fig. 16.8, the L band is observed at excitation wavelengths of 488, 532, 633, and 671 nm, indicating that the long linear carbon chains formed in the CNTs possess the corresponding band gaps of 1.85–2.54 eV, which are approximately consistent with the reported values [21, 22]. The distribution of the band gaps may be attributed to the length and the encapsulating environment of the long linear carbon chains.

In the present study, the presence of the long linear carbon chains was found in the damaged regions of the SWCNT films after electrical discharge in the field electron experiments. Although the electrical discharge plays an important role in the formation of the long linear carbon chains, the formation mechanism is still unclear and a subject for the future researches.

16.4 Conclusions

Field electron emission experiments were carried out using the SWCNT films until electrical discharge occurred. Raman spectroscopy and TEM of the SWCNT films after electrical discharge revealed the formation of the long linear carbon chains confined within the SWCNTs as well as the DWCNTs and TWCNTs. The present results can potentially lead to a novel way to produce a bulk production of the long linear carbon chains encapsulated in CNTs by field electron emission accompanied with electrical discharge of the SWCNT films.

References

1. Casari, C. S., Tommasini. M., Tykwinski, R. R. and Milani, A. (2016). Carbon-atomwires: 1-D systems with tunable properties, *Nanoscale*, **8**, pp. 4414–4435.
2. Yang, X., Huang, Y., Cao, B. and To, A. C. (2017). Length and temperature dependence of the mechanical properties of finite size carbine, *Phys. E*, **93**, pp. 124–131.
3. Wang, M. and Lin, S. (2015). Ballistic thermal transport in carbyne and cumulene with micron-scale spectral acoustic phonon mean free path, *Sci. Rep.*, **5**, 18122.

4. Shen, L., Zeng, M., Yang, S.-W., Zhang, C., Wang, X. and Feng, Y. (2010). Electron transport properties of atomic carbon nanowires between graphene electrodes, *J. Am. Chem. Soc.*, **132**, pp. 11484–11486.
5. Akdim, B. and Pachter, R. (2011). Switching behavior of carbon chains bridging graphene nanoribbons: Effects of uniaxial strain, *ACS Nano*, **5**, pp. 1769–1774.
6. Zeng, M. G., Shen, L., Cai, Y. Q., Sha, Z. D. and Feng, Y. P. (2010). Perfect spin-filter and spin-valve in carbon atomic chains, *Appl. Phys. Lett.*, **96**, 042104.
7. Sorokin, P. B., Lee, H., Antipina, L. Yu., Singh, A. K. and Yakobson, B. I. (2011). Calcium-decorated carbyne networks as hydrogen storage media, *Nano Lett.*, **11**, pp. 2660–2665.
8. Zhao, X., Ando, Y., Liu, Y., Jinno, M. and Suzuki, T. (2003). Carbon nanowire made of a long linear carbon chain inserted inside a multiwalled carbon nanotube, *Phys. Rev. Lett.*, **90**, 187401.
9. Scuderi, V., Scalese, S., Bagiante, S., Compagnini, G., D'Urso, L. and Privitera, V. (2009). Direct observation of the formation of linear C chain/carbon nanotube hybrid systems, *Carbon*, **47**, pp. 2112–2142.
10. Scalese, S., Scuderi, V., Bagiante, S., Simone, F., Russo, P., D'Urso, L., Compagnini, G. and Privitera, V. (2010). Controlled synthesis of carbon nanotubes and linear C chains by arc discharge in liquid nitrogen, *J. Appl. Phys.*, **107**, 014304.
11. Andrade, N. F., Vasconcelos, T. L., Gouvea, C. P., Archanjo, B. S., Achete, C. A., Kim, Y. A., Endo, M., Fantini, C., Dresselhaus, M. S. and Filho, A. G. S. (2015). Linear carbon chains encapsulated in multiwall carbon nanotubes: Resonance Raman spectroscopy and transmission electron microscopy studies, *Carbon*, **90**, pp. 172–180.
12. Wang, Z., Ke, X., Zhu, Z., Zhang, F., Ruan, M. and Yang, J. (2000). Carbon-atom chain formation in the core of nanotubes, *Phys. Rev. B*, **61**, pp. R2472–R2474.
13. Cazzanelli, E., Castriota, M., Caputi, L. S., Cupolillo, A., Giallombardo, C. and Papagno, L. (2007). High-temperature evolution of linear carbon chains inside multiwalled nanotubes, *Phys. Rev. B*, **75**, 121405(R).
14. Zhao, C., Kitaura, R., Hara, H., Irle, S. and Shinohara, H. (2011). Growth of linear carbon chains inside thin double-wall carbon nanotubes, *J. Phys. Chem. C*, **115**, pp. 13166–13170.
15. Zhang, J., Feng, Y., Ishiwata, H., Miyata, Y., Kitaura, R., Dahl. J. E. P., Carlson, R. M. K., Shinohara, H. and Tománek, D. (2012). Synthesis

and transformation of linear adamantane assemblies inside carbon nanotubes, *ACS Nano*, **6**, pp. 8674–8683.

16. Shi, L., Rohringer, P., Suenaga, K., Niimi. Y., Kotakoski, J., Meyer, J. C., Peterlik, H., Wanko, M., Cahangirov, S., Rubio, A., Lapin, Z. J., Novotny, L., Ayala, P. and Pichler, T. (2016). Confined linear carbon chains as a route to bulk carbine, *Nat. Mater.*, **15**, pp. 634–640.
17. Toma, S., Asaka, K., Irita, M. and Saito, Y. (2019). Bulk synthesis of linear carbon chains confined inside single-wall carbon nanotubes by vacuum discharge, *Surf. Interface Anal.*, **51**, pp. 131–135.
18. Asaka, K., Toma, S. and Saito, Y. (2019). Raman spectroscopy and transmission electron microscopy of long linear atomic carbon chains generated by field electron emission accompanied with electrical discharge of carbon nanotube films, *SN Appl. Sci.*, **1**, 493.
19. Spindt, C. A., Brodie, I., Humphrey, L. and Westerberg, E. R. (1976). Physical properties of thin-film field emission cathodes with molybdenum cones, *J. Appl. Phys.*, **47**, pp. 5248–5263.
20. Asaka, K., Toma, S. and Saito, Y. (2020). Raman spectroscopy and TEM of long linear carbon chain formed in CNT field emission cathode, *e-J. Surf. Sci. Nanotech.*, **18**, pp. 159–163.
21. Shi, L., Rohringer, P., Wanko, M., Rubio, A., Wasserroth, S., Reich, S., Cambre, S., Wenseleers, W., Ayala, P. and Pichler, T. (2017). Electronic band gaps of confined linear carbon chains ranging from polyyne to carbyne, *Phys. Rev. Mater.*, **1**, pp. 075601–075607.
22. Heeg, S., Shi, L., Pichler, T. and Novotny, L. (2018). Raman resonance profile of an individual confined long linear carbon chain, *Carbon*, **139**, pp. 581–585.

Chapter 17

Emission of C_{20}^{+} by Field Evaporation from CNT

Yahachi Saito
Toyota Physical and Chemical Research Institute, Yokomichi 41-1, Nagakute, Aichi 480-1192, Japan
ysaito@toyotariken.jp

Carbon forms various structures in a small assembly of carbon atoms, e.g., linear chain, cage structures like C_{60} fullerene. Under extremely high electric field, carbon atoms and clusters are emitted as ions via field evaporation from the surface of carbon emitters. Among the field-evaporated carbon clusters from carbon nanotubes, C_{20}^{+} was observed as a prominent species in mass spectra.

17.1 Introduction

In 1995, Rinzler et al. [1] showed in their field-emission study that an open-edged multiwall carbon nanotube (MWCNT) gave high field-emission currents even at a low applied voltage owing to an extremely high field enhancement, and in order to explain this high field enhancement they postulated that a linear carbon chain (an atomic

Nanostructured Carbon Electron Emitters and Their Applications
Edited by Yahachi Saito

ISBN 978-981-4877-62-6 (Hardcover), 978-1-003-14199-0 (eBook)
www.jennystanford.com

wire) was unraveled from an open-edge of graphene comprising the MWCNT by a strong electrostatic force. However, the formation of atomic chains at open-ended carbon nanotubes (CNTs) under electric field was not verified. Field-emission microscopy (FEM) is usually inappropriate to the observation in atomic resolution, but field ion microscopy (FIM) affords to observe individual carbon atoms. Using carbon polyacrylonitrile-based fibers, Mazilova et al. [2, 3] reported FIM studies on formation of linear carbon chains and their mechanical behaviors under high electric fields.

FIM of CNTs gave images that rapidly changed from clusters consisting of bright spots arranged in a pentagon to streaks radially arranged in a daisy-shape, as has been described in Chapter 1. Such a variation in FIM images is ascribed to a change in the atomic structure of a CNT tip induced by a high electric field. Under a very high positive electric field, surface atoms of a solid can be removed from their lattice sites by the field induced on the solid surface. This phenomenon is called field evaporation, which emits surface atoms as ion species. In order to study the field evaporation of nanotubes, carbon species emitted from MWCNTs and SWCNTs have been investigated by mass spectrometry [4].

Interestingly, C_{20}^{+} ions were found to be the most prominent species in mass spectra of the field evaporated carbon from CNTs. C_{20} clusters are predicted to have several metastable structures (isomers), among which the smallest fullerene structure consisting of only pentagons (dodecahedron) is included. The preferential formation of C_{20} clusters is not limited to field evaporation method, but is reportedly found in other experimental techniques such as spark arc discharge of carbon solid [5] and substitution reaction of perhydrogenated dodecahedrane ($C_{20}H_{20}$) [6].

17.2 Field Evaporation of Carbon Ions from CNT Under High Electric Field

The field strength required for field evaporation depends on emitter materials, being highest for refractory metals such as W and Mo and low for soft metals such as Ni. The activation energy of the field evaporation is mainly governed by the binding energy of the surface atoms, the ionization energy, the work function, and the applied electric field. Onset field strengths of field evaporation for various forms of carbon (graphite) reported in literatures are listed together with metals in Table 17.1. The evaporation field of carbon distributes from 20 to 80 V/nm, depending on the form of carbon. Graphite and highly graphitized carbon seem to have high stability against field evaporation, their onset field strength being in a range of refractory metals.

Table 17.1 Field strength of field evaporation for carbon and metals

Emitter materials	Strength of evaporation field (V/nm)	References
Carbon fiber	70–80 (at 77 K)	V. A. Ksenofontov et al., *J. Phys.: Condens. Matter* 19 (2007) 466204
C (graphite)	>45 (at low temp.)	J. Liu and T. T. Tsong, *Phys. Rev. B* 38 (1988) 8490
C (glassy carbon)	23 (at liq. N_2 temp.)	M. Futamoto et al., *Surf. Sci.* 86 (1979) 718
MWCNT, SWCNT	≈10 (at RT)	Ref. [4]
W	57 (at low temp.)	T. T. Tsong, *Surf. Sci.* 70 (1978) 211
Mo	46 (at low temp.)	Ibid.
Ni	35 (at low temp.)	Ibid.

The field strength at the tip of CNT can be estimated from the geometrical field factor β (= E_{loc}/V_{app}, where E_{loc} is the local field at the tip and V_{app} the applied voltage), which can be calculated from the slope of the Fowler–Nordheim (F–N) plot because the slope of F–N plot is proportional to $-\phi^2/\beta$, where ϕ is the work function [7]. The threshold field strength for observing of carbon ions by field evaporation was reported to be about 10 V/nm for both MWCNT and SWCNT, and the ion intensity increased rapidly with the increase of the field strength [4]. The threshold field strength for CNTs is lower than those reported for carbon fiber, graphite, and glassy carbon (Table 17.1). This may be due to the difference in temperature of field emitters; the CNT emitters were at room temperature, while the other carbon emitters were at low temperatures (e.g., liq. N_2 temp.). Since the field evaporation is thermally activated events, at higher temperature the evaporation rate is higher and thus the threshold field required for the observation become lower.

17.3 Magic Cluster Ion, C_{20}^+, in Field Evaporation Mass Spectra

Both a lump of MWCNTs-produced arc discharge and a purified SWCNT bundle were used to investigate the field evaporation of carbon atoms from CNTs. Field emitted carbon ions emitted from CNT emitters, which were biased by positive high electric potentials (from 3 to 9 kV) relative to an extractor electrode, were mass analyzed by a single-focusing magnetic mass spectrometer. Figure 17.1 shows a mass spectrum of carbon cluster ions field-evaporated from a limp of MWCNTs [4]. The intensity of C_{20}^+ ions is prominently high, indicating either the high stability or a preferred formation of C_{20}^+ ions. The inset of Fig. 17.1, which is a high-resolution mass spectrum in a region around 240 u corresponding to C_{20}^+, shows satellite peaks due to the combination of carbon isotopes ^{12}C and ^{13}C. The relative intensities of the resolved peaks at 240, 241, and 242 u are well reproduced by the binomial distribution calculated from the natural abundance of carbon isotopes, 98.9% for ^{12}C and 1.1% for ^{13}C, substantiating the assignment of the peak at 240 u as C_{20}^+. Purified SWCNT bundles also gave similar mass spectra by field evaporation, showing the C_{20}^+ ions as prominent clusters [4].

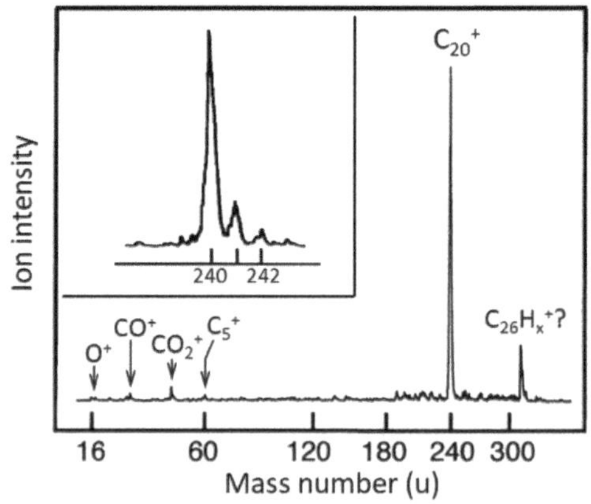

Figure 17.1 Mass spectrum of carbon cluster ions field-evaporated from MWCNTs. Inset shows a high-resolution mass spectrum around 240 u. The local field strength at the emitter tip was estimated ≈17 V/nm, and the emitter temperature was room temperature [4].

Liu and Tsong [8] investigated pulsed laser-stimulated field evaporation from a polycrystal graphite tip by a time-of-flight mass spectrometer. They observed singly charged C_3, C_5, and C_7 as magic clusters in addition to multiply charged (+2 and +3) clusters, but no information on C_{20}^+.

The mechanism of the selective formation of C_{20} is not obvious. If the C_{20}^+ ions were formed via atomic wires at a nanotube tip, then it would be difficult to explain why the wires snap off when the number of carbon atoms reaches 20.

Von Helden et al. [9] have experimentally showed that carbon clusters in the gas phase develop from linear chains to planar rings and cage structures (fullerenes) as their size increases. When carbon clusters possess the linear chain or ring structures, it is difficult to explain the uniqueness of 20 atoms which constitute a cluster; Arbitrary numbers of atoms (though even–odd effects may exist) are allowed to make a cluster with the linear and ring structures. Among the various isomers of carbon clusters (Fig. 17.2), on the other hand, the cage structure (fullerene) restricts the number of constituent atoms. From the magic nature of the cluster, it is postulated that the C_{20} cluster has the dodecahedron structure possessing equivalent 20 corners on which carbon atoms are placed. The dodecahedron consisting of only pentagons is the smallest fullerene structure with the I_h symmetry.

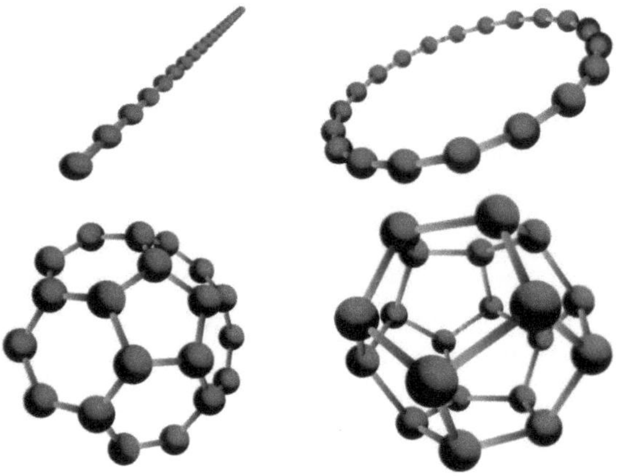

Figure 17.2 Isomers of C_{20} clusters: linear chain, ring, bowl, and dodecahedron.

17.4 Other Evidences of the Existence of C_{20} Clusters in Gas Phase

Besides the field evaporation from CNT emitters described above, several experimental and theoretical studies showing the selective and preferential formation of C_{20} clusters are reported. Kurokawa et al. formed carbon deposit on metal surfaces by using an arc-plasma gun for carbon evaporation and found chains of fused C_{20} clusters by scanning tunneling microscopy [5]. They considered that the C_{20} fullerenes are deformed from the ideal I_h symmetry to the D_{2h} (D_2 or C_{2v}) symmetry to attain twofold symmetry, enabling the formation of one-dimensional chains of the C_{20} molecules on the metal surfaces. The C_{20} fullerenes were probably formed either in carbon plasma generated by the arc-plasma gun before the deposition, or on the metal surfaces during diffusion and coagulation of deposited carbon clusters. Anyway, the observation of the C_{20} molecules suggests their selective formation and/or high stability.

Prinzbach et al. tried to produce the cage-structured C_{20} from its perhydrogenated dodecahedrane ($C_{20}H_{20}$) via substitution reactions [6], and characterized the product by mass-selective anion photoelectron spectroscopy for comparison with other known isomers, bowl (dehydrated corannulene), and monocyclic ring. The spectroscopic features of the C_{20} produced from the perhydrogenated dodecahedrane were significantly different from the other isomers, suggesting the existence of cage-structured isomer.

Theoretical studies have shown that the cage-structured C_{20} is less energetically stable than the bowl and ring isomers consisting of 20 carbon atoms [10]. Although it is metastable, its stability with respect to the decay to a lower energy configuration has been simulated, suggesting that the C_{20} fullerene retains its chemical structure under heating to very high temperatures ($\approx$3000 K) [11]. Therefore, once C_{20} fullerenes are formed, it is expected that the isomer survives at least room temperature.

17.5 Conclusion

Mass spectrometry of carbon ions field-evaporated from MWCNT and SWCNT emitters, to which high positive electric voltages were applied, showed that the C_{20}^{+} cluster ions were emitted as the most dominant species with exclusively high intensity compared with other carbon ions. The overwhelming high intensity of C_{20}^{+} indicates its preferential formation and/or high stability.

The number of atoms, 20, is a special one which can form the smallest fullerene, dodecahedron. There are several experimental and theoretical reports suggesting the formation and stability of the C_{20} fullerene.

References

1. Rinzler, A. G., Hafner, J. H., Nikolaev, P., Lou, L., Kim, S. G., Tomanek, D., Nordlander, P., Colbert, D. T. and Smalley, R. E. (1995). Unraveling nanotubes: Field emission from an atomic wire, *Science*, **269**, pp. 1550–1553.
2. Mazilova, T. I., Mikhailovskij, I. M., Ksenofontov, V. A. and Sadanov, E. V. (2009). Field-ion microscopy of quantum oscillations of linear carbon atomic chains, *Nano Lett.*, **9**, pp. 774–778.
3. Mikhailovskij, I. M., Sadanov, E. V., Kotrechko, S., Ksenofontov, V. A. and Mazilova, T. I. (2013). Measurement of the inherent strength of carbon atomic chains, *Phys. Rev. B*, **87**, pp. 269045410-1–045410-7
4. Hata, K., Ariff, M., Tohji, K. and Saito, Y. (1999). Selective formation of C cluster ions by field evaporation from carbon nanotubes, *Chem. Phys. Lett.*, **308**, pp. 343–346.
5. Kurokawa, S., Yamamoto, D., Hirashige, K. and Sakai, A. (2016). Possible formation of one-dimensional chains of C_{20} fullerenes observed by scanning tunneling microscopy, *Appl. Phys. Express*, **9**, pp. 045102-1–045102-4
6. Prinzbach, H., Weiler, A., Landenberger, P., Wahl, F., Wörth, J., Scott, L. T., Gelmont, M., Olvano, D. v., and Issendorff, B. (2000). Gas-phase production and photoelectron spectroscopy of the smallest fullerene, C_{20}, *Nature*, **407**, pp. 60–63.
7. Gomer, R. (1961). *Field Emission and Field Ionization* (Harvard University Press, Cambridge, MA), Chapter 1 "Theory of Field Emission", pp. 1–31.
8. Liu, J. and Tsong, T. T. (1988). Kinetic-energy and mass analysis of carbon cluster ions in pulsed-laser-stimulated field evaporation, *Phys. Rev. B*, **38**, pp. 8490–8493.

9. Von Helden, G., Gotts, N. G. and Bowers, M. T. (1993). Experimental evidence for the formation of fullerenes by collisional heating of carbon rings in the gas phase, *Nature*, **363**, pp. 60–63.

10. Lu, J., Re, S., Choe, Y., Nagase, S., Zhou, Y., Han, R., Peng, L., Zhang, X. and Zhao, X. (2003). Theoretical identification of C_{20} carbon clusters: Prevalence of the monocyclic isomer and existence of the smallest fullerene and bowl isomer, *Phys. Rev. B*, **67**, pp. 125415-1–125415-7.

11. Davydov, I. V., Podlivaev, A. I. and Openov, L. A. (2005). Anomalous thermal stability of metastable C20 fullerene, *Phys. Solid State*, **47**, pp. 778–784.

Index